Translational Bioinformatics

Volume 8

Series editor

Aims and Scope

The Book Series in Translational Bioinformatics is a powerful and integrative resource for understanding and translating discoveries and advances of genomic, transcriptomic, proteomic and bioinformatic technologies into the study of human diseases. The Series represents leading global opinions on the translation of bioinformatics sciences into both the clinical setting and descriptions to medical informatics. It presents the critical evidence to further understand the molecular mechanisms underlying organ or cell dysfunctions in human diseases, the results of genomic, transcriptomic, proteomic and bioinformatic studies from human tissues dedicated to the discovery and validation of diagnostic and prognostic disease biomarkers, essential information on the identification and validation of novel drug targets and the application of tissue genomics, transcriptomics, proteomics and bioinformatics in drug efficacy and toxicity in clinical research.

The Book Series in Translational Bioinformatics focuses on outstanding articles/chapters presenting significant recent works in genomic, transcriptomic, proteomic and bioinformatic profiles related to human organ or cell dysfunctions and clinical findings. The Series includes bioinformatics-driven molecular and cellular disease mechanisms, the understanding of human diseases and the improvement of patient prognoses. Additionally, it provides practical and useful study insights into and protocols of design and methodology.

Series Description

Translational bioinformatics is defined as the development of storage-related, analytic, and interpretive methods to optimize the transformation of increasingly voluminous biomedical data, and genomic data in particular, into proactive, predictive, preventive, and participatory health. Translational bioinformatics includes research on the development of novel techniques for the integration of biological and clinical data and the evolution of clinical informatics methodology to encompass biological observations. The end product of translational bioinformatics is the newly found knowledge from these integrative efforts that can be disseminated to a variety of stakeholders including biomedical scientists, clinicians, and patients. Issues related to database management, administration, or policy will be coordinated through the clinical research informatics domain. Analytic, storage-related, and interpretive methods should be used to improve predictions, early diagnostics, severity monitoring, therapeutic effects, and the prognosis of human diseases.

Recently Published and Forthcoming Volumes

Bioinformatics for Diagnosis, Prognosis and Treatment of Complex Diseases
Editor: Bairong Shen
Volume 4

Single Cell Sequencing and Systems Immunology
Editors: Xiangdong Wang, Xiaoming Chen, Zhihong Sun, Jinglin Xia
Volume 5

Genomics and Proteomics for Clinical Discovery and Development
Editor: György Marko-Varga
Volume 6

Computational and Statistical Epigenomics
Editor: Andrew E. Teschendorff
Volume 7

More information about this series at http://www.springer.com/series/11057

Ailin Tao · Eyal Raz
Editors

Allergy Bioinformatics

Editors
Ailin Tao
The Second Affiliated Hospital
of Guangzhou Medical University
Guangzhou
China

Eyal Raz
The Second Affiliated Hospital
of Guangzhou Medical University
Guangzhou
China

and

Department of Medicine
University of California
San Diego
USA

ISSN 2213-2775 ISSN 2213-2783 (electronic)
Translational Bioinformatics
ISBN 978-94-017-7442-0 ISBN 978-94-017-7444-4 (eBook)
DOI 10.1007/978-94-017-7444-4

Library of Congress Control Number: 2015952056

Springer Dordrecht Heidelberg New York London

Printed on acid-free paper

Springer Science+Business Media B.V. Dordrecht is part of Springer Science+Business Media (www.springer.com)

Foreword

Allergies have been challenging human beings all over the world for many, many years, and, while the prevalence of allergic diseases has been increasing, there have been continuous breakthroughs in allergy research.

In recent years, the computational classification of proteins has accelerated research in the field of allergy and guided experimental design. Our book entitled "Allergy Bioinformatics" is one of the first to systematically describe the application of bioinformatics in allergy. This book makes an outstanding contribution in the areas of allergen number reduction, bioinformatic classifiers for allergen sequence discrimination, allergenicity evaluation and modification, and allergen database and allergen discriminating software, in addition to redefining the terms *antigenicity*, *immunogenicity*, and *allergenicity*. This book is based on the accumulation and incorporation of the authors' vast knowledge and experience and includes increasingly important in-depth allergy studies, as well as homology prediction, reduction of number of allergens for cross-reactivity, development of an allergen database, and vaccine and epitope design. This book also provides an overview of the development of bioinformatics in allergy research over the past decade and provides a complete and reliable reference for researchers in related fields.

Bioinformatics, one of the great frontiers of contemporary sciences, is an essential part of the natural sciences, allergy bioinformatics is a new and fresh resource for many clinical and preclinical scientists working on allergy and immune diseases and has great reference potential, yet it also presents challenges to the current state of allergy research. Most professional databases are lacking important information and need to be integrated with sequence data and the scientific literature, in addition to containing impeccable data analyses. Of the currently available databases, the allergen database ALLERGENIA (http://allergenia.gzhmu.edu.cn) is one of the most comprehensive thanks to the significant contributions and efforts of Professor Ailin Tao.

Professor Tao is one of the leading scientists in global allergology research and has been a pioneer in allergy bioinformatics. He is a well-recognized expert in this field and has developed a large number of bioinformatic software, e.g., SORTALLER (http://sortaller.gzhmu.edu.cn) for allergen discrimination that displays high specificity (98.4 %) and sensitivity (98.6 %), as well as a Matthews correlation coefficient up to 0.970. At present, SORTALLER outperforms all other allergen prediction software and has been widely used by researchers all over the world. Professor Tao has also built an allergen database ALLERGENIA that has several advantages over other databases, such as good coverage, better non-redundancy, excellent astringency and accuracy, and user-friendly analyzing functions. I was impressed when I visited Professor Tao's laboratory and communicated with his colleagues regarding the achievements Professor Tao and his team have reached in allergenicity assessment and modification based on bioinformatics methods that they have developed. They established a number of animal and cell models for the evaluation of allergenicity, which they were then able to use to modify the allergenicity of allergens and immunotoxins in order that they are safely used for targeted immunotherapy. Professor Tao is a scientific thinker and active researcher in the field of allergy. He has proposed a number of new concepts, including "Representative Major Allergens," "Broad-spectrum Immunomodulator," and the hypothesis of "Balanced Stimulation by Whole Antigens." I do believe that this book, edited by Professor Tao and Eyal Raz, another giant immunologist, will contribute much more to allergology in the future. Please join me in congratulating Professor Tao and his co-authors for such an excellent and successful book and wish them much success in all their future endeavors.

August 2015 Xiangdong Wang

Preface

At a meeting three years ago, I bumped into Distinguished Professor Xiang-dong Wang, from the Zhongshan Affiliated Hospital of Fudan University. According to his seniority, I respectfully call him ShiShu. Because his research also involves bioinformatics methods, we talked more about this topic. I told him about our work in allergy bioinformatics and, in particular, about the allergen discriminating software SORTALLER and the allergen database ALLERGENIA that we have constructed. These are currently some of the best software and database systems available that have been established for allergenicity assessment and modification and substantive progress has been made in our laboratory. ShiShu's face lit up upon hearing this. He invited me to become editor-in-chief of the book "Allergy Bioinformatics" and to introduce our achievements to the booming community of Allergology. Many thanks to him for his wisdom, his tenderness, and his recommendations that have brought us this opportunity to reframe international research in allergy bioinformatics.

I and my 15-year friend Dr. Eyal Raz, professor from the University of California at San Diego, also a Distinguished Professor at Guangzhou Medical University, began to discuss and decide together the contents, chapters, and candidate contributors to this book. In addition, we discussed immunological theory, particularly a novel understanding of Antigenicity and a redefinition Allergenicity, which will facilitate the evaluation of all antigens in the future. We believe this to be a highlight of this book, and thus, it is hard to imagine that this book could be published without his encouragement, support, and urging. It was our honor to have also invited Professor Rudolf Valenta, from Medical University of Vienna, Wien, Vienna, Austria, to be a contributor to this book. Unfortunately, his busy schedule and our failure to follow-up prevented his contribution to the text, but we adopted some of his suggestions regarding the chapter setup. Our sincere wishes of peace and happiness always to Professor Rudolf Valenta.

With the help of Professor Hong-Der Shen, from Taipei Veterans General Hospital, Taiwan, China, we invited Professor Wayne Thomas, an allergological expert from Telethon Kids Institute, University of Western Australia, to contribute to this book. He presented us with a wonderful and timely book chapter. Pingchang Yang, formerly an associate professor at Canada McMaster University and currently Distinguished Professor at Shenzhen University Medical College, has also happily contributed to this book. In addition, Taiwan expert Professor Jiu-yao Wang (Department of Pediatrics, Allergy and Clinical Immunology Research Center, College of Medicine, National Cheng Kung University, Tainan, Taiwan, China) and Professors Chih-hsing Hung and Chang-Hung Kuo (Kaohsiung Medical University Hospital, Kaohsiung, Taiwan, China), who hold important positions worldwide in the field of allergy for their academic achievements and are worthy of teaching China's mainland scholars, gladly joined our writing team and made excellent contributions. Some authors (including professors Zehong Zou, Jianguo Zhang, and He Lai; Drs Wen Li, Xueting Liu, Junyan Zhang, Shan Wang, Zhaoyu Liu, Junshu Liu, Ying He, Yuyi Huang, Huifang Chen, and Juan Wang) are my colleagues working in the allergy clinic or engaged in allergenicity evaluation and modification, bioinformatics software and/or database construction, and I have much gratitude to them for their diligence and professionalism in their contributions to this book.

Sincere appreciation goes to the students who have worked in Guangdong Provincial Key Laboratory of Allergy and Clinical Immunology, the State Key Clinical Specialty in Allergy, the State Key Laboratory of Respiratory Disease, and the Second Affiliated Hospital of Guangzhou Medical University. They have been dedicated to allergen data testing, online software research and development, and experimental allergology. However, their names have not often been listed as authors for publication. I would like to acknowledge them here. They are Yanfang Li, Honglin Xia, Jierong Gao, Jiangli Cheng, Junling Peng, Huifeng Dong, Gui Xu, Jiang Li, Yuhe Guo, Huiyan Wu, Linmei Li, Chenxia Xu, Shufen Cao, Zhi'ao Guan, Zhiheng Ji, Bizhou Li, Renren Shi, Juan Wu, Lanyan Xiao, Miaolian Zhang, Jiandong Zhu, Baotong Wu, etc. I wish them all much success and enjoyment in their present positions.

After opening this book, in addition to the professional knowledge in allergy bioinformatics, you will also enjoy the American English! Beyond the main contributions by the authors, my old friend Lucinda G. Beck from UCSD has repeatedly edited the language and provided professional revisions for all parts of this book in preparation for publication. She was busy with her family during this time and was always eager to help me. I love her language talent, earnest, hard-working attitude, and meticulousness. The existence of this book will forever record my gratitude and benediction to her.

Because of the many interesting contributions from named and unnamed authors, we believe this book will be beneficial to the readers.

Fourteen years ago, I was lucky enough to become a postdoctoral student of the Yangtze River Scholar Professor Shao-Heng He. Since then, I have had a never-ending enthusiasm for allergology. Professor He's 60th birthday celebration will be held next April 2016 and I dedicate this book to him as a birthday gift.

Many many thanks to my wife Shulan Wang and my daughter Tianyu Tao for keeping the lights on the living room waiting for me to return home from my office at midnight. Their patience and support have allowed me to be devoted to my allergy research and to writing and revising this book.

Thank you all!

August 2015 Ailin Tao

Contents

Chapter 1
Introduction

Shan Wang, Ailin Tao and Eyal Raz

Abstract Allergy is a mistargeted immune reaction that occurs after the body has been primed by a certain antigen known as allergen and is subsequently restimulated by the same antigen to generate clinical symptoms and transient or chronic organ dysfunction. Allergic diseases affect mainly the skin and mucosal tissues such as sinuses, lung and the gut. Under certain conditions the hypersensitivity reaction is systemic and in this case it can be fatal due to systemic reaction. Over the last few decades the prevalence of allergic diseases in Western countries has been increasing across age, sex and racial groups and its annual healthcare costs have risen substantially. It is estimated that 20 % of the US population and approximately 36 % of the Chinese population suffer from some type of allergic disease. The study of the biochemical and immunological properties of allergens has been an important focus of allergy in addition to the bioinformatics of allergens, which consists of computer science, statistics, mathematics, and engineering aiming to dissect the biochemistry and immune properties of allergens at multiple levels. We envision that the use of bioinformatics will play a more important role in biomedical research of allergy in the future and promote the development of related disciplines and industries that will be of great practical importance.

Keywords Allergens · Immunotherapy · Bioinformatic · Cross-reactivity · IgE

S. Wang · A. Tao (✉) · E. Raz
Guangdong Provincial Key Laboratory of Allergy and Clinical Immunology, The State Key Clinical Specialty in Allergy, The State Key Laboratory of Respiratory Disease, The Second Affiliated Hospital of Guangzhou Medical University, 250# Changgang Road East, Guangzhou 510260, People's Republic of China
e-mail: AerobiologiaTao@163.com

S. Wang
e-mail: eraz@ucsd.edu

E. Raz (✉)
Department of Medicine, University of California, San Diego, La Jolla, CA, USA
e mail: craz@ucsd.edu

A. Tao and E. Raz (eds.), *Allergy Bioinformatics*, Translational Bioinformatics 8,
DOI 10.1007/978-94-017-7444-4_1

1.1 Overview

Allergy is a hypersensitivity immune response. Allergic disease a mis-targeted immune reaction that occurs after the body has been primed by a certain antigen known as allergen and is subsequently re-stimulated by the same antigen to generate clinical symptoms and transient or chronic organ dysfunction. Allergic diseases affect mainly the skin and mucosal tissues such sinuses, lungs, and the gut. Under certain conditions, the hypersensitivity reaction is systemic and in this case it can be fatal due to systemic reaction (Ruffoni et al. 2015; Yang et al. 2015). Over the last few decades the prevalence of allergic diseases in Western countries has been increasing across age, sex, and racial groups and its annual healthcare costs have risen substantially. It is estimated that 20 % of the US population and approximately 36 % of the Chinese population suffer from some type of allergic disease (Wang and Zhang 2012). The study of the biochemical and immunological properties of allergens has been an important focus of allergy in addition to the bioinformatics of allergens, which consists of computer science, statistics, mathematics, and engineering aiming to dissect the biochemistry and immune properties of allergens at multiple levels and which is the topic of this book.

1.1.1 Identification of New Allergenic Determinants

The diagnosis and treatment of allergic diseases is inseparable from allergens, thus, the preparation of crude allergenic extracts was the initial step in allergen characterization and are mainly made from pollens, fungi, mites, and certain food ingredients. The crude extracts can be used for different diagnostic tests or for in vitro assays. Currently, Beijing Union Medical College Hospital has isolated and purified more than 100 kinds of allergens, and the Second Affiliated Hospital of Guangzhou Medical University has also purified more than 60 kinds of allergens.

1.1.2 Allergen Component Identification

Traditional diagnostic tests for allergies usually use natural extracts that can result in false positives because they contain pan-allergens. In order to improve the accuracy of diagnosis and treatment, the identification of specific allergens is imperative. Over the last 20 years, the development of cloning and expression systems, as well as novel techniques in molecular biology and immunology, propelled allergen discovery and characterization. In addition, various allergenic components have been identified, and at present more than 100 types of related allergenic components are available (Kazemi-Shirazi et al. 2002). Furthermore, in terms of specific immunotherapy, European guidelines emphasize the importance of treating multi-allergen-sensitized patients with a single dominant allergen, and clinical trials have confirmed that certain major allergens can be used for both diagnosis (Bublin et al. 2010) and for allergen-specific immunotherapy. Interestingly, this approach results in the same effects as immunotherapy with an allergen extract (Mobs et al. 2010).

1.1.3 Limiting the Number of Major Representative Allergens

Numerous allergenic sources and major allergens have been condensed into fewer major representative molecules (He et al. 2014; Jenkins et al. 2005; Radauer and Breiteneder 2006; Radauer et al. 2008), thereby allowing the use of a smaller number of allergens for the conducting of comprehensive allergen testing and immunotherapy. Currently, the major allergens have been reduced to 21 representative allergens, which have been further divided into seven structural classes, each of which contains similar structural components (He et al. 2014).

1.1.4 Allergen Classifiers and Software

Sequence comparison of novel proteins for similarity to known allergens is used to ascertain the safety of the gene(s) to be transformed in several genetically modified crops. Over the past 10 years, several sophisticated bioinformatic methods have recorded substantial progress in allergen prediction (Barrio et al. 2007; Fiers et al. 2004; Muh et al. 2009; Saha and Raghava 2006; Zhang et al. 2007, 2012). Among the allergen prediction methods, SORTALLER predicts allergenic properties by using a novel algorithm on AFFPs (Allergen Family Featured Peptides). AFFPs are allergen-specific peptides panned from nonredundant allergens and harbor perfect information with noise fragments eliminated because of their similarity to non-allergens (Zhang et al. 2012). AFFPs are substantially optimized on most of the SVM (support vector machines)-based classifier parameters (Webb-Robertson et al. 2008). Thus, SORTALLER performs significantly better than other existing software and attains an optimal balance of high specificity (98.4 %) and sensitivity (98.6 %) for discriminating allergenic proteins from several independent datasets of protein sequences of diverse sources, and it is also highlighted by a Matthews correlation coefficient (MCC) as high as 0.970, fast running speed and the ability to rapidly predict a batch of amino acid sequences with a single click (Zhang et al. 2012). In other words, the more efficiently the classifier behaves, the more accurate the allergen predicting method is.

1.1.5 *Identification of Cross-Reactivity among Different Allergens*

Cross-reactivity among different allergens occurs on several levels. First, the primary amino acid structure of different allergens may be homologous (Jenkins et al. 2005). Based on the similarity of their amino acid sequence, major antigens from different species can be divided into clusters with obvious differences and then further subdivided into 21 representative allergens with extremely low amino acid sequence similarity (He et al. 2014). Second, pan-allergens, such as profilin and polcalcin, are conserved in the whole evolutionary tree and play an important role in cross-reactivity (Mas et al. 2014).

1.1.6 *Classification of Allergens*

As the prevalence of allergic disease is on the rise, the current use of recombinant allergens for diagnostic tests and immunotherapy is in high demand. Most allergen genes/amino acid sequences can be obtained from public databases. Bioinformatics cluster analysis of allergens downloaded from public databases showed that allergens could be grouped into seven or eight large clusters according to their kinship distance (He et al. 2014; Tao et al. 2007; Tao and He 2004; Wu et al. 2009). In addition, it has been shown that numerous allergenic sources and major allergens can be condensed into fewer major representative ones, thus allowing for the use of a smaller number of allergens when conducting comprehensive allergen testing and immunotherapy treatments (He et al. 2014; Tao et al. 2007). Much clinical evidence has also proven that specific immunotherapy with one type of allergenic reagent, or allergen-component resolved diagnosis and immunotherapy can prevent both the progression of allergies and the acquisition of new allergic sensitizations (Bublin et al. 2010; Mobs et al. 2010; Zhang et al. 2012).

When their 3D structure was assessed, whole allergens interestingly fell into seven or eight structural classes (He et al. 2014; Radauer et al. 2008), each of which contained similar structural components (He et al. 2014). This result most likely explains why similar clinical symptoms are triggered by different allergens.

Recently, with the rapid developments in bioinformatics, computational classification of proteins has revolutionized allergen research by guiding experimental design and, therefore, bioinformatic classifiers for allergen sequence discrimination should be a focus of allergy research.

1.2 Bioinformatics Application in Allergy

Over the last 20 years, information technology generated by computer software and network analysis has been increasingly used and has spawned the emerging interdisciplinary field of bioinformatics. Bioinformatics uses computation and biological theories to analyze a variety of proteins, nucleic acid sequences and even scientific literature through collection, processing, storage, transmission and retrieval analyses, and interpretation (Medigue and Moszer 2007; Peitz et al. 2002). With the rapid increase of raw data and associated allergens, the importance of bioinformatics in allergy has become increasingly valuable and is now in use for the tasks described below.

1.2.1 Homology Prediction

Sequence similarity analysis using bioinformatics is the simplest and fastest way to predict possible sensitization by comparing the homology of the product expressed with the genes of known allergens. FASTA and BLAST are the most commonly used tools for sequence comparison (Fasoli et al. 2009; Hileman et al. 2002; Platt et al. 2014). FASTA is the best for small local matching searches, and the local database is easy to use. Therefore, the FASTA program is used more often to analyze similarity to known allergens.

The bioinformatics section for the determination of protein sensitization formulated by the United Nations Food and Agriculture Organization, the World Health Organization (FAO/WHO), and the International Life Sciences Institute (ILSI) contains the following two criteria: First, at least six contiguous amino acids of the unknown protein are identical to the amino acids sequence of known allergens (Fiers et al. 2004; Kleter and Peijnenburg 2003). Second, the homology between them has more than 35 % shared identity in a sliding window of 80 amino acids. Meeting either of these criteria indicates potential for cross-reactivity. The allergenicity of the newly expressed protein in the genetically modified organisms can be predicted from the positive results of these two criteria, according to probability studies. However, when judging protein allergenicity by homology with eight or more contiguous amino acids there are less false positives (Silvanovich et al. 2006). And sequence similarity analysis does not include conformational epitope analysis. Thus, to determine the allergenicity of the encoded protein more accurately, it is necessary to conduct further serologic tests to verify the binding of specific IgE to the protein encoded by the candidate gene.

1.2.2 Reducing the Number of Allergens for Cross-Reactivity

Cross sensitivity between some plant food allergens such as vegetables and fruits often exists among patients with allergies to pollen (Hirano et al. 2013; van Rhijn et al. 2013). The results from bioinformatic analyses show that the allergy symptoms caused by most plant proteins are very similar (Lin et al. 2012). These results show that the allergens' structures are highly similar to each other and suggest that the conserved structure and biological activity may play an important role in the characteristics of the allergic symptoms seen with any particular allergen. Structural bioinformatics analysis showed that the conserved sequence of the 3D structure should be included in the assessment of any potential cross-allergenicity because even a single conserved region on the surface of the protein structure can generate a cross-reaction (Ginalski et al. 2003; Oezguen et al. 2008). In addition, through the assessment of the conserved 3D structure of a newly developed food protein, we can predict whether there are cross-reactions with other allergens (Ivanciuc et al. 2002). This could be of great significance in the prevention of severe food allergies.

It is difficult to find a new allergen that has less than 50 % homology with other known allergens. This has been one of the greatest achievements of international allergy academia. It also shows that the identification of new allergens has come to an end. The existence of cross reactive allergens or pan-allergens is considered to be the main reason why patients are allergic to a variety of allergens (Pfiffner et al. 2012; Sicherer 2001). Cross-reactivity occurs extensively at three levels, which provides the theoretical basis for the "redundant" data: (1) A lot of allergens have homology in their amino acid structures; (2) Some proteins (such as the pan-allergen profilin) are quite conserved throughout the whole evolutionary tree from plants to animals and play an important role in the cross-reactivity of allergens; (3) Allergens from different species form similar antigenic/allergenic epitopes because they share the same spatial structure characteristics. More and more literature has shown that, when compared with desensitization therapy using preparations containing all allergens of the species, simple desensitization therapy using only the major allergen can have the same effect. Therefore, we can improve allergen-specific immunotherapy treatment by reducing the number of allergens through bioinformatics analysis to obtain the main representative allergens.

1.2.3 Development of Allergen Database

Presently, there are several comprehensive allergen databases in use in the world. These include: (1) the WHO/IUIS allergen database (World Health Organization and International Union of Immunological Societies), which was established by the World Health Organization and the International Union of Immunological Societies Allergen Nomenclature Committee, (2) the Allergome database, which is a database website that was established in order to provide appropriate information about allergens, (3) the AllergenOnline database, which provides manual reviewing of the allergen list and a database with sequence searching capability and (4) the ALLERGENIA database, which contains a large number and type of allergens, has good nonredundancy, correctness, and astringency, and has friendly analyzing functions that make it easy to use.

1.3 Allergenicity Evaluation and Modification

Predicting antigen allergenicity is essential in molecular biology experiments, such as choosing a candidate gene for transforming, developing synthetic peptide vaccines, preparing of diagnostic reagents, and screening of monoclonal antibodies (Radauer et al. 2008). It is now possible to modify the high allergenicity of a protein encoded by gene(s) with fancy medical/agronomic traits (Fiers et al. 2004). After differentiating *Allergenicity* from *Antigenicity* and *Immunogenicity*, this book describes procedures and methods for the evaluation and modification of allergenicity.

1.3.1 Antigenicity, Allergenicity, and Immunogenicity

This book proposes that antigenicity includes two types of immune phenomena, allergenicity and immunogenicity. Allergenicity refers to the ability of an antigen to trigger a hypersensitivity immune response and, in most cases, without any protective/prophylactic effect. Allergenicity involves mast cells, basophils, lymphocytes, and other cellular reactions. Immunogenicity refers to the ability of an antigen to trigger protective immune responses that include immune defense, immune homeostasis, and immune surveillance. The future goal of specific immunotherapy is to reduce the allergenicity of allergens without diminishing their immunogenicity, as this encourages the development of immune tolerance while reducing any adverse reactions during the desensitization treatment so that the immunotherapy treatment has the best chance of success.

1.3.2 Predicting Binding Affinity of MHC with Allergen Peptides

The key step in the initiation of the adaptive immune response is recognition and binding to the antigen peptide by MHC molecules, i.e., T cells do not recognize unbound antigens and can only recognize the antigen when presented by MHC (Zinkernagel and Doherty 1974). The binding affinity of MHC with allergen peptides can be used to predict allergenicity, however, the traditional screening for and synthesis of dominant peptides with higher allergenicity is cost- and time-consuming. Currently, using bioinformatic models, it is possible to predict the epitope(s) of allergen peptides and the definitive MHC molecules by identifying large numbers of peptide sequences that bind to MHC molecules with high affinity (Knapp et al. 2009). In this way, the special epitopes with high/low binding affinity can be selected.

1.3.3 Designing Vaccines and Epitopes

Epitopes, also known as antigenic determinants, are immunologically active elements that exist on the surface of an antigen and have a special structure and immunological activity. They can stimulate the production of antibodies or T cell immunity. Epitopes can be divided into immuno-dominant epitopes, sub-dominant epitopes, and recessive epitopes. Epitopes are also divided into B cell epitopes and T cell epitopes. B cell epitopes are often hydrophilic peptides and are generally located at the folds of long chain amino acids of 3D macromolecular antigens (Sharon et al. 2014). T cell epitopes are usually hydrophobic peptides that are buried in the internal structure of the protein (Hoze et al. 2013).

Using molecular biology techniques, recombinant allergens that have the same immunological properties as their natural counterparts have been created, and many experiments have demonstrated that most IgE epitopes are the same in both recombinant and natural allergens. Recombinant allergen vaccines of high purity can be mass-produced. Their allergenicity can be reduced but they still retain favorable immunogenicity after modification. These recombinant allergen vaccines are comparable to natural allergen vaccines in diagnosis and treatment (Focke-Tejkl et al. 2015; Niederberger et al. 2014). Immunotherapy with modified recombinant allergens can safely achieve efficacy by reducing the formation of IgE.

Epitope-based vaccines are a new type of vaccines developed in recent years (Ghochikyan et al. 2014; Zhao et al. 2013). An epitope of a pathogenic microorganism is expressed either in vitro or by genetic engineering methods and then used as a vaccine. Epitope vaccines can be divided into B cell epitope vaccines, T cell epitope vaccines, or multi-epitope vaccines that combine these two kinds of epitopes. Epitope vaccines are also divided into synthetic peptide vaccines, recombinant epitope vaccines, and epitope DNA vaccines (Smith et al. 2014).

In the process of epitope vaccine development, the first and most important task is the screening and identification of the various epitopes and then selecting those that provide the required immune response.

1.4 Conclusion

Bioinformatics is an important frontier in contemporary life sciences and natural sciences and will be an essential part of the natural sciences in the twenty-first century. Bioinformatics has been widely used in various areas of research such as genome, immunology, and drug development, has generated considerable economic and social benefits, and has played a very important role in allergy research. Because it is fast and efficient, bioinformatics is necessary in any in-depth allergy study and has played an important role in the evaluation of some allergens. But from another perspective, as a new discipline, bioinformatics still has some problems and deficiencies. The database is the necessary foundation for any bioinformatic analysis and a general database cannot meet the needs of professionals and their disciplines. There is an especially severe lack of professional databases that have been integrated with sequence data and scientific literature, and that contain impeccable analysis modules. Furthermore, the scientific literature is scattered in diverse databases of different integrities. Researchers must collect and analyze the related literature in order to determine the allergenicity of the candidate protein(s) along with sequence-based allergenicity evaluation (Zou et al. 2012). The use of bioinformatics will play a more important role in biomedical research in future and promote the development of a number of related disciplines and industries that will be of great practical importance and strategic significance. However, validation of bioinformatics-related hits using traditional biological and immunological methods is still necessary to increase meaningful results.

References

Barrio AM, Soeria-Atmadja D, Nister A, Gustafsson MG, Hammerling U, Bongcam-Rudloff E. EVALLER: a web server for *in silico* assessment of potential protein allergenicity. Nucleic Acids Res. 2007;35(suppl_2):W694–700.

Bublin M, Pfister M, Radauer C, Oberhuber C, Bulley S, Dewitt A, Lidholm J, Reese G, Vieths S, Breiteneder H, et al. Component-resolved diagnosis of kiwifruit allergy with purified natural and recombinant kiwifruit allergens. J Allergy Clin Immunol. 2010;125(3):687–94 (94.e1).

Fasoli E, Pastorello EA, Farioli L, Scibilia J, Aldini G, Carini M, Marocco A, Boschetti E, Righetti PG. Searching for allergens in maize kernels via proteomic tools. J Proteomics. 2009;72(3):501–10.

Fiers M, Kleter G, Nijland H, Peijnenburg A, Nap J, van Ham R. Allermatch, a webtool for the prediction of potential allergenicity according to current FAO/WHO Codex Alimentarius Guidelines. BMC Bioinf. 2004;5:133.

Focke-Tejkl M, Weber M, Niespodziana K, Neubauer A, Huber H, Henning R, Stegfellner G, Maderegger B, Hauer M, Stolz F, et al. Development and characterization of a recombinant, hypoallergenic, peptide-based vaccine for grass pollen allergy. J Allergy Clin Immunol. 2015;135(5):1207–17 (e1–e11).

Ghochikyan A, Petrushina I, Davtyan H, Hovakimyan A, Saing T, Davtyan A, Cribbs DH, Agadjanyan MG. Immunogenicity of epitope vaccines targeting different B cell antigenic determinants of human alpha-synuclein: feasibility study. Neurosci Lett. 2014;560:86–91.

Ginalski K, Elofsson A, Fischer D, Rychlewski L. 3D-Jury: a simple approach to improve protein structure predictions. Bioinformatics. 2003;19(8):1015–8.

He Y, Liu X, Huang Y, Zou Z, Chen H, Lai H, Zhang L, Wu Q, Zhang J, Wang S, et al. Reduction of the number of major representative allergens: From clinical testing to 3-dimensional structures. Mediat Inflamm. 2014;2014:291618.

Hileman RE, Silvanovich A, Goodman RE, Rice EA, Holleschak G, Astwood JD, Hefle SL. Bioinformatic methods for allergenicity assessment using a comprehensive allergen database. Int Arch Allergy Immunol. 2002;128(4):280–91.

Hirano K, Hino S, Oshima K, Okajima T, Nadano D, Urisu A, Takaiwa F, Matsuda T. Allergenic potential of rice-pollen proteins: expression, immuno-cross reactivity and IgE-binding. J Biochem. 2013;154(2):195–205.

Hoze E, Tsaban L, Maman Y, Louzoun Y. Predictor for the effect of amino acid composition on CD4+ T cell epitopes preprocessing. J Immunol Methods. 2013;391(1–2):163–73.

Ivanciuc O, Schein CH, Braun W. Data mining of sequences and 3D structures of allergenic proteins. Bioinformatics. 2002;18(10):1358–64.

Jenkins JA, Griffiths-Jones S, Shewry PR, Breiteneder H, Mills EN. Structural relatedness of plant food allergens with specific reference to cross-reactive allergens: an in silico analysis. J Allergy Clin Immunol. 2005;115(1):163–70.

Kazemi-Shirazi L, Niederberger V, Linhart B, Lidholm J, Kraft D, Valenta R. Recombinant marker allergens: diagnostic gatekeepers for the treatment of allergy. Int Arch Allergy Immunol. 2002;127(4):259–68.

Kleter GA, Peijnenburg AA. Presence of potential allergy-related linear epitopes in novel proteins from conventional crops and the implication for the safety assessment of these crops with respect to the current testing of genetically modified crops. Plant Biotechnol J. 2003;1(5):371–80.

Knapp B, Omasits U, Bohle B, Maillere B, Ebner C, Schreiner W, Jahn-Schmid B. 3-Layer-based analysis of peptide-MHC interaction: in silico prediction, peptide binding affinity and T cell activation in a relevant allergen-specific model. Mol Immunol. 2009;46(8–9):1839–44.

Lin J, Bruni FM, Fu Z, Maloney J, Bardina L, Boner AL, Gimenez G, Sampson HA. A bioinformatics approach to identify patients with symptomatic peanut allergy using peptide microarray immunoassay. Journal Allergy Clin Immunol. 2012;129(5):1321–8 (e5).

Mas S, Garrido-Arandia M, Batanero E, Purohit A, Pauli G, Rodriguez R, Barderas R, Villalba M. Characterization of profilin and polcalcin panallergens from ash pollen. J Investig Allergol Clin Immunol. 2014;24(4):257–66.

Medigue C, Moszer I. Annotation, comparison and databases for hundreds of bacterial genomes. Res Microbiol. 2007;158(10):724–36.

Mobs C, Slotosch C, Loffler H, Jakob T, Hertl M, Pfutzner W. Birch pollen immunotherapy leads to differential induction of regulatory T cells and delayed helper T cell immune deviation. J Immunol. 2010;184(4):2194–203.

Muh H, Tong J, Tammi M. AllerHunter: a SVM-pairwise system for assessment of allergenicity and allergic cross-reactivity in proteins. PLoS ONE. 2009;4(6):e5861.

Niederberger V, Eckl-Dorna J, Pauli G. Recombinant allergen-based provocation testing. Methods. 2014;66(1):96–105.

Oezguen N, Zhou B, Negi SS, Ivanciuc O, Schein CH, Labesse G, Braun W. Comprehensive 3D-modeling of allergenic proteins and amino acid composition of potential conformational IgE epitopes. Mol Immunol. 2008;45(14):3740–7.

Peitz M, Pfannkuche K, Rajewsky K, Edenhofer F. Ability of the hydrophobic FGF and basic TAT peptides to promote cellular uptake of recombinant Cre recombinase: a tool for efficient genetic engineering of mammalian genomes. Proc Natl Acad Sci USA. 2002;99(7):4489–94.

Pfiffner P, Stadler BM, Rasi C, Scala E, Mari A. Cross-reactions vs co-sensitization evaluated by in silico motifs and in vitro IgE microarray testing. Allergy. 2012;67(2):210–6.

Platt M, Howell S, Sachdeva R, Dumont C. Allergen cross-reactivity in allergic rhinitis and oral-allergy syndrome: a bioinformatic protein sequence analysis. Int Forum Allergy Rhinology. 2014;4(7):559–64.

Radauer C, Breiteneder H. Pollen allergens are restricted to few protein families and show distinct patterns of species distribution. J Allergy Clin Immunol. 2006;117(1):141–7.

Radauer C, Bublin M, Wagner S, Mari A, Breiteneder H. Allergens are distributed into few protein families and possess a restricted number of biochemical functions. J Allergy Clin Immunol. 2008;121(4):847–52 (e7).

Ruffoni S, Barberi S, Bernardo L, Ferrara F, Furgani A, Tosca MA, Schiavetti I, Ciprandi G. Anaphylaxis in pediatric population: a 1-year survey on the Medical Emergency Service in Liguria, Italy. Int J. Immunopathol Pharmacol. 2015.

Saha S, Raghava GPS. AlgPred: prediction of allergenic proteins and mapping of IgE epitopes. Nucleic Acids Res. 2006;4(suppl_2):W202–9.

Sharon J, Rynkiewicz MJ, Lu Z, Yang CY. Discovery of protective B-cell epitopes for development of antimicrobial vaccines and antibody therapeutics. Immunology. 2014;142(1):1–23.

Sicherer SH. Clinical implications of cross-reactive food allergens. J Allergy Clin Immunol. 2001;108(6):881–90.

Silvanovich A, Nemeth MA, Song P, Herman R, Tagliani L, Bannon GA. The value of short amino acid sequence matches for prediction of protein allergenicity. Toxicol Sci Off J Soc Toxicol. 2006;90(1):252–8.

Smith HA, Rekoske BT, McNeel DG. DNA vaccines encoding altered peptide ligands for SSX2 enhance epitope-specific CD8+ T-cell immune responses. Vaccine. 2014;32(15):1707–15.

Tao A, Zou Z, Lai H, Huang B, Zhang J, Li Y. Species distribution and progressive clustering of the major allergens worldwide (in Chinese with English abstract). Chin J Microbiol Immunol. 2007;27(5):451–6.

Tao AL, He SH. Bridging PCR and partially overlapping primers for novel allergen gene cloning and expression insert decoration. World J Gastroenterol. 2004;10(14):2103–8.

van Rhijn BD, van Ree R, Versteeg SA, Vlieg-Boerstra BJ, Sprikkelman AB, Terreehorst I, Smout AJ, Bredenoord AJ. Birch pollen sensitization with cross-reactivity to food allergens predominates in adults with eosinophilic esophagitis. Allergy. 2013;68(11):1475–81.

Wang RQ, Zhang HY. Two hundred thousands results of allergen specific IgE detection. Chin J Allergy Clin Immunol. 2012;6(1):18–23.

Webb-Robertson B-JM, Cannon WR, Oehmen CS, Shah AR, Gurumoorthi V, Lipton MS, Waters KM. A support vector machine model for the prediction of proteotypic peptides for accurate mass and time proteomics. Bioinformatics. 2008;24(13):1503–9.

Wu Q, Zhang J, Wang S, Zhang J, Tao A, Sun B, Ivanciuc O, Garcia T, Torres M, Schein CH, et al. Characteristic motifs for families of allergenic proteins. Mediators Inflamm. 2009;46(4):559–68.

Yang L, Clements S, Joks R. A retrospective study of peanut and tree nut allergy: Sensitization and correlations with clinical manifestations. Allergy Rhinol (Providence, R.I.). 2015;6(1):39–43.

Zhang L, Huang Y, Zou Z, He Y, Chen X, Tao A. SORTALLER: predicting allergens using substantially optimized algorithm on allergen family featured peptides. Bioinformatics. 2012;28(16):2178–9.

Zhang ZH, Koh JLY, Zhang GL, Choo KH, Tammi MT, Tong JC. AllerTool: a web server for predicting allergenicity and allergic cross-reactivity in proteins. Bioinformatics. 2007;23(4):504–6.

Zhao L, Zhang M, Cong H. Advances in the study of HLA-restricted epitope vaccines. Human Vaccines Immunotherapeutics. 2013;9(12):2566–77.

Zinkernagel RM, Doherty PC. Restriction of in vitro T cell-mediated cytotoxicity in lymphocytic choriomeningitis within a syngeneic or semiallogeneic system. Nature. 1974;248(5450):701–2.

Zou Z, He Y, Ruan L, Sun B, Chen H, Liu S, Yang X, Tao A. A bioinformatic evaluation of potential allergenicity of 85 candidate genes in transgenic organisms. Chin Sci Bull. 2012;57(15):1824–32.

Author's Biography

Dr. Shan Wang is a Docent and Junior Research Scientist at the Guangdong Provincial Key Laboratory of Allergy & Clinical Immunology, the State Key Clinical Specialty in Allergy, the State Key Laboratory of Respiratory Disease, the Second Affiliated Hospital of Guangzhou Medical University. Email: april0860@163.com.

Dr. Wang received her PhD from South China University of Technology, working on the interaction of immunocytes and tumor cells as well as novel cell death pathways. She is currently working at Guangzhou Medical University. Her main research interests focus on disease associations, including the relationship between food allergy and autoimmune diseases, the pathogenesis of autoimmune diseases, and also the phenomenon of cell-in-cell in inflammation and cancer. Dr. Wang has published several scientific papers in internationally well-recognized journals. She has participated in several projects of the Science Foundation of National and Guangdong Provincial and presided over a project from the National Natural Science Foundation.

Ailin Tao is a Professor at Guangzhou Medical University, Director of Guangdong Provincial Key Laboratory of Allergy and Clinical Immunology, Principal Investigator of the State Key Laboratory of Respiratory Disease, Deputy Director of the State Key Clinical Specialty in Allergy of the Second Affiliated Hospital of Guangzhou Medical University, Member of the State Committee for Transgenic Safety Assessment, Standing Committee Member of Allergy Branch of Guangdong Medical Association, Member of Guangdong Provincial Committee for Transgenic Safety Assessment.

Prof. Ailin Tao earned his doctorate degree from the State Key Laboratory of Crop Genetic Improvement of Huazhong Agricultural University in 2002, followed by a postdoctoral training at Postdoctoral Station of Basic Medicine in Shantou University Medical College, majoring in allergen proteins. His most recent research has been on allergy bioinformatics, allergy and clinical immunology, and disease models such as allergic asthma, allergic rhinitis, infection and inflammation induced by allergy, inflammatory and protracted diseases caused by antigens or superantigens. He has gained experience in the field of allergology including the mechanisms of immune tolerance, allergy-triggering factors and chronic inflammation pathways and allergenicity evaluation and modification for food and drugs. He proposed some new concepts including "Representative Major Allergens," "Allergenicity Attenuation" of immunotoxin and allergens, "broad-spectrum immunomodulator" as well as the theoretical hypothesis of "Balanced Stimulation by Whole Antigens." Prof. Tao's laboratory focuses on the diagnosis of allergic disease and the medical evaluation of food and drug allergenicity and its modification. Prof. TAO has now constructed a system for the prediction, quantitative assessment, and simultaneous modification of epitope allergenicity, which has been applied to more than 20 allergens, and he also developed a bioinformatics software program for allergen epitope prediction, SORTALLER (http://sortaller.gzhmu.edu.cn), which performed significantly better than other existing software, reaching a perfect balance of high specificity (98.4 %) and sensitivity (98.6 %) for discriminating allergenic proteins from several independent datasets of protein sequences of diverse sources. Furthermore, this program has a Matthews correlation coefficient as high as 0.970, a fast running speed and can rapidly predict a set of amino acid sequences with a single click. The software has been frequently used by researchers from many institutions in China and over 30 countries worldwide, thus becoming the number one allergen epitope prediction software program. Prof Tao has set up an allergen database ALLERGENIA (http://ALLERGENIA.gzhmu.edu.cn) that has several advantages over other databases such as a wide selection of nonredundant allergens, excellent astringency, and accuracy, and friendly and usable analytical functions.

Eyal Raz, MD, Professor Eyal Raz is an Allergist and Immunologist whose work is focused on the fundamentals of innate immunity and the crosstalk between innate and adaptive immunity. His research group studies the TLR and the non-TLR (e.g., TRP) signaling pathways by which the host recognizes and responds to microbial agents and environmental cues, especially at mucosal surfaces. They study the principles of host-commensal interaction in the G-I tract and explore its impact on various inflammatory conditions and their transformation to cancer as well as the outcome of this interaction on the development and differentiation of CD4 T cells. Their work further examines the consequence of this activation on antigen presentation by dendritic cells and its impact on priming of naïve CD4 and CD8 T cells, as well as the generation of memory T cell responses. Using a similar strategy, they have investigated the role of transient receptor potential (TRP) family members in inflammation. This basic science approach helps in the development of vaccines against various infectious agents (e.g., HIV), tumors and allergic diseases (e.g., asthma). Some of the therapeutic approaches that have been developed are already in use in humans. Similarly, Dr. Raz's laboratory explores the role of innate immunity in the inhibition of certain inflammatory conditions. In particular, they study the principles of host-commensal interaction in the G-I tract and explore how this interaction affects various inflammatory intestinal conditions and their human analogs such as inflammatory bowel disease and colitis-associated cancer. Dr. Raz has published more than 10 papers in Nature, Science, Nature Medicine, Nature Cell Biology, Nature Biotechnology and Nature Immunology and many others in internationally well-recognized journals, and during 2002–2012 achieved a total impact factor of 400 points.

Chapter 2
Allergic Disease Epidemiology

Juan Wang, Junshu Wu and He Lai

Abstract Allergic disease represents a spectrum of disorders characterized by abnormal sensitivity mediated by IgE. Approximately, 25 % of the population in industrialized countries suffers from some form of allergic disease such as allergic rhinitis (AR) or hay fever, allergic asthma, food allergy, allergic skin inflammation, and anaphylaxis, particularly in children and young adults. The sequelae of allergies may present in many organ systems and the manifestations of allergic disease are often associated with symptoms at multiple sites. In order to facilitate the implementation of effective treatment and prevention strategies, it is important and necessary for clinicians and other involved personnel to understand the epidemiology of allergic disease. This chapter focuses on the epidemiology of allergies causing asthma, AR, chronic urticaria (CU), eczema, drug allergies, IgE-mediated food allergies, allergic conjunctivitis, Henoch–Schönlein purpura, and eosinophilic gastroenteritis.

Keywords Allergic disease · Allergen · Allergic rhinitis · Epidemiology · Meta-analysis

J. Wang · H. Lai (✉)
Guangdong Provincial Key Laboratory of Allergy and Clinical Immunology, The State Key Clinical Specialty in Allergy, The State Key Laboratory of Respiratory Disease, The Second Affiliated Hospital of Guangzhou Medical University, 250# Changgang Road East, Guangzhou, China
e-mail: 1102796746@qq.com

J. Wang
e-mail: wangjuan20110808@163.com

J. Wu
The Ophthalmic Department, The Second Affiliated Hospital of Guangzhou Medical University, 250# Changgang Road East, Guangzhou, China
e-mail: wujs78@126.com

A. Tao and E. Raz (eds.), *Allergy Bioinformatics*, Translational Bioinformatics 8,
DOI 10.1007/978-94-017-7444-4_2

2.1 Introduction

Allergic diseases are common and their incidence has been continuously rising with the developments in technology and increasingly severe environmental pollution. Allergic diseases have a large social and economic impact that include the costs of health care, lost work and school hours, and lower quality of life. This occurs not only in industrialized and developed countries but also in the vast impoverished areas around the globe (De Sario et al. 2013).

Statistics show that respiratory diseases have been increasing all over the world, especially allergic respiratory disease whose incidence is rising at an alarming rate (Hjern 2012). The World Allergy Organization (WAO) reports that approximately one of five people suffer from some form of allergic disease such as allergic rhinitis (AR), asthma, conjunctivitis, eczema, food allergies, drug allergies, and other severe allergic reactions. The incidence of allergic diseases among on-duty soldiers in Switzerland has increased nearly three times in the past 30 years (Braback 2012). In short, incidences of allergic diseases have been substantially rising around the world in both developed and developing countries (Dimitrov et al. 2014). Allergies, as an increasingly severe health problem, have become a great concern of both governments and individuals. Allergy has been called "the twenty-first century disease."

2.2 The Ubiquitous Allergens

A significant risk factor for the increasing incidence of allergic diseases is the pervasive presence of allergens (Hernandez-Cadena et al. 2015). Humans are exposed to 8500 kinds of compounds and approximately 2800 of them, including certain cosmetics, are contact allergens (Yue et al. 2009). That is, a large number, but only a few of these contact allergens, actually cause allergic reactions in clinical trials and reactions to inhalant allergens and important food allergens are actually rare.

Inhalant allergens and important food allergens in China are consistent with those reported in other countries (Wang and Zhang 2012). The common inhalant allergens in China include *Dermatophagoides pteronyssinus* (D.p.), *Dermatophagoides farina* (D.f.), Artemisia pollen, *Humulus* pollen, *Alternaria*, ash pollen, cypress pollen, ragweed pollen, birch pollen, cockroaches, *Platanus Hispanica* pollen, cocklebur pollen, house dust, *Cladosporium herbarum*, dog dander, cat dander, feather grass pollen, *Aspergillus fumigatus*, etc. The important food allergens include eggs, milk, peanuts, soybeans, shrimp, crab, and some nuts, grains, and fruits. A six-year survey on allergic diseases in Guangzhou (Sun and Zheng 2014), a city in south China, revealed that at least 15 types of allergens are common in Guangzhou: D.p. and D.f. are the most common among inhalant allergens, while egg and milk are the most common food allergens. The most common allergens vary among different age groups; 9–18 year olds are most allergic to

D.p., D.f., and *Blomia tropicalis*, 3–6 year olds are most allergic to eggs and those younger than 3 years of age are most allergic to milk (Sun and Zheng 2014). The rate of positive sIgE against D.p. among patients with positive sIgE against eggs or milk increases with their age (Sun and Zheng 2014). It has been reported that the most common inhalant allergens in Changsha are flour mite and D.p. and the most common food allergen is shrimp (Lu et al. 2011).

A few new allergens have been found in recent years. Smith et al. identified two cat-derived allergens: the taste gland protein Fel d 7 and the latherin-like protein Fel d 8 (Smith et al. 2011); Ma et al. first identified two IgE binding proteins: Tab a 1 and Tab a 2 (Ma et al. 2011; An et al. 2013) identified eight D.f. allergens; Ayuso et al. identified two novel shrimp allergens (Yue et al. 2009; Zhang and Zhang 2014) and a tropomyosin-like allergen of sea urchin (*Strongylocentrotus purpuratus*) (XP_001192266), which showed only 22 % sequence identity and 35 % similarity with Lit v 1 (ACB38288.1) (Bergon-Sendin et al. 2014). A new major allergen was isolated from dog urine and identified as prostatic kallikrein (Begum et al. 2012) and a closely related or identical protein was detected in dog dander. The recombinant form of prostatic kallikrein displayed similar immunologic and biochemical properties to those of the natural protein and bound IgE antibodies from 70 % of the subjects with dog allergy. The dog allergen kallikrein was also found to cross-react with human prostate-specific antigen, a key culprit in IgE-mediated vaginal reactions to semen (Begum et al. 2012). Thus, allergen exposure is very common (Torres et al. 2014).

2.3 The Epidemiology of Common Allergic Diseases

Allergic diseases occur in people of all ages, from newborns to the elderly, and often in those with a genetic predisposition. Allergic diseases are often characterized by immediate allergic reactions and are mainly manifested as respiratory allergies, skin allergies, digestive tract allergies and anaphylactic shock. Common clinical allergic diseases include asthma, AR, allergic dermatitis, food allergies, allergic conjunctivitis, allergic purpura, and eosinophilic gastroenteritis. Understanding the epidemiology of common allergic diseases would provide a reference for their prevention and treatment.

2.3.1 Asthma

The incidence of asthma has been growing rapidly since the 1960s: from 2 to 10 % in Switzerland; from 7.3 to 8.2 % in the US during 2001–2009 with 17,700,000 cases of adult asthma and 7,000,000 cases of child asthma recorded in 2010; and by 38 % in Italy in the 20 years from 1991 to 2010 (de Marco et al. 2012). According to WHO estimates, 235 million people in the world currently suffer from asthma (Bergon-Sendin et al. 2014), with over 50 % of the adult cases and over 80 % of the infant cases caused by house dust mite, with 250,000 reported deaths (Begum et al. 2012).

Due to the variations in diagnostic criteria among the different epidemiological studies, the incidence of confirmed and suspected asthma varies among different countries. In addition, the incidence of asthma in some population groups may be significantly higher than in others. Statistics have shown that the risk of asthma increases with age and is negatively correlated with the education level of the parents but seems not to be correlated with household income (Hammer-Helmich et al. 2014). One-year observational study on 3761 Taiwanese children less than 12 years old indicated that the correlation between the incidence of asthma and a family member smoking plus family socioeconomic status is gender-specific (Strong and Chang 2014). Multivariate logistic regression analysis revealed that family member smoking might predict the risk of asthma in girls but not boys; girls in low-income households are more likely to have asthma (Strong and Chang 2014). In addition, the asthma risk in both boys and girls is negatively correlated with the education level of the father in Taiwan (Strong and Chang 2014). According to 2008 and 2010 statistics provided by National Health Interview Survey (Table 2.1), National Center for Health Statistics, the incidence of asthma increased in the black and white non-Spanish population in the US but decreased in other race and skin-color populations (http://www.cdc.gov/asthma/nhis/default.htm).

An epidemiological investigation of adult asthma in Asia indicated a similar increasing trend as seen on other continents (Song et al. 2014). The incidence of asthma in adult residents was < 5 % lower than that seen in European adults, while the incidence of asthma among elderly Asians was about 1.3–15.3 %, which is relatively high. This could be attributed to the aging of this population (extended life expectancy and/or reduced birth rate) and more attention should be paid to the problem of asthma among the elderly people in Asia. In China, the National Pediatric Asthma Collaborative Group carried out an asthma epidemiology survey from September 2009 to August 2010 on 463,982 children from 27 provinces, autonomous regions, and four municipalities. The result indicated that the total asthma incidence rate was 3.02 % (95 %CI, 2.97–3.06 %) in the major cities of China and the prevalence in two years (2009–2010) was 2.32 % (95 %CI, 2.28–2.37 %) (National Cooperative Group on Childhood, etc. 2013). The prevalence of asthma was significantly different among regions, cities, ages, and genders; it was higher in male children (3.51 %) than in female children (2.29 %), highest in preschool children (3–5 years old), highest in East China and lowest in Northeast China, and highest in Shanghai and lowest in Lasa. Nearly one-third of children with asthma were not diagnosed at an early stage or not diagnosed correctly, thus, treatment and management of asthma in children also awaits improvement (National Cooperative Group on Childhood, etc. 2013).

The causes of the increasing incidence of asthma are not clear but several hypotheses have been proposed. Western researchers believe that the rapid rise in asthma over the past three decades in Western societies has been attributed to

Table 2.1 Current prevalence of asthma among children and adults by sex, race/ethnicity, region, and family income

	Prevalence rate (%)[a]					
	All ages		Children		Adults	
	Total		Age 18		Age 18+	
Characteristic	2010	2006–2008	2010	2006–2008	2010	2006–2008
Total	8.5 ± 0.18	7.8 ± 0.2	9.4 ± 0.35	9.3 ± 0.4	8.2 ± 0.21	7.3 ± 0.25
Male	7 ± 0.23	6.9 ± 0.3	10.5 ± 0.52	10.7 ± 0.65	5.8 ± 0.26	5.5 ± 0.35
Female	9.9 ± 0.26	8.6 ± 0.35	8.2 ± 0.42	7.8 ± 0.6	10.4 ± 0.31	8.9 ± 0.35
White non-hispanic	8.1 ± 0.22	7.8 ± 0.3	8.2 ± 0.48	8.2 ± 0.65	8.1 ± 0.26	7.7 ± 0.35
Male	6.5 ± 0.29	6.8 ± 0.45	9.2 ± 0.75	9.5 ± 0.95	5.7 ± 0.32	5.9 ± 0.45
Female	9.7 ± 0.33	8.7 ± 0.35	7.3 ± 0.59	6.9 ± 0.85	10.3 ± 0.39	9.3 ± 0.5
Black non-hispanic	12.1 ± 0.55	9.5 ± 0.55	15.9 ± 1.07	14.6 ± 1.45	10.7 ± 0.64	7.8 ± 0.6
Male	10.8 ± 0.77	8.5 ± 0.8	18 ± 1.48	16.5 ± 1.65	7.6 ± 0.91	5.7 ± 0.85
Female	13.3 ± 0.71	10.3 ± 0.8	13.7 ± 1.49	12.7 ± 1.75	13.2 ± 0.8	9.5 ± 0.8
Other non-hispanic	8.2 ± 0.66	14.8 ± 2.15	9.5 ± 1.35	13.6 ± 2.75	7.6 ± 0.72	15.1 ± 2.65
Male	6.8 ± 0.75	12.1 ± 2.95	8.8 ± 1.55	14.6 ± 4.3	5.8 ± 0.97	11.2 ± 3.65
Female	9.5 ± 0.99	17.4 ± 3.15	10.1 ± 1.94	12.6 ± 3.85	9.2 ± 1.08	19.1 ± 4
Hispanic	7.3 ± 0.4	14.2 ± 1.85	8.1 ± 0.57	18.4 ± 3.8	6.9 ± 0.49	12.8 ± 2
Male	6.6 ± 0.56	11.3 ± 2.5	9.8 ± 0.82	23.6 ± 1.1	4.9 ± 0.7	7.0 ± 2.65
Female	8.1 ± 0.5	16.9 ± 2.4	6.3 ± 0.68	13.0 ± 3.8	9.1 ± 0.72	18.2 ± 2.85
Puerto Rican[b]	18.5 ± 1.88		19.5 ± 2.6		18.1 ± 2.55	
Male	18.9 ± 3.09		23.1 ± 4.17		16.8 ± 4.23	
Female	18.2 ± 1.77		15.9 ± 2.57		19.3 ± 2.55	
Mexican/ Mexican–American[b]	6.3 ± 0.43		6.9 ± 0.62		6 ± 0.55	
Male	5.7 ± 0.58		8.5 ± 0.86		4.1 ± 0.69	

continued

Table 2.1 (continued)

	Prevalence rate (%)[a]					
	All ages		Children		Adults	
	Total		Age 18		Age 18+	
Female	6.9 ± 0.64		5.2 ± 0.74		7.9 ± 0.95	
Region						
Northeast	8.8 ± 0.5		9.5 ± 0.74		8.6 ± 0.57	
Midwest	8.7 ± 0.4		10.2 ± 0.82		8.2 ± 0.45	
South	8.3 ± 0.29		9.9 ± 0.62		7.8 ± 0.33	
West	8.3 ± 0.36		7.8 ± 0.58		8.4 ± 0.44	
Ratio of family income to poverty threshold[c]						
0–0.99	11.2 ± 0.51		12.1 ± 0.92		10.7 ± 0.56	
1.00–2.49	8.8 ± 0.32		9.6 ± 0.59		8.5 ± 0.38	
2.50–4.49	8.2 ± 0.38		8.8 ± 0.63		8.1 ± 0.45	
4.50 and above	6.9 ± 0.31		6.9 ± 0.54		6.9 ± 0.36	

Source National Health Interview Survey, National Center for Health Statistics, CDC from 2006 to 2008 and March 1, 2012

[a]95 % confidence interval

All relative standard errors are < 30 % unless otherwise indicated

[b]As a subset of hispanic

[c] Missing responses imputed

numerous diverse factors including increased awareness of the disease, altered lifestyle and activity patterns, and ill-defined changes in environmental exposures (Gilmour et al. 2006). In addition, smoking and obesity have some correlation with asthma (Eriksson et al. 2015); children of smoking parents show a higher incidence of asthma than those of non-smoking parents and severe obesity is closely correlated with adult-female asthma (Jackson et al. 2013). Furthermore, the type and timing of microbial exposure also play an important role in the development of asthma (Wildfire et al. 2014).

The pathogenesis of asthma has not been fully clarified. Recent studies tend to consider asthma a multifactorial airway disease that arises from a relatively common genetic background interfaced with exposures to allergens and airborne irritants (Gilmour et al. 2006). It is believed to be correlated with allergy, airway inflammation, increased airflow resistance, and airway hyperresponsiveness (Gu and Zhao 2011) With the development of molecular biology in recent years, studies on asthma-related genes have made great progress. Asthma-related gene mutations have been found on chromosomes 2, 3, 5, 6, 7, 9, 11, 12, 13, 14, 17, and 19, but the specific relationship has yet to be defined (Gu and Zhao 2011). In recent years, a Genome-Wide Association Study (GWAS) has been performed on the studies of the pathogenesis of asthma. A correlation has been identified between mutations in Chromosome 17q21 and asthma in Chinese Han people (Li et al. 2012). It was also found that mutations in Chromosome 17q21 were associated with primary asthma in Northeast China Han children (Yu et al. 2014). Using the GWAS network analysis platform "Identify Candidate Causal SNPs and Pathway" (ICSNPathway), it has also been identified that four candidate Single Nucleotide Polymorphisms (SNP) sites (rs7192, rs20541, rs1058808, and rs17350764), four genes (*HLA-DRA, IL-13, ERBB2,* and *OR52J3*), and 21 related metabolic pathways correlate with incidences of asthma (Song and Lee 2013). A novel SNP (rs10044254) was identified correlating with down-regulation of *FBXL7* and increased sensitivity of asthmatic patients to glucocorticoids but emphasized that this is an important regulatory mechanism of sensitivity to glucocorticoids in children, however, not in adults (Park et al. 2014). Recent studies identified some genes that may correlate with total IgE in asthma patients (*CRIM1, ZNF71, TLN1,* and *SYNPO2*) and demonstrated a correlation between mite-specific IgE and SNPs near OPPK1 (may be related to D.p.-induced asthma) and *LOC730217* (may be related to D.f.-related asthma) (Kim et al. 2013). Understanding the pathogenesis of asthma and the reasons for its increased incidence may suggest more effective strategies for the prevention and treatment of asthma.

2.3.2 Allergic Rhinitis (AR)

AR is a common disease among children with an incidence of 15–25 % (Adamia et al. 2014) but is often overlooked, misdiagnosed, or mistreated. AR may lead to severe rhinitis and asthma and is a global health problem (Chiang et al. 2012). This disease not only causes nasal and non-nasal inflammation of the respiratory

Table 2.2 The prevalence of allergic rhinitis in adults and children in different cities in China

	Prevalence of allergic rhinitis (%)	
Cities in China	Adults	Children
Beijing	8.7	14.46
Shihezi		12.56
Urumqi	24.1	10.1
Hohhot		4.5
Xi`An	9.1	3.9
Chengdu	34.3	10.1
Chongqing	32.3	20.42
Wuhan	19.3	8.3
Changsha	16.1	
Guangzhou	7.83	14.1
Shenzhen		20.1
Harbin		4.9
Changchun	11.2	
Shenyang	15.7	
Shanghai	13.6	13.1
Nanjing	13.3	
Hangzhou	8.9	

system but also results in fatigue of the affected individual and hindered cognitive ability. Genetic factors, asthma, upper respiratory tract infection, use of antibiotics in the first year after birth, living in a grassy environment and exposure to dust, certain gas, or smoke are all risk factors for AR (Eriksson et al. 2012; Galffy et al. 2014). The most common allergens that cause AR are dust mite, pollen, herbs, *Alternaria solani* and, German cockroach (Yang et al. 2011).

Epidemiological data have shown a rapid increase in the prevalence of AR in the past decades (Yang et al. 2013). The incidence of AR was lower than 1 % in the 1920s and began to increase after the industrial revolution, slowly in the 1950s–1980s, but sharply since at least 1990 (de Marco et al. 2012). AR in wealthy African countries, Taiwan, and some Middle Eastern countries is even higher than that of Western Europe and North America (Katelaris et al. 2012). The prevalence of AR is strongly associated with asthma and the incidences of both are on the rise in both developed and developing countries (de Marco et al. 2012; Pesce et al. 2012; Sanjana et al. 2014). Regions with a high incidence of AR often show a high incidence of asthma as well (Khan 2014; Wang et al. 2012b); the incidence of asthma is < 2 % among AR-free individuals but is as high as 10–40 % among AR patients (Ozdoganoglu and Songu 2012). Also, the prevalence of comorbid allergic diseases decreased with age (Hong et al. 2012).

The incidence of AR is about 10–25 % around the world, 10–20 % in US and Europe combined, or 12–13 % of Americans and 23–30 % of Europeans (Ozdoganoglu and Songu 2012; Zhang and Zhang 2014). Eriksson et al. reported in 2012 that the incidence of AR in Sweden increased to 28 % (lower in males than in females, 26.6 versus 29.1 %) and the incidence was 33.6 % for those aged 30–40 (Eriksson et al. 2012). The incidence of AR was reported as 8.7–24.1 % in China (Zhang and Zhang 2014) and another study more specifically reported a 9.1 % incidence in Northern China (Wang et al. 2012b). Available data indicated that despite variations in the prevalence of AR in different regions of China, the prevalence of AR has increased in both adults and children over the past two decades (Zhang and Zhang 2014) (Table 2.2). The incidence of AR in Batumi of Adjara was 15.3 % and higher in boys than in girls. This difference in the incidence of AR between genders was in consistent with results reported by Chiang WC et al. (Chiang et al. 2012). According to the 2008 Nutrition and Health Survey in the Philippines, the incidence of AR in this country was 20 %, similar between males and females, higher in rural areas than in urban areas, highest in the 40–49 age group, more prevalent in May and June of the year, and similar between coastal and inland areas (Abong et al. 2012). Understanding the epidemiology of AR in different areas would provide references for further study and prevention of this disease.

2.3.3 Allergic Dermatitis

According to the differences in manifestations, inducing factors and prognosis, common atopic dermatitis is divided into chronic urticaria (CU), eczema, cold urticaria, solar dermatitis, skin scratch disease, contact dermatitis, angioedema, drug allergy, and so on. In this section, we will introduce the epidemiology of some of the most common types of atopic dermatitis.

(1) **Chronic Urticaria (CU)**

Urticaria is a common inflammatory skin disease that affects approximately 20 % of the general population (Papadopoulos et al. 2014). According to the current EAACI/GA^{2}LEN/EDF/WAO guideline, urticaria can be classified into the following two subtypes: spontaneous urticaria (SU, including acute spontaneous urticarial and chronic spontaneous urticaria) and inducible urticaria (including cold urticaria, delayed pressure urticaria, heat urticaria, solar urticaria, symptomatic dermographism, vibratory angioedema, aquagenic urticaria, cholinergic urticaria, and contact urticaria) (Zuberbier et al. 2014). Chronic spontaneous urticaria (CSU), usually called CU, is the most common subtype of all forms of non-acute urticaria, and accounts for 25 % of the cases of urticaria (Losol et al. 2014). Although accurate data on the prevalence of urticaria are unavailable, it is estimated that 15–25 % of the US population are affected at some time of their lives with urticaria and that 33 % of all urticaria cases are considered to be chronic (Rance and Goldberg 2013).

CU is defined as a clinical course over more than six weeks, with an average disease duration between two and five years (Rance and Goldberg 2013) that mainly affects the skin and is caused by degranulation of cutaneous mast cells and/or basophils and the release of histamine and other inflammatory mediators such as arachidonic acid metabolites, leukotrienes (LTC4, D4 and E4), prostaglandin D2, serotonin, acetylcholine, platelet activating factor, heparin, codeine, anaphylatoxins C3, C5a, quinones, and neurotransmitters released from cutaneous nerve endings (Criado et al. 2013). Food/food additives, drugs, psychological conditions, mosquito bites, autoreactivity, and alcohol consumption were the predisposing factors for CU (Wildfire et al. 2014). Genetic factors also play a role in occurrence of this disease, since 31.4 % of the patients with urticaria had a family history of this disease. Note that aspirin and other nonsteroidal anti-inflammatory drugs are the most common drugs that cause chronic urticarial (Losol et al. 2014). Molecular genetic mechanisms of CU have been studied in recent years. Losol P et al. (Losol et al. 2014) showed that genes involved in CU pathogenesis were those related to mast cell activation and histamine (including *FcεRI, HNMT, HRH1, HRH2, TNF-α, TGFβ1, ADORA3*, and *IL-10*), the arachidonic acid pathway (including *ALOX5, CysLTR1, LTC4S*, and *PTGER4*), HLA class I and II alleles, and other genes (*UGT1A6, CYP2C9, NAT2, ACE*, and *PTPN22*). Candidate genes and GWAS were used to further reveal the molecular mechanism of urticaria in order to provide molecular markers for the different types of urticaria and to find new therapeutic targets.

The majority of studies of CU show that it can occur in populations of all ages, with a typical onset in the third to fourth decade of life, and females are affected nearly twice as often as males (Losol et al. 2014). The incidence of CU is now up to 5 % in the general population and over 10 % in people that have certain allergies or skin diseases. 45–90 % of people with CU suffer from itching with no known cause, with women being four times more likely to suffer from itching than men (Rance and Goldberg 2013). About 40 % of CU patients experience concurrent angioedema (Losol et al. 2014). CU has a profound impact on quality of life and causes immense distress to patients, necessitating effective treatment.

(2) **Eczema**

Atopic dermatitis, also known as eczema, is a common childhood atopic disease associated with chronicity and impaired quality of life. Eczema affects about 10–20 % of children (mostly before 5 years old) and about 1–3 % of adults in the UK and the incidence of eczema has increased as much as threefold in the past 40 years (Adams et al. 2013). The International Study of Asthma and Allergies in Childhood (ISAAC) hase III study on the global variations in the prevalence of eczema symptoms in children indicated that for 6–7 year olds (data on 385,853 participants from 143 centers in 60 countries), the prevalence of current eczema ranged from 0.9 % in India to 22.5 % in Ecuador, with new data showing high rates in Asia and Latin America (Odhiambo et al. 2009). For 13–14 year olds (data on 663,256 participants from 230 centers in 96 countries), the prevalence ranged from 0.2 % in China to 24.6 % in Columbia, with the highest rates in Africa and Latin America. In both age groups, the current prevalence of eczema was lower for boys than girls (Odhiambo et al. 2009).

Eczema is a complex disease caused by multiple genetic and environmental factors. Genetic factors play a significant role in eczema. We postulated that eczema is also related to specific SNPs. The first GWAS reported noncoding rs7927894 on 11q13 to be associated with eczema in German children (Esparza-Gordillo et al. 2009). It has been reported that 11q13 and the gene encoding filaggrin (*FLG*) are important in the pathogenesis of childhood eczema (Wang et al. 2012a; Ziyab et al. 2014). Filaggrin molecules and their metabolites are crucial for skin barrier functions as they organize the keratin filaments and maintain skin hydration. Filaggrin molecules are metabolized to their constituent amino acids when the relative humidity drops below 80 % and *FLG* mutation carriers display reduced synthesis of the osmolytic "natural moisturizing factor" intended to protect the skin from drying (Thyssen 2012). A systematic study pointed out that exposure to antibiotics in the first year of life, but not prenatally, was more common in children with eczema (Tsakok et al. 2013). It has been reported that eczema was associated with high parental educational level but there was no association with household income (Hammer-Helmich et al. 2014).

The incidence of eczema is still on the rise in both developed and developing countries and approximately one-third of children with severe atopic eczema also suffer from a food allergy, whereas food allergies are rare in adult patients (Wassmann and Werfel 2015). Eczema has been correlated with food allergies, however, this correlation is often over-emphasized. The prevalence of food allergies in children with eczema is estimated to be between 33 and 63 % (Santiago 2015) but it does not mean that the food allergies are the cause of the eczema. Food allergies seem to play a key role in eczema flares, especially in those children with moderate-to-severe disease. It is true that many foods, including eggs, peanuts, cow's milk, soy, tree nuts, fish, and shellfish, may aggravate eczema in infants and children, especially in the first two years of life (Ben-Shoshan et al. 2015) and elimination diets will likely be part of the treatment (Santiago 2015).

2.3.4 Drug Allergy

Drug allergy is any reaction caused by a drug with clinical features indicating an immunological mechanism. Each year, approximately 62,000 people in England are admitted in the hospital after experiencing a serious allergic reaction to a drug and up to 15 % of inpatients have a prolonged hospital stay as a result of an adverse drug reaction (Dworzynski et al. 2014). Analysis of patient safety incidents reported to the National Reporting and Learning System between 2005 and 2013 identified 18,079 incidents involving drug allergy including six deaths, 19 "severe harms," 4980 "other harms," and 13,071 "near-misses" (Dworzynski et al. 2014). In China, where there is a long history of traditional Chinese medicine useage, herb-induced adverse reactions are also increasing, especially herb-induced allergies. However, the true incidence of drug allergic reactions in China is not known, probably due to under-reporting. Some people are never offered referral to specialist services and instead stay in primary care while others have their drug allergy managed by other health care disciplines. Therefore, only a small proportion of people are treated in specialist allergy centers.

Penicillin allergy remains the most common reported drug allergy affecting 10 % of the general population depending on the specific population evaluated (Albin and Agarwal 2014). Nonsteroidal anti-inflammatory drugs (NSAIDs) are the second most common cause of drug-induced hypersensitivity reactions and account for 21–25 % of adverse drug events (Karakaya et al. 2013). Allergic reactions to NSAIDs such as ibuprofen, diclofenac, naproxen, and aspirin are common (Nascimento-Sampaio et al. 2015). In particular, 21 % of people with asthma are affected by NSAIDs (Morales et al. 2014). About 35 % of people with CU have severe reactions to NSAIDs, involving angioedema and anaphylaxis after administration of NSAIDs (Karakaya et al. 2013). Anaphylaxis during general anesthesia occurs in one in 13,000 in France and one in 20,000–30,000 in Australia, and it is estimated by extrapolation that the incidence of anaphylaxis is 175–1000 reactions per annum in the United Kingdom (Krishna et al. 2014).

2.3.5 Food Allergy

Food allergy is a serious public health problem and studies of food allergy began in 1905. It is a reproducible adverse event that elicits a pathologic IgE-mediated or non-IgE-mediated reaction, leading to a variety of clinical symptoms that affect the patient's quality of life such as runny nose, itchy eyes, dry throat, rash, and breathing difficulties, or even fatal anaphylaxis (Dyer and Gupta 2013; Zukiewicz-Sobczak et al. 2013).

Although up to 170 kinds of foods have been proven to be able to cause allergic reactions, in reality, only a small number are responsible for most food allergies. Approximately, 90 % of all food allergies are caused by only eight types of foods: milk, eggs, peanuts, other nuts, fish, shellfish, soy, and cereals sensitizing at different frequencies (Zukiewicz-Sobczak et al. 2013). Among them, peanuts, tree nuts, finned fish, crustaceans, fruit, and vegetables account for 85 % of the food-allergic reactions in adults (O'Neil et al. 2011), while milk, wheat, fish, soy, and peanuts are the most common food allergens in children. Prevention of possible life-threatening allergies is not only an important medical issue but also a responsibility of the food industry and related supervisors. However, currently, there are no effective therapies for food allergies (Burks et al. 2012). The standard therapy for the treatment of food allergies is allergen avoidance and prompts treatment of allergic reactions that occur on accidental exposure (Umetsu et al. 2015). However, in reality, the problem of food allergies is a difficult one to control due to the increasing size of the food-allergic population.

Due to the differences in diagnostic methods, the reported rate of food allergies differs greatly among different countries. In recent years, the incidence of food allergies has been on the rise for unclear reasons but is likely due to a complex interplay between biologic, genetic, and environmental factors (Ezell et al. 2014) and may be due to the use of genetically modified products (Zukiewicz-Sobczak et al. 2013). Over the past 15 years, the prevalence of food allergy appears to have doubled or even quadrupled in the US, UK, and China (Umetsu et al. 2015). It is estimated that approximately 220–250 million people worldwide have suffered from food allergies (Umetsu et al. 2015). Food allergy affects 4–8 % of children and 5 % of adults in the US, with the prevalence of peanut allergy alone being 1.8 % of adults in the US, 2 % of 8 year old children in the UK, and 3 % of young children in Australia (Umetsu et al. 2015). One survey suggested that there was a significant association between the incidence of food allergy and race, age, income, and geographic region (Gupta et al. 2011).

2.3.6 Allergic Conjunctivitis

Allergic conjunctivitis is the most common form of ocular allergy. Approximately, 40 % of individuals in developed countries suffer from allergic conjunctivitis (Chigbu and Coyne 2015). In the United States, allergic conjunctivitis constitutes over 90 % of all ocular allergies and approximately 50 % of Americans have had at least one allergic reaction in their lives, with approximately one-third of them having allergies of the eye (Chigbu and Coyne 2015). Allergic conjunctivitis is caused by exposure of conjunctiva to allergens leading to an immune response and a series of tissue- and cell-responsive diseases that are manifested to varying degrees with symptoms including red eye (the most common sign of allergy conjunctivitis), watery eyes (88 %), itching (88 %), redness (78 %), soreness (75 %), swelling (72 %), or stinging (65 %) (Almaliotis et al. 2013), and can affect school performance and work productivity (Chigbu and Coyne 2015). According to the pathological mechanism and clinical symptoms of the allergic response on the ocular surface, ocular allergies present with a variety of types such as seasonal allergic conjunctivitis (SAC), perennial allergic conjunctivitis (PAC), vernal catarrh keratoconjunctivitis (VKC), giant papillary conjuntivitis (GPC), and atopic keratoconjunctivitis (AKC) (Almaliotis et al. 2013; Chigbu and Coyne 2015).

The incidence of allergic conjunctivitis has been on the rise in recent years, especially in developing countries, and can be partially attributed to increasingly severe environmental pollution, increased number of pets, the wearing of contact lenses, eye cosmetic use, and other factors (La Rosa et al. 2013). Due to different geographical environments and medical conditions, the incidence of ocular allergy in various regions is significantly different, from 5 to 22 %, and more recent studies report rates as high as 40 % (Almaliotis et al. 2013; Kumah et al. 2015). One questionnaire revealed that 396 12–13 year old children of a population-based sample in Sweden showed a 19.1 % incidence of allergic conjunctivitis and a 17.6 % incidence of AR with a 92 % correlation (Hesselmar et al. 2001). But in reality, the overall incidence is greater than reported.

In China, there is not a lot of data on the epidemiology of allergic conjunctivitis. Investigation on 6179 patients with moderate or severe allergic conjunctivitis showed that the central regions of China have the highest prevalence of allergic conjunctivitis (45.1 %) and in other areas including the north, south, southwest, and central-north the prevalence was 8.0 %, 18.1 %, 1.06 %, and 18.1 %, respectively (Li et al. 2008). SAC, PAC, and VKC are the most common forms of allergic conjunctivitis in China, with SAC and PAC being the major types (74.4 %). The incidence of SAC is 22.3 % in children, significantly greater than in adults (8.3 %), while the prevalence of PAC and GPC is less than 10 % in children (Li et al. 2008).

(1) **SAC, PAC, and GPC**

SAC and PAC are the most common types of ocular allergy and are self-limited (Almaliotis et al. 2013). SAC, also called hay-fever conjunctivitis and is a seasonal variant of allergic conjunctivitis, usually occurs in the spring and summer and generally abates during the winter months (La Rosa et al. 2013). SAC is the most prevalent type of allergic conjunctivitis and is usually due to outdoor aeroallergens including pollen, grass, mold spores, and other outdoor seasonal antigens (Almaliotis et al. 2013). In the US, the most common pollen allergen is ragweed, while in China it is *Artemisia* pollen (Almaliotis et al. 2013).

Generally, PAC, which can occur throughout the year with exposure to perennial allergens, is the perennial variant of allergic conjunctivitis due to indoor airborne allergens (Almaliotis et al. 2013). The incidence and clinical symptoms of PAC are lower (3.5:10,000) and milder than that of SAC (Dart et al. 1986). Aeroallergens associated with PAC patients are dust mites, pet dander, feathers, and mold (Chigbu and Coyne 2015). Compared with SAC, PAC patients are reactive to house dust (42 vs. 0 %) and the clinical symptoms are more closely aligned with perennial AR (Goldberg et al. 1998). SAC and PAC originate mostly in childhood, with males being affected more often. The reactivity can gradually disappear after puberty but about 50 % of patients relapse at the ages of 18–35. At this stage, the incidence between males and females is similar.

In 1950, GPC was reported for the first time. This disease can be caused by limbal sutures, contact lenses, ocular prostheses, and limbal dermoids (La Rosa et al. 2013) and, thus, is classified as an iatrogenic disease not, an allergic disease. Data showed that 1–5 % of patients wearing rigid gas permeable contact lenses and 15 % of patients wearing soft contact lenses suffer from GPC, which is similar to VKC. Patients from 10 to 50 years old, especially men, have suffered prolonged courses of AKC but the most severe cases occur in 30–50 year olds (Rich and Hanifin 1985).

(2) **AKC and VKC**

AKC and VKC are characterized by chronic immune inflammation with T cell infiltration and may be sight threatening. The symptoms of AKC are more serious than other ocular allergies and can be exacerbated in summer and winter (La Rosa et al. 2013). AKC is considered the ocular counterpart of allergic dermatitis, or eczema (La Rosa et al. 2013). The main symptoms of AKC are itchy eyes, dry eyes, and chronic eyelid eczema, usually with superficial punctate keratopathy or keratohelcosis. A small number of patients may have complications including conjunctival scarring, trantas dots, posterior capsular opacification or uveitis, with severe complications potentially causing blindness (La Rosa et al. 2013). AKC can persist throughout life.

VKC accounts for approximately 0.5 % of allergic eye diseases (Sehgal and Jain 1994), generally has a course of 5–10 years, and is more common in warm climates and warm weather months (La Rosa et al. 2013). VKC commonly appears in spring, summer, and autumn, with the spring plant pollens being particularly allergenic. But some patients are affected all year round. VKC occurs more often in the tropics than in northern climates and rarely in frigid zones, thus, the disease is more prevalent in sub-Saharan Africa and the Middle East (La Rosa et al. 2013). Young people are typically affected, with onset at ages 10–20 in 60 % of VKC patients and those patients typically also have a history of seasonal allergies, eczema, or asthma (La Rosa et al. 2013). In Europe, the prevalence of VKC ranges from 1.2 to 10.6 cases per 10,000, although the prevalence of associated corneal complications is much lower (0.3–2.3 per 10,000) (La Rosa et al. 2013).

2.3.7 Henoc–Schönlein Purpura (HSP)

Henoch–Schönlein purpura (HSP), also known as anaphylactoid purpura or allergic purpura, is a common autoimmune and multisystem allergic disease and is a type III hypersensitivity mediated small vessel vasculitis characterized by a clinical tetrad of specifically palpable purpura, arthralgia, abdominal pain, and renal disease, with gastrointestinal and renal involvement more prevalent in older adults (Kurdi et al. 2014). Leukocytoclastic vasculitis accompanied by immunoglobulin A (IgA) immune complexes within affected organs is the cause of these manifestations (Gupta et al. 2015).

Despite these years of study on HSP, the underlying pathogenesis of HSP remains unclear. Although some cases lack a clear precipitating event, streptococcal infections, staphylococcal infections, vaccinations, medications, and insect bites have been implicated as possible triggers (Ha et al. 2015). In adults, medications are the most common triggers and the most frequently involved medications include antibiotics, angiotensin-converting enzyme inhibitors (ACEI), angiotensin-converting enzyme II receptor antagonist (AGIIR), and NSAIDs (Gonzalez et al. 2009).

The incidence of childhood HSP is on the rise with an annual incidence of 13–20 cases per 100,000 children < 17 years old (Yang et al. 2015). Severe cases are becoming more frequent and can lead to serious consequences if not treated timely and correctly. HSP predominantly affects children and is rare in the adult population (Gupta et al. 2015). Up to 90 % of cases are in children between 4 and 11 years of age (Gupta et al. 2015). The incidence of HSP in this age group is 8–20 cases per 100,000 children annually. Approximately, 50 % of patients are < 5 years of age, 75 % of patients are < 8 years of age, and 90 % are < 10 years of age (Bluman and Goldman 2014; Lim et al. 2015), compared to 1.3 cases per 100,000 adults annually (Kang and Park 2014). Boys are affected more often than girls (male-to-female ratio is 1.5:1) and the incidence was the highest in Caucasian people and the lowest in African–American people (Bluman and Goldman 2014).

HSP most commonly occurs in October, November, January, February, and March and is relatively rare in July, August, and September and, in 90 % of cases, HSP is most frequently associated with recent respiratory tract infections (Kurdi et al. 2014).

2.3.8 Eosinophilic Gastroenteritis (EG)

Eosinophilic gastroenteritis (EG) was first reported by Kaijser in 1937 (Ingle and Hinge Ingle 2013). EG is an uncommon gastrointestinal disease affecting both children and adults, and is characterized by varying degrees of focal or diffuse infiltration of eosinophils in the gastrointestinal tract in the absence of secondary causes (Martillo et al. 2015), with the percent of peripheral blood eosinophils increasing up to 15–70 % in 80 % of the patients. Because the ranges of eosinophil numbers seen in normal and abnormal gastric and intestinal mucosa are not standardized, it is difficult to define eosinophilic gastroenteritis (Martillo et al. 2015). EG is a progressive disease occurring in industrial countries with an incidence that has been rising over the last decade even though the overall prevalence of allergy-mediated gastrointestinal disease is rare (Raithel et al. 2014). It occurs at any age but most commonly presents between the third and fifth decades of life, with the incidence of EG in the US being about 2.5 cases per 100,000 adults (Verheijden and Ennecker-Jans 2010).

However, the exact etiology of EG is still unclear. Frequently, EG patients have a personal or family history of asthma or other allergic disorders. Data show that in some cases, EG has an association with parasitic infections, allergic mechanisms, or medications such as enalapril, rifampicin, and indomethacin (Hepburn et al. 2010; Raithel et al. 2014). More than half of EG patients have a history of asthma, AR, urticaria, or eczema and up to 62 % of cases are sensitized to foods such as milk, eggs, mutton, and sea shrimp (Raithel et al. 2014; Rodriguez Jimenez et al. 2011).

Eosinophilic gastroenteritis presents with nonspecific gastrointestinal symptoms and in almost one-third of cases has concomitant esophageal or colonic involvement (Reed et al. 2015). It remains difficult to treat, with high rates of endoscopy. Current treatment for EG lacks direct evidence but after dietary measures, i.e., avoid eating or having contact with potential allergens, some patients may become cured (Assa'ad 2009; Hommel et al. 2012). The symptoms of EG may be alleviated without extra treatment, indicating that EG may be either an allergic or nonallergic disorder (Ekunno et al. 2012). A few studies have reported that low-dose steroid therapy may be an effective treatment for chronic relapsing EG and this has become the preferred treatment for preventing grave complications like ascites and intestinal obstruction (Ingle and Hinge Ingle 2013). In most patients, the response to corticosteroids is rapid, however, immuno-suppressants, sodium cromoglycate, or leukotriene-receptor antagonists may be needed by patients treated long-term with corticosteroids (Famularo et al. 2011).

The increase in the incidence of these allergic diseases, which underlie the phenomena associated with allergic hypersensitivity to allergens, and the progress of knowledge in new fields such as immunology, molecular biology, and genetics in the past few years have led to an unprecedented increase in interest in the difficult field of allergy. Consequently, it is hoped that this knowledge will continue to be further expanded in the coming years.

2.4 Meta-analysis

In a narrow sense, meta-analysis is a systematic quantitative analysis of previous published studies and is essentially a summarization of multiple studies that have the same goals. It is an evaluation of the combined significance of multiple results, namely, it is a series of procedures to obtain a quantitative average conclusion and to answer certain questions by combining the results of several studies.

The idea of meta-analysis first emerged in the early 1930s and was initially applied to social sciences like pedagogy and psychology. In 1955, this method was first applied to medical science (Floyd and Ohnmeiss 2000). In 1976, Gene Glass was the first to name this method "Meta-analysis" and established a series of procedures and methods to obtain a representative conclusion, namely, to use statistical concepts and methods to collect, sort, and analyze multiple previous empirical studies so as to get a clear understanding of a certain issue of concern and complementing traditional review articles in this area of research (Sena et al. 2014). In the late 1970s, as medical science began to integrate some concepts of social and behavioral sciences, meta-analysis was introduced into medical science and subsequently emerged in the medical literature.

Given the importance of meta-analysis in evidence-based clinical decision-making, Chinese clinicians and scientific researchers began to pay attention to the results of meta-analysis based research, however, the overall quality is relatively low in both methodology and reports. Many problems are encountered during writing of a meta-analysis such as a low rate of complete document recall, failure to list excluded trials, unclear patient characteristics, diagnostic criteria and therapeutic ranges, lack of quality control before combining data, lack of monitoring and control of potential biased data, unstandardized statistical analysis, lack of quality assessment of original research data, lack of sensitivity analysis with alternative methods, lack of assessment of publication bias, and lack of evaluation of the application potential. The reasons for such deficiencies are due to a lack of sufficient understanding of this method, not knowing when and how to make a meta-analysis, and failing to follow the standardized procedure and writing format.

However, not all problems require a meta-analytical approach. Performing a meta-analysis only for the purpose of publishing a paper can hardly yield results of practical value. Early in 1987, Chalmers TC summarized the four aims of meta-analysis (Chalmers 1988): (1) to deal with inconsistencies between studies that may be caused by different levels of research, different subjects, and various interfering factors; (2) to enhance statistical power. Some randomized studies may have too small a control group to yield a solid conclusion. Using meta-analysis, the statistical power of these studies can be enhanced; (3) to enhance specificity and accuracy. When the results of several similar studies do not agree with each other, meta-analysis can combine studies to obtain an average effect and generate a more definite conclusion from controversial or even contradictory findings, resulting in a conclusion that is more specific and accurate; (4) to answer new questions. Through meta-analysis, certain unclarified issues due to deficiencies in a single study can be revealed, which may guide future studies.

2.4.1 Meta-analysis Procedures

Meta-analysis follows certain procedural steps in order to come to a conclusion that is objective and convincing, and they are described as follows:

First, the problem to be solved shall be clearly proposed and a research program shall be developed. This is the beginning step of every meta-analysis; a detailed research program shall clarify the objectives of the meta-analysis and problems to be solved, establish a proper set of inclusion and exclusion criteria, plan ways to select databases for literature and key word searches and decide which statistical indicators will be used for data merging.

Second, related literature shall be collected. According to the purpose of the study, proper databases or other data sources shall be selected. An online and manual search shall be combined to comprehensively and unambiguously collect related data. Comprehensiveness is an important feature of meta-analysis, which means, besides published papers, the author shall also collect unpublished data through all possible means such as conference papers, abstracts, and unpublished clinical trials, as well as personally obtained information. Attention shall be paid to references of the collected literature: if some of the literature meets the requirements of the searcher but the data or some other content were not clear, the original author shall be contacted for important information so as to reduce publication bias. For clarity, the process of retrieving relevant literature shall be described in the most detailed way possible. Data source, keywords, and search strategy shall be explained to allow others to repeat the study and to ensure objectivity and reliability of the meta-analysis.

Third, all research results that are collected shall meet the inclusion criteria. All experimental studies that may potentially meet the requirements shall be evaluated for relevance to the content. Generally, determination of the studies to be included shall be performed independently by two persons and, when inconsistency occurs, a consensus shall be made through discussion or by referral to a third party.

Fourth, the quality of the included studies shall be assessed. Quantitative and qualitative assessment shall be made based on proper standards. To assess the quality of the included research, an evaluation of how well system errors and bias were prevented during design and implementation of the experiments shall be performed. Rigorous experiments generate more reliable results, while low-quality studies may exaggerate the effect of interventions or have false negative results. There is currently no golden standard for assessing the quality of experimental studies, however, some criteria and scales may be useful in evaluation of randomized trials such as the Jadad scale that is often used for quality assessment by Cochrane systematic review. Using this scale, at least two researchers can independently evaluate the selected studies to avoid inclusion of research results of different qualities that may lead to inaccurate analysis results.

Fifth, data shall be extracted from the included literature. After comprehensively selecting the appropriate literature, related information shall be extracted. The characteristics of each study shall be noted such as experimental design (whether it is a randomized trial or not), research background, research methods, sample size, data measurement, and statistical analysis, allowing the included studies to be classified into several categories for comparison or analysis. It is important to design a data collection strategy with a proper design so that common information and characteristics of each study can be best represented. To ensure reliability of data collection, two or more researchers shall independently assess the selected data, and when inconsistency occurs, a consensus shall be made by discussion or by referral to a third party.

Sixth, the data shall be statistically analyzed. The five steps above are the preliminary work of meta-analysis and they are time-consuming but have a great influence on the results. Statistical analysis of meta-analysis is relatively easy and can be done with the help of the Review Manager (Revman) software developed by the Cochrane Collaboration group, which developed and maintains the Cochrane systematic review system. This system calculates the weighted average of included data and statistical values in a comprehensive manner. Effect variables can be a continuous variable of measurement data or a dichotomous variable of count data. The effect of a certain intervention is generally represented by the average value for continuous variables and by rate difference (RD), odds ratio (OR), and relative risk (RR) for dichotomous variables.

For different statistical values and different statistical assumptions, the method of meta-analysis varies. In the case of different statistical values, the methods of meta-analysis can be grouped into three categories: (1) The first category uses effect size as the statistical value and is well suited for independent studies that use continuous measurement data. Currently, this type of method is mainly used in social science (pedagogy, psychology, etc.), clinical medicine, and ecology and represents the size and direction of an effect or phenomenon. (2) The second category is mainly used in epidemiology, etiology, public science, and other medical sciences and the statistical value is the relative risk, hazard ratio, and risk difference. (3) The third category is the regression method that emerged in medical science in the late 1980s. The statistical value is the dose-response slope and this method is well suited for independent studies whose results are disaggregated data.

Before combining statistics, the included studies must be first tested for heterogeneity. If the results of the independent studies share a common effect size, namely, the results are in good agreement, a fixed effects model shall be used such as the Mantel–Haenszel method, Peto method, or General Variance-Based method for statistical analysis. If heterogeneity exists among the results of different studies, namely, the size of the effect varies, aggregation of the data shall be performed with great caution. If the aggregated data is still of clinical significance despite heterogeneity, a random effects model shall be used for data aggregation such as the Dersimonian and Laird model for statistical analysis. Since the random effects model is more in line with reality, it is well recognized by analysts and has been widely used. If the heterogeneity is severe, its source should be identified and generally includes clinical heterogeneity, methodological heterogeneity, and statistical heterogeneity. Researchers may perform subgroup analysis based on the source of the heterogeneity, or sensitivity analysis or multiple regression analysis, in order to select the appropriate model for statistical analysis.

Meta-analysis is the aggregation of multiple studies but the results of such aggregation may be greatly affected by a single study whereby inclusion or exclusion of such a study would generate a completely different conclusion. In this situation, a sensitivity analysis to identify the key factors that affected the conclusion should be performed in order to find the reasons why different studies generate a different conclusion. Stratified analysis is the most commonly used method for sensitivity analysis, that is, the included studies are categorized into different subgroups according to certain characteristics of a study such as sample size, quality of methodology, inclusion of unpublished studies, etc. Combining of data can be first performed in each subgroup. Then, the overall effects and that of each subgroup can be compared.

Besides RevMan, many other software programs can be used for data processing of meta-analysis such as the free MIX software, the business software Comprehensive Meta-Analysis, Meta-Win, etc., and some statistical software also integrate programs for meta-analysis such as STATA, SAS, and WinBUGs.

Seventh, an analysis report can be made (possibly by diagrams). The analysis shall be evaluated and a conclusion can be made, namely, the significance of the results for related clinical practice and future studies shall be summarized. In essence, meta-analysis is an observational study so that interpretation of the results should be made with great caution. Many of those interested in meta-analysis pay the most attention to this final part of meta-analysis or even just go directly to this part, however, the actual clinical problems are very complex and a meta-analysis should not be expected to provide an absolute solution, thus, more information should be provided in this report. For example, the methodological quality and disadvantages of the included studies shall be explained and methodology of the meta-analysis itself, applicability of the results, and other information related to medical decision-making (pros and cons and cost of the intervention) shall be clarified. Results analysis shall also include assessment of the effect and explain the deficiencies in experimental design and data analysis, so as to guide future studies.

As can be seen, meta-analysis has a rigorous design and clear standard for literature inclusion: it systematically considers the effect of research method, outcome measure, categorization, and conclusion; it provides the measurement index (aggregated statistic value) and a mechanism to quantitatively estimate the effect size so that the analysis results are highly objective and scientific; it improves the overall statistical capacity of the literature; and it considers the quality of independent studies.

Besides the above seven steps, many other details should be included to generate a high-quality meta-analysis. A series of international standards are gradually being established to evaluate and help improve the quality of meta-analysis such as the recently proposed MOOSE, QUOROM, PRISMA, etc. (Bello et al. 2015). By adhering to such uniform reporting formats, the author can not only improve the clarity of the meta-analysis but also avoid missing important information. In addition, such a format would make it easier for editors and reviewers to control the quality of meta-analyses so that meta-analyses can be more scientific and standardized and become a true high-level reference for clinical decision-making.

References

Abong J, Kwong S, Alava H, Castor M, De Leon J. Prevalence of allergic rhinitis in Filipino adults based on the National Nutrition and Health Survey 2008. Asia Pac Allergy. 2012;2(2):129–35.

Adamia N, Manjavidze N, Ghughunishvili M, Ubiria I, Saginadze L, Katamadze N, Zerekidze V, Gigauri T, Arakhamia T, Gogodze N, et al. PO-0990 Prevalence Of Allergic Rhinitis In Children's Population—Adjara Region. Arch Dis Child. 2014;99(Suppl_2):A574.

Adams JD, Filho ORF, Boutemy A, Adams JD, Filho ORF, Boutemy A. Treatments for atopic dermatitis. J Pharm Drug Develop. 2013;1(1).

Albin S, Agarwal S. Prevalence and characteristics of reported penicillin allergy in an urban outpatient adult population. Allergy Asthma Proc. 2014;35(6):489–94.

Almaliotis D, Michailopoulos P, Gioulekas D, Giouleka P, Papakosta D, Siempis T, Karampatakis V. Allergic conjunctivitis and the most common allergens in Northern Greece. World Allergy Organ J. 2013;6(1):12.

An S, Chen L, Long C, Liu X, Xu X, Lu X, Rong M, Liu Z, Lai R. *Dermatophagoides farinae* allergens diversity identification by proteomics. Mol Cell Proteomics. 2013;12(7):1818–28.

Assa'ad A. Eosinophilic gastrointestinal disorders. Allergy Asthma Proc. 2009;30(1):17–22.

Begum K, Haque S, Islam T, Nesa JU, Sarker ZH, Begum K, Begum K, Haque S, Islam T, Nesa JU. Factors related to severe acute asthma attack and treatment. Int J Pharm Sci Res. 2012;3(8).

Bello A, Wiebe N, Garg A, Tonelli M. Evidence-based decision-making 2: Systematic reviews and meta-analysis. Methods Mol Biol. 2015;1281:397–416.

Ben-Shoshan M, Soller L, Harrington DW, Knoll M, La Vieille S, Fragapane J, Joseph L, St Pierre Y, Wilson K, Elliott SJ, et al. Eczema in early childhood, sociodemographic factors and lifestyle habits are associated with food allergy: a nested case-control study. Int Arch Allergy Immunol. 2015;166(3):199–207.

Bergon-Sendin E, Perez-Grande C, Lora-Pablos D, De la Cruz-Bertolo J, Moral-Pumarega M, Pallas-Alonso C. PO-0613 Real time safety auditing of monitoring and respiratory support in a tertiary neonatal intensive care unit. Arch Dis Child. 2014;99(Suppl_2):A455.

Bluman J, Goldman RD. Henoch-Schönlein purpura in children: limited benefit of corticosteroids. Can Fam Physician. 2014;60(11):1007–10.

Braback L. Allergic diseases: health in Sweden: The National Public Health Report 2012. Chapter 14. Scand J Public Health. 2012;40(9_suppl):268–74.

Burks AW, Tang M, Sicherer S, Muraro A, Eigenmann PA, Ebisawa M, Fiocchi A, Chiang W, Beyer K, Wood R, et al. ICON: food allergy. J Allergy Clin Immunol. 2012;129(4):906–20.

Chalmers TC. Meta-analysis in clinical medicine. Trans Am Clin Climatol Assoc. 1988;99:144–50.

Chiang W, Chen Y, Tan H, Balakrishnan A, Liew W, Lim H, Goh S, Loh W, Wong P, Teoh O, et al. Allergic rhinitis and non-allergic rhinitis in children in the tropics: prevalence and risk associations. Pediatr Pulmonol. 2012;47(10):1026–33.

Chigbu DI, Coyne AM. Update and clinical utility of alcaftadine ophthalmic solution 0.25 % in the treatment of allergic conjunctivitis. Clin Ophthalmol. 2015;9:1215–25.

Criado PR, Antinori LC, Maruta CW, Reis VM. Evaluation of D-dimer serum levels among patients with chronic urticaria, psoriasis and urticarial vasculitis. An Bras Dermatol. 2013;88(3):355–60.

Dart JK, Buckley RJ, Monnickendan M, Prasad J. Perennial allergic conjunctivitis: definition, clinical characteristics and prevalence. A comparison with seasonal allergic conjunctivitis. Trans Ophthalmol Soc U K. 1986;105(Pt 5):513–20.

de Marco R, Cappa V, Accordini S, Rava M, Antonicelli L, Bortolami O, Braggion M, Bugiani M, Casali L, Cazzoletti L, et al. Trends in the prevalence of asthma and allergic rhinitis in Italy between 1991 and 2010. Eur Respir J. 2012;39(4):883–92.

De Sario M, Katsouyanni K, Michelozzi P. Climate change, extreme weather events, air pollution and respiratory health in Europe. Eur Respir J. 2013;42(3):826–43.

Dimitrov I, Naneva L, Doytchinova I, Bangov I. AllergenFP: allergenicity prediction by descriptor fingerprints. Bioinformatics. 2014;30(6):846–51.

Dworzynski K, Ardern-Jones M, Nasser S, Guideline Development Group; National Institute for Health and Care Excellence. Diagnosis and management of drug allergy in adults, children and young people: summary of NICE guidance. BMJ. 2014;349:g4852.

Dyer AA, Gupta R. Epidemiology of childhood food allergy. Pediatr Ann. 2013;42(6):91–5.

Ekunno N, Munsayac K, Pelletier A, Wilkins T. Eosinophilic gastroenteritis presenting with severe anemia and near syncope. J Am Board Fam Med. 2012;25(6):913–8.

Eriksson J, Ekerljung L, Bossios A, Bjerg A, Wennergren G, Ronmark E, Toren K, Lotvall J, Lundback B. Aspirin-intolerant asthma in the population: prevalence and important determinants. Clin Exp Allergy. 2015;45(1):211–9.

Eriksson J, Ekerljung L, Ronmark E, Dahlen B, Ahlstedt S, Dahlen S, Lundback B. Update of prevalence of self-reported allergic rhinitis and chronic nasal symptoms among adults in Sweden. Clin Respir J. 2012;6(3):159–68.

Esparza-Gordillo J, Weidinger S, Folster-Holst R, Bauerfeind A, Ruschendorf F, Patone G, Rohde K, Marenholz I, Schulz F, Kerscher T, et al. A common variant on chromosome 11q13 is associated with atopic dermatitis. Nat Genet. 2009;41(5):596–601.

Ezell JM, Ownby DR, Zoratti EM, Havstad S, Nicholas C, Nageotte C, Misiak R, Enberg R, Johnson CC, Joseph CL. Using a physician panel to estimate food allergy prevalence in a longitudinal birth cohort. Ann Epidemiol. 2014;24(7):551–3.

Famularo G, Prantera C, Nunnari J, Gasbarrone L. Eosinophilic gastroenteritis in a young man. CMAJ. 2011;183(1):E65–7.

Floyd T, Ohnmeiss D. A meta-analysis of autograft versus allograft in anterior cervical fusion. Eur Spine J. 2000;9(5):398–403.

Galffy G, Hirshberg A, Katona G, Sultesz M. The 2007-2013 trends in the prevalence of allergic rhinitis and asthma in primary schoolchildren in Budapest. Eur Respir J. 2014;44(Suppl_58):P4098.

Gilmour MI, Jaakkola MS, London SJ, Nel AE, Rogers CA. How exposure to environmental tobacco smoke, outdoor air pollutants, and increased pollen burdens influences the incidence of asthma. Environ Health Perspect. 2006;114(4):627–33.

Goldberg A, Confino-Cohen R, Waisel Y. Allergic responses to pollen of ornamental plants: high incidence in the general atopic population and especially among flower growers. J Allergy Clin Immunol. 1998;102(2):210–4.

Gonzalez LM, Janniger CK, Schwartz RA. Pediatric Henoch-Schönlein purpura. Int J Dermatol. 2009;48(11):1157–65.

Gu M, Zhao J. Mapping and localization of susceptible genes in asthma. Chin Med J (Engl). 2011;124(1):132–43.

Gupta N, Kim J, Njei B. Spontaneous bacterial peritonitis and Henoch-Schönlein purpura in a patient with liver cirrhosis. Case Rep Med. 2015;2015:340894.

Gupta RS, Springston EE, Warrier MR, Smith B, Kumar R, Pongracic J, Holl JL. The prevalence, severity, and distribution of childhood food allergy in the United States. Pediatrics. 2011;128(1):e9–17.

Ha SE, Ban TH, Jung SM, Bae KN, Chung BH, Park CW, Choi BS. Henoch-Schönlein purpura secondary to infective endocarditis in a patient with pulmonary valve stenosis and a ventricular septal defect. Korean J Intern Med. 2015;30(3):406–10.

Hammer-Helmich L, Linneberg A, Thomsen SF, Glumer C. Association between parental socioeconomic position and prevalence of asthma, atopic eczema and hay fever in children. Scand J Public Health. 2014;42(2):120–7.

Hepburn I, Sridhar S, Schade R. Eosinophilic ascites, an unusual presentation of eosinophilic gastroenteritis: a case report and review. World J Gastrointest Pathophysiol. 2010;1(5):166–70.

Hernandez-Cadena L, Zeldin DC, Barraza-Villarreal A, Sever ML, Sly PD, London SJ, Escamilla-Nunez MC, Romieu I. Indoor determinants of dustborne allergens in Mexican homes. Allergy Asthma Proc. 2015;36(2):130–7.

Hesselmar B, Aberg B, Eriksson B, Aberg N. Allergic rhinoconjunctivitis, eczema, and sensitization in two areas with differing climates. Pediatr Allergy Immunol. 2001;12(4):208–15.

Hjern A. Migration and public health: Health in Sweden: The National Public Health Report 2012. Chapter 13. Scand J Public Health. 2012;40(9 Suppl):255–67.

Hommel K, Franciosi J, Gray W, Hente E, Ahrens A, Rothenberg M. Behavioral functioning and treatment adherence in pediatric eosinophilic gastrointestinal disorders. Pediatr Allergy Immunol. 2012;23(5):494–9.

Hong S, Son D, Lim W, Kim S, Kim H, Yum H, Kwon H. The prevalence of atopic dermatitis, asthma, and allergic rhinitis and the comorbidity of allergic diseases in children. Environ Health Toxicol. 2012;27:e2012006.

Ingle SB, Hinge Ingle CR. Eosinophilic gastroenteritis: an unusual type of gastroenteritis. World J Gastroenterol. 2013;19(31):5061–6.

Jackson TL, Gjelsvik A, Garro A, Pearlman DN. Correlates of smoking during an economic recession among parents of children with asthma. J Asthma. 2013;50(5):457–62.

Kang Y, Park JS. Differences in clinical manifestations and outcomes between adult and child patients with Henoch-Schönlein purpura. 2014;29(2):198–203.

Karakaya G, Celebioglu E, Kalyoncu AF. Non-steroidal anti-inflammatory drug hypersensitivity in adults and the factors associated with asthma. Respir Med. 2013;107(7):967–74.

Katelaris CH, Lee BW, Potter PC, Maspero JF, Cingi C, Lopatin A, Saffer M, Xu G, Walters RD. Prevalence and diversity of allergic rhinitis in regions of the world beyond Europe and North America. Clin Exp Allergy. 2012;42(2):186–207.

Khan DA. Allergic rhinitis and asthma: epidemiology and common pathophysiology. Allergy Asthma Proc. 2014;35(5):357–61.

Kim JH, Cheong HS, Park JS, Jang AS, Uh ST, Kim YH, Kim MK, Choi IS, Cho SH, Choi BW, et al. A genome-wide association study of total serum and mite-specific IgEs in asthma patients. PLoS ONE. 2013;8(8):e71958.

Krishna MT, York M, Chin T, Gnanakumaran G, Heslegrave J, Derbridge C, Huissoon A, Diwakar L, Eren E, Crossman RJ, et al. Multi-centre retrospective analysis of anaphylaxis during general anaesthesia in the United Kingdom: aetiology and diagnostic performance of acute serum tryptase. Clin Exp Immunol. 2014;178(2):399–404.

Kumalı DB, Lartey SY, Yemanyi F, Boateng EG, Awualı E. Prevalence of allergic conjunctivitis among basic school children in the Kumasi Metropolis (Ghana): a community-based cross-sectional study. BMC Ophthalmol. 2015;15(1):69.

Kurdi MS, Deva RS, Theerth KA. An interesting perioperative rendezvous with a case of Henoch-Schönlein purpura. Anesth Essays Res. 2014;8(3):404–6.

La Rosa M, Lionetti E, Reibaldi M, Russo A, Longo A, Leonardi S, Tomarchio S, Avitabile T, Reibaldi A. Allergic conjunctivitis: a comprehensive review of the literature. Ital J Pediatr. 2013;39:18.

Li F, Tan J, Yang X, Wu Y, Wu D, Li M. Genetic variants on 17q21 are associated with asthma in a Han Chinese population. Genet Mol Res. 2012;11(1):340–7.

Li Y, Zhang X, Lv L, Yuan J, G-q Sun. Preliminary results of olopatadine hydrochloride in treating allergic conjunctivitis: a multicentral open trial (in Chinese with English abstract). Ophthalmol China. 2008;17(3):166–70.

Lim Y, Yi BH, Lee HK, Hong HS, Lee MH, Choi SY, Park JO. Henoch-Schönlein purpura: ultrasonography of scrotal and penile involvement. Ultrasonography. 2015;34(2):144–7.

Losol P, Yoo HS, Park HS. Molecular genetic mechanisms of chronic urticaria. Allergy Asthma Immunol Res. 2014;6(1):13–21.

Lu Y, Xie Z, Zhao S, Zhang H, Liu Y, Chen X, Jiang W. Prevalence of allergens for Changsha patients with allergic rhinitis. Lin Chung Er Bi Yan Hou Tou Jing Wai Ke Za Zhi. 2011;25(11):491–4.

Ma D, Li Y, Dong J, An S, Wang Y, Liu C, Yang X, Yang H, Xu X, Lin D, et al. Purification and characterization of two new allergens from the salivary glands of the horsefly, *Tabanus yao*. Allergy. 2011;66(1):101–9.

Martillo M, Abed J, Herman M, Abed E, Shi W, Munot K, Mankal PK, Gurunathan R, Ionescu G, Kotler DP. An atypical case of eosinophilic gastroenteritis presenting as hypovolemic shock. Case Rep Gastroenterol. 2015;9(2):142–51.

Morales DR, Lipworth BJ, Guthrie B, Jackson C, Donnan PT, Santiago VH. Safety risks for patients with aspirin-exacerbated respiratory disease after acute exposure to selective nonsteroidal anti-inflammatory drugs and COX-2 inhibitors: meta-analysis of controlled clinical trials. J Allergy Clin Immunol. 2014;134(1):40–5.

Nascimento-Sampaio FS, Leite Mdos S, Leopold DA, Silva SG, Schwingel PA, Mendes CM, Souza-Machado A, Campos Rde A. Influence of upper airway abnormalities on the control of severe asthma: a cross-sectional study. Int Forum Allergy Rhinol. 2015;5(5):371–9.

National Cooperative Group on Childhood A, Institute of Environmental H, Related Product Safety CCfDC, Prevention, Chinese Center for Disease C, and Prevention. Third nationwide survey of childhood asthma in urban areas of China. Zhonghua Er Ke Za Zhi. 2013;51(10):729–35.

O'Neil CE, Zanovec M, Nicklas TA. A review of food allergy and nutritional considerations in the food-allergic adult. Am J Lifestyle Med. 2011;5(1):49–62.

Odhiambo J, Williams H, Clayton T, Robertson C, Asher M. Global variations in prevalence of eczema symptoms in children from ISAAC Phase Three. J Allergy Clin Immunol. 2009;124(6):1251–8 (e23).

Ozdoganoglu T, Songu M. The burden of allergic rhinitis and asthma. Therapeutic Adv Resp Dis. 2012;6(1):11–23.

Papadopoulos J, Karpouzis A, Tentes J, Kouskoukis C. Assessment of Interleukins IL-4, IL-6, IL-8, IL-10 in acute urticaria. J Clin Med Res. 2014;6(2):133–7.

Park H, Dahlin A, Tse S, Duan Q, Schuemann B, Martinez F, Peters S, Szefler S, Lima J, Kubo M. Genetic predictors associated with improvement of asthma symptoms in response to inhaled corticosteroids. J Allergy Clin Immunol. 2014;133(3):664–9 (e5).

Pesce G, Marchetti P, Girardi P, Marcon A, Cazzoletti L, Accordini S, Antonicelli L, Bugiani M, Casali L, Cerveri I, et al. Latitude variation in the prevalence of asthma and allergic rhinitis in Italy: results from the GEIRD study. Eur Respir J. 2012;40(Suppl_56):P4777.

Raithel M, Hahn M, Donhuijsen K, Hagel AF, Nagel A, Rieker RJ, Neurath MF, Reinshagen M. Eosinophilic gastroenteritis with refractory ulcer disease and gastrointestinal bleeding as a rare manifestation of seronegative gastrointestinal food allergy. Nutr J. 2014;13:93.

Rance K, Goldberg P. The management and treatment of chronic urticaria: a clinical update and review. J Asthma Allergy Educators. 2013;4(6):305–9.

Reed C, Woosley JT, Dellon ES. Clinical characteristics, treatment outcomes, and resource utilization in children and adults with eosinophilic gastroenteritis. Dig Liver Dis. 2015;47(3):197–201.

Rich LF, Hanifin JM. Ocular complications of atopic dermatitis and other eczemas. Int Ophthalmol Clin. 1985;25(1):61–76.

Rodriguez Jimenez B, Dominguez Ortega J, Gonzalez Garcia JM, Kindelan Recarte C. Eosinophilic gastroenteritis due to allergy to cow's milk. J Investig Allergol Clin Immunol. 2011;21(2):150–2.

Sanjana JM, Mahesh PA, Jayaraj BS, Lokesh KS. Changing trends in the prevalence of asthma and allergic rhinitis in children in Mysore, South India. Eur Respir J. 2014;44(Suppl_58): P1187.

Santiago S. Food allergies and eczema. Pediatr Ann. 2015;44(7):265–7.

Sehgal VN, Jain S. Atopic dermatitis: ocular changes. Int J Dermatol. 1994;33(1):11–5.

Sena ES, Currie GL, McCann SK, Macleod MR, Howells DW. Systematic reviews and meta-analysis of preclinical studies: why perform them and how to appraise them critically. J Cereb Blood Flow Metab. 2014;34(5):737–42.

Smith W, O'Neil SE, Hales BJ, Chai TL, Hazell LA, Tanyaratsrisakul S, Piboonpocanum S, Thomas WR. Two newly identified cat allergens: the von Ebner gland protein Fel d 7 and the latherin-like protein Fel d 8. Int Arch Allergy Immunol. 2011;156(2):159–70.

Song G, Lee Y. Pathway analysis of genome-wide association study on asthma. Hum Immunol. 2013;74(2):256–60.

Song W-J, Kang M-G, Chang Y-S, Cho S-H. Epidemiology of adult asthma in Asia: toward a better understanding. Asia Pacific Allergy. 2014;4(2):75–85.

Strong C, Chang L-Y. Family socioeconomic status, household tobacco smoke, and asthma attack among children below 12 years of age: Gender differences. J Child Health Care. 2014;18(4):388–98.

Sun B, Zheng P. Prevalence of allergen sensitization among patients with allergic diseases in Guangzhou, Southern China: a six-year observational study. Eur Respir J. 2014;44(Suppl_58):P4040.

Thyssen J. The association between filaggrin mutations, hand eczema and contact dermatitis: a clear picture is emerging. Br J Dermatol. 2012;167(6):1197–8.

Tsakok T, McKeever TM, Yeo L, Flohr C. Does early life exposure to antibiotics increase the risk of eczema? A systematic review. Br J Dermatol. 2013;169(5):983–91.

Umetsu DT, Rachid R, Schneider LC. Oral immunotherapy and anti-IgE antibody treatment for food allergy. World Allergy Organ J. 2015;8(1):20.

Verheijden NA, Ennecker-Jans SA. A rare cause of abdominal pain: eosinophilic gastroenteritis. Neth J Med. 2010;68(11):367–9.

Wang R, Zhang H. Two hundred thousands results of allergen specific IgE detection (in Chinese with English abstract). Chin J Allergy Clin Immunol. 2012;01:18–23.

Wang SS, Hon KL, Sy HY, Kong AP, Chan IH, Tse LY, Lam CW, Wong GW, Chan JC, Leung TF. Interactions between genetic variants of FLG and chromosome 11q13 locus determine susceptibility for eczema phenotypes. J Invest Dermatol. 2012a;132(7):1930–2.

Wang ZH, Lin WS, Li SY, Zhao SC, Wang L, Yang ZG, Chen J, Zhang ZF, Yu JZ. Analysis of the correlation of prevalence in allergic rhinitis and other allergic diseases. Zhonghua Er Bi Yan Hou Tou Jing Wai Ke Za Zhi. 2012b;47(5):379–82.

Wassmann A, Werfel T. Atopic eczema and food allergy. Chem Immunol Allergy. 2015;101:181–90.

Wildfire J, Calatroni A, Lynch SV, Jr HAB, Fujimura K, Rauch M, Lynn H. Interactive exploration of microbial exposure, asthma and allergy using a web-based tool. J Allergy Clin Immunol. 2014;133(2):AB240.

Yang G, Liu ZQ, Yang PC. Treatment of allergic rhinitis with probiotics: an alternative approach. N Am J Med Sci. 2013;5(8):465–8.

Yang Y, Zhao Y, Wang CS, Wang XD, Zhang L. Prevalence of sensitization to aeroallergens in 10030 patients with allergic rhinitis. Zhonghua Er Bi Yan Hou Tou Jing Wai Ke Za Zhi. 2011;46(11):914–20.

Yang YH, Tsai IJ, Chang CJ, Chuang YH, Hsu HY, Chiang BL. The interaction between circulating complement proteins and cutaneous microvascular endothelial cells in the development of childhood Henoch-Schönlein purpura. PLoS ONE. 2015;10(3):e0120411.

Yu X, Yu C, Ren Z, Deng Y, Song J, Zhang H, Zhou H. Genetic variants of 17q21 are associated with childhood-onset asthma and related phenotypes in a northeastern Han Chinese population: a case–control study. Tissue Antigens. 2014;83(5):330–6.

Yue X, Zhang H, Fong X, Fan X. Clinical observation of Velvetfeeling treating contact dermatitis (in Chinese with English abstract). Guide China Med. 2009;7(13):69–70.

Zhang Y, Zhang L. Prevalence of allergic rhinitis in china. Allergy Asthma Immunol Res. 2014;6(2):105–13.

Ziyab AH, Karmaus W, Zhang H, Holloway JW, Steck SE, Ewart S, Arshad SH. Association of filaggrin variants with asthma and rhinitis: is eczema or allergic sensitization status an effect modifier? Int Arch Allergy Immunol. 2014;164(4):308–18.

Zuberbier T, Aberer W, Asero R, Bindslev-Jensen C, Brzoza Z, Canonica GW, Church MK, Ensina LF, Gimenez-Arnau A, Godse K, et al. The EAACI/GA(2) LEN/EDF/WAO Guideline for the definition, classification, diagnosis, and management of urticaria: the 2013 revision and update. Allergy. 2014;69(7):868–87.

Zukiewicz-Sobczak WA, Wroblewska P, Adamczuk P, Kopczynski P. Causes, symptoms and prevention of food allergy. Postepy Dermatol Alergol. 2013;30(2):113–6.

Author's Biography

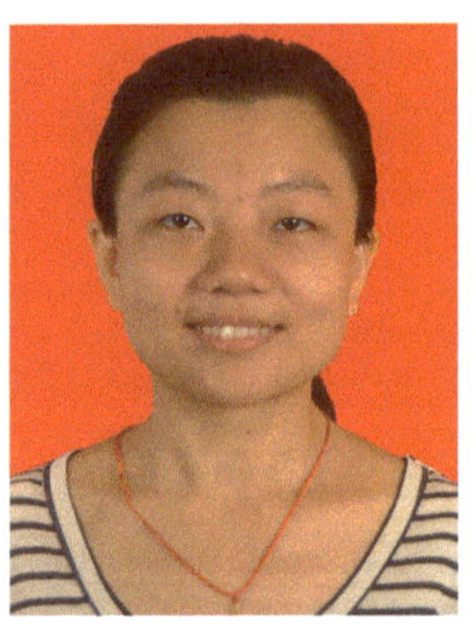

Juan Wang received her Bachelor's Degree in Clinical Medicine from Nanchang University and her Master's Degree in Immunology and Molecular Biology from Guangzhou Medical University in 2009 and 2012, respectively.

Currently, she is working at the Guangdong Provincial Key Laboratory of Allergy and Clinical Immunology, The Second Affiliated Hospital of Guangzhou Medical University. E-mail: wangjuan20110808@163.com.

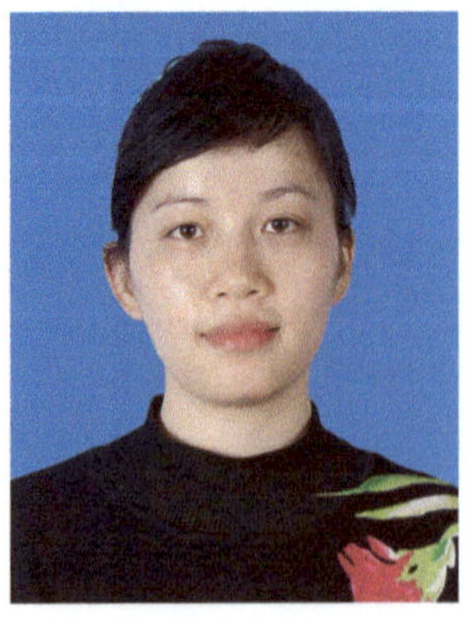

Dr. Junshu Wu is working in the Ophthalmic Department as an Associate Chief Doctor in Ophthalmology at the Second Affiliated Hospital of Guangzhou Medical University. E-mail: wujs78@126.com.

She received her PhD at Sun Yat-San University in 2004, studying the mechanism of experimental myopia. Her current main research interests focus on eye surface diseases and refractive surgery, including the epidemiology of allergic eye diseases. Dr. Wu has a total of 11 first-author articles that have been published in high-level academic journals. In the past few years, she has presided over several projects for the Guangdong Natural Science Foundation, the Guangdong Medical Research Foundation, and the State Key Laboratory Foundation of Ophthalmology.

Professor He Lai is the Director of Allergy Department of the Second Affiliated Hospital of Guangzhou Medical University, Vice-Chairman of the Chinese Medical Association Guangzhou Branch, President of the Guangdong Medical Association Allergy Branch, Council Member of the Society of Allergology, Chinese Medical Association. After graduating in July 1986, Prof. Lai has been working for 30 years as an allergist and Otolaryngology physician at the Second Affiliated Hospital of Guangzhou Medical University. She has been continuously engaged in clinical research of allergic rhinitis and other allergic diseases, and has presided over many key projects of the Science and Technology Bureau of Guangzhou and Guangdong Province. Her research has focused mainly on immunotherapy and its mechanisms, etc., especially regarding the strict control of indications for the use of immunotherapy and optimization of its treatment.

Chapter 3
The Pathogenesis of Allergy: A Brief Introduction

Zhi-Qiang Liu and Ping-Chang Yang

Abstract Allergic diseases are mainly IgE-mediated diseases that include allergic rhinitis, allergic asthma, allergic dermatitis, and food allergy, etc. The precise causative factors of allergic diseases are not clear. It has been proposed that some people are genetically prone to develop allergic diseases. When these people encounter foreign antigens, dendritic cells capture and process the antigens, transfer the antigen information to T helper (T_H) cells, and a skewed immune response is induced that resuts in a T_H2 polarization status in the body. These antigen-specific T_H2 cells present the antigen information to B cells and drive these cells to become plasma cells that produce antigen-specific IgE. This antigen-specific IgE sensitizes mast cells by binding to the high-affinity IgE receptors on their surface. When the sensitized mast cells are re-exposed to the specific antigens, they are activated to release allergic mediators such as histamine, tryptase, leukotrines, and serotonin, etc. that initiate allergic attacks.

Keywords Pathogenesis · Immunoglobulin E · Cytokines · Predisposition

3.1 Introduction

Allergy, formally called type I hypersensitivity, is described as an immediate reaction of the immune system to certain antigens that is largely mediated by IgE antibodies (Uzzaman and Cho 2012). Broadly, allergy is a mistargeted immune response that occurs after the body has been primed with certain exposed antigen(s) that then triggers an abnormal immune response seen mainly as physiological dysfunction or tissue damage. Allergens are usually external, innocuous

Z.-Q. Liu · P.-C. Yang (✉)
Allergy and Immunology Branch, College of Medicine, Shenzhen University, Nanhai Avenue 3688, Shenzhen, Guangdong 518060, People's Republic of China
e-mail: pcy2356@163.com

A. Tao and E. Raz (eds.), *Allergy Bioinformatics*, Translational Bioinformatics 8,
DOI 10.1007/978-94-017-7444-4_3

substances that include: (1) airborne particles such as dust and pollen, which are the causes of asthma symptoms; (2) foods, insect stings, and some medications such as penicillin that may result in allergies; and (3) substances such as latex coming in contact with the skin that can cause contact dermatitis or eczema. The allergic reactions can sometimes be severe and result in life-threatening reactions called anaphylaxis. Some people have a predisposed and exaggerated tendency to become sensitized to a variety of common environmental allergens, which is termed atopy. Two things, host and environment, are generally responsible for this predisposition. The internal host factors are mainly heredity, sex, race and age. Exposure to infectious diseases in early childhood, environmental pollution, allergen levels, and dietary changes are some of the external environmental factors. In recent decades, the prevalence of allergy has increased greatly, while the pathogenic processes that initiate a skewed T_H2 cell polarization and antigen-specific IgE antibody production still need further study.

3.2 Mechanisms of Allergy

The immunological mechanisms involved in allergic reactions can be generally categorized into three types: IgE mediated, mixed (IgE/Non-IgE), and non-IgE mediated (cellular, delayed type hypersensitivity) (Ho et al. 2014). Because it is one of the most common allergies, food allergy has been the most extensively studied. Both immunological and nonimmunological mechanisms can cause adverse food reactions, with the most common immunological responses being IgE mediated. Non-IgE-mediated hypersensitivity to foods can also occur, such as enterocolitis, benign eosinophilic colitis, allergic eosinophilic gastroenteritis, and celiac disease. The same foods that cause food allergies can also be responsible for non-IgE-mediated hypersensitivity (Fodey 2009). Different from food allergies, food intolerance is defined as an adverse food reaction that includes metabolic, toxic, psychological, and pharmacological reactions, but without the involvement of immunological mechanisms (Ja et al. 2010).

3.3 IgE-Dependent Mechanism of Allergy

The pathophysiology of IgE-dependent type I hypersensitivity generally consists of two phases: the acute response and the late-phase response. In the early stage, antigen is processed and presented by antigen-presenting cells, mostly by dendritic cells, in the form of a MHC II-antigen peptide complex. The antigenic peptide is then recognized by naïve $CD4^+$ T cells on a specific T cell receptor (TCR), resulting in their proliferation and differentiation into T_H2 lymphocytes. This response also includes the release of cytokines and costimulatory factors from stimulated dendritic cells. The expression of costimulatory molecules is essential for the activation of naïve T cells (Ruiter and Shreffler 2012). T_H2 cells produce interleukin (IL)-4, IL-5, IL-13, which are important in responses against extracellular, multicellular pathogens such as parasites. With the help of T_H2 cells, antigen-specific B lymphocytes differentiate into plasma cells that undergo immunoglobulin class switching to IgE production. Secreted IgE circulates in the blood and binds to the surface of mast cells and basophils on IgE-specific receptors (FcεRI), resulting in the sensitization of these cells to specific allergens. Later, when re-exposed to the same allergens, cross-linking of the allergen bound IgE and FcεRI activates the sensitized cells to undergo degranulation, during which they release histamine and other inflammatory chemical mediators that cause allergic symptoms. With the subsidence of the acute response, the late-phase response may occur, which begins with the migration of other leukocytes such as neutrophils, eosinophils, and macrophages to the initial site of allergen exposure (Grimbaldeston et al. 2006).

3.4 IgE-Independent Mechanism of Allergy

For non-IgE-mediated allergy, the mechanism may involve immune complexes formed by food antigens and antibodies to them, such as IgG, IgM, and IgA, as well as cell-mediated immunity. Non-IgE-mediated allergy to foods causes a wide range of diseases from atopic dermatitis to celiac disease (Pasha and Saeed 2013). Studies suggest that non-IgE-mediated gastrointestinal food allergy may be a nonimmediate type allergic reaction in which antigen-presenting cells, T cells, epithelial cells, and eosinophils are involved, but the precise mechanism of non-IgE-mediated allergy remains elusive (Nomura et al. 2012). Non-IgE-mediated and irritant-induced occupational rhinitis is caused by work exposure to drugs, wood dust, chemicals, etc., and is different than allergic occupational rhinitis, which is an IgE-mediated hypersensitivity reaction. Non-IgE-mediated occupational rhinitis may be due to contact with low molecular weight chemicals. The allergic mechanism of this disorder has not yet been characterized and needs to be studied further (Siracusa et al. 2013).

3.5 Understanding IgG Involvement in Allergy

Allergic diseases are usually defined as a group of allergic reactions driven by IgE-dependent mechanisms and affect approximately 22 % of the world population. In some allergy-like reactions, the serum IgE levels are normal, suggesting that the IgE-mediated mechanism might not be the only mechanism causing allergic disorders. IgG may also play a role in the pathogenesis of allergy (He 2013). The role of IgG and FcγR signaling in humans with allergic disease is still controversial, but allergen-specific IgG could contribute to the pathogenesis of allergy. IgG may interact with the Fc-gamma receptor (FcγR) on antigen-presenting cells and cause the induction of allergic inflammation. Studies showed that IgE-deficient mice were able to develop airway hyperactivity, and specific IgG may have played a role. High-affinity human IgG receptors FcγRIIA and FcγRI may be involved in an IgG-mediated allergic response and anaphylaxis (Hofmaier et al. 2014). It has been reported that FcγRI contributes to IgG2a-induced type I hypersensitivity and passive systemic anaphylaxis, while FcγRIIB negatively regulates hypersensitivity reactions in mice (Bruhns 2012). In addition, some studies found increases in allergen-specific IgG antibodies in humans after specific immunotherapy. One of the IgG subclasses, IgG4, may be able to block IgE antigens from binding to mast cells and, thus, inhibit the allergic response (Williams 2012). The role of antigen-specific IgG in the pathogenesis of allergic diseases remains obscure and varies according to the level of exposure to specific antigens.

3.6 Innate and Adaptive Immunity in Allergy

Innate immunity is a primitive system for host defense that occurs at the time of the initial encounter with antigen and does not require prior exposure. The development of an adaptive immune response requires gene rearrangement, which evolved specifically for antigen recognition. Innate immune responses are characterized by rapid and transient reactions, while adaptive immunity develops slowly but is more powerful and rapid upon re-exposure to the same antigens due to immunological memory (Vighi et al. 2008). The T_H2 promoting milieu is essential; however, the causes and the origin of this milieu are unclear (Wambre 2012). In fact, innate immunity and adaptive immunity are two undifferentiable parts of the immune system; both of them play an important role in the T_H2 response promoting milieu (Finn and Bigby 2009). Airway epithelium or skin barrier tissue can be stimulated by allergens or viral infection to secret IL-25, IL-33 and TSLP and these cytokines act on dendritic cells to bias the response towards T_H2 differentiation (Saenz 2008; Schleimer et al. 2007).

The gastrointestinal system is a major site of the body that is in contact with external stimuli and commensal flora. It is a prominent part of the immune system, with gut-associated lymphoid tissue (GALT) orchestrating the generation of immune tolerance, eliminating luminal antigens, and also damaging the intestinal mucosa through innate and adaptive immune responses. Food allergy is a specific immune response that repeatedly occurs on exposure to a given food (Burks et al. 2012). Why this occurs is still obscure (Sicherer 2011). Immune adjuvants play an important role in the generation of a skewed T_H2 response and have the capacity to speed and amplify the immune response to an antigen through pattern recognition receptors (PRRs), resulting in proinflammatory cytokine release, innate immune cell activation, and initiation of an adaptive immune response. The structural and functional properties of food proteins play an important role in their allergenicity. However, in order to elicit immune responses, food allergens must survive or bypass digestion, as only intact allergens can be recognized by specific IgE. Protein glycosylation may contribute to protein stability and some proteins in both food and inhalant allergens bind lipids that protect the antigens from digestion and enhance absorption. Environmental exposure to peanut through an impaired skin barrier increases allergy risk and maternal early intake of peanut reduces the infant's risk of peanut allergy. The offspring of peanut-sensitized mice are prone to developing peanut allergy. A diverse and healthy infant diet has a protective effect in inhibiting the development of atopic disease (Sicherer and Leung 2015).

In summary, allergy is an immediate immune reaction of the body to allergens and is mostly mediated by IgE antibodies. It may occur in certain individuals with genetic susceptibility under certain environmental influences. The incidence of allergic diseases has increased rapidly in recent decades; thus, it is important that the pathogenesis of allergy continue to be further elucidated.

References

Bruhns P. Properties of mouse and human IgG receptors and their contribution to disease models. Blood. 2012;119(24):5640–9.

Burks AW, Tang M, Sicherer S, Muraro A, Pa Eigenmann, Ebisawa M, et al. ICON: food allergy. J Allergy Clin Immunol. 2012;129(4):906–20.

Finn PW, Bigby TD. Innate immunity and asthma. Proc Am Thorac Soc. 2009;6(3):260–5.

Fodey TL, Greer NM, Crooks SR. Antibody production: Low dose immunogen vs. low incorporation hapten using salmeterol as a model. Anal Chim Acta. 2009;637(1–2):328–32.

Grimbaldeston MA, Metz M, Yu M, Tsai M, Galli SJ. Effector and potential immunoregulatory roles of mast cells in IgE-associated acquired immune responses. Curr Opin Immunol. 2006;18(6):751–60.

He SH, Zhang HY, Zeng XN, Chen D, Yang PC. Mast cells and basophils are essential for allergies: mechanisms of allergic inflammation and a proposed procedure for diagnosis. Acta Pharmacol Sin. 2013;34(10):1270–83.

Ho MH, Wong Wh, Chang C. Clinical spectrum of food allergies: a comprehensive review. Clin Rev Allergy Immunol. 2014;46(3):225–40.

Hofmaier S, Comberiati P, Matricardi PM. Immunoglobulin G in IgE-mediated allergy and allergen-specific immunotherapy. Eur Ann Allergy Clin Immunol. 2014;46(1):6–11.

Ja Boyce, Assa'ad A, Aw Burks, Sm Jones, Ha Sampson, Ra Wood, et al. Guidelines for the diagnosis and management of food allergy in the United States: report of the NIAID-sponsored expert panel. J Allergy Clin Immunol. 2010;126(6 Suppl):007.

Nomura I, Morita H, Ohya Y, Saito H, Matsumoto K. Non-IgE-mediated gastrointestinal food allergies: distinct differences in clinical phenotype between Western countries and Japan. Curr Allergy Asthma Rep. 2012;12(4):297–303.

Pasha I, Saeed F, Sultan MT, Batool R, Aziz M, Ahmed W. Wheat allergy & intolerence; recent updates and perspectives. Crit Rev Food Sci Nutr. 2013;2:2.

Ruiter B, Shreffler WG. Innate immunostimulatory properties of allergens and their relevance to food allergy. Semin Immunopathol. 2012;34(5):617–32.

Saenz SA, Taylor BC, Artis D. Welcome to the neighborhood: epithelial cell-derived cytokines license innate and adaptive immune responses at mucosal sites. Immunol Rev. 2008;226:172–90.

Schleimer RP, Kato A, Kern R, Kuperman D, Avila PC. Epithelium: at the interface of innate and adaptive immune responses. J Allergy Clin Immunol. 2007;120(6):1279–84.

Sicherer SH. Epidemiology of food allergy. J Allergy Clin Immunol. 2011;127(3):594–602.

Sicherer SH, Leung DY. Advances in allergic skin disease, anaphylaxis, and hypersensitivity reactions to foods, drugs, and insects in 2014. J Allergy Clin Immunol. 2015;135(2):357–67.

Siracusa A, Folletti I, Moscato G. Non-IgE-mediated and irritant-induced work-related rhinitis. Curr Opin Allergy Clin Immunol. 2013;13(2):159–66.

Uzzaman A, Cho SH. Chapter 28. Classification of hypersensitivity reactions. Allergy Asthma Proc. 2012;33(Suppl 1):S96–9.

Vighi G, Marcucci F, Sensi L, Di Cara G, Frati F. Allergy and the gastrointestinal system. Clin Exp Immunol. 2008;1:3–6.

Wambre E, James EA, Kwok WW. Characterization of $CD4^{+}T$ cell subsets in allergy. Curr Opin Immunol. 2012;24(6):700–6.

Williams JW, Tjota MY, Sperling AI. The contribution of allergen-specific IgG to the development of th2-mediated airway inflammation. J Allergy. 2012;236075(10):21.

Authors' Biography

Dr. Zhi-Qiang Liu PhD is an Associate Professor and Senior Research Scientist at Longgang Central Hospital and ENT Institute of Shenzhen. He is working on the pathogenesis of food allergy, allergic rhinitis, and the immune mechanisms of inflammatory bowel diseases. Dr. Zhi-Qiang Liu has published more than 20 scientific papers in internationally well-recognized journals and has written one book chapter.

Professor Ping-Chang Yang PhD is the Director of the Allergy and Immunology Center at Shenzhen University School of Medicine in Shenzhen, China. Dr. Yang's research focuses on the pathogenesis of allergy, specifically immune regulation and immune tolerance. One of Dr. Yang's research findings is that intestinal epithelial cells play a critical role in initiating the development of antigen-specific immune tolerance and that T_H2 cytokines, such as IL-13, can compromise established immune tolerance. Using several immune-competent molecules, Dr. Yang has found that the compromised immune tolerance can be restored in an allergic environment. Dr. Yang's next goal is to elucidate the mechanism by which the IgE isotype switch is initiated in B cells.

Chapter 4
Genetics and Epigenetic Regulation in Allergic Diseases

Chang-Hung Kuo and Chih-Hsing Hung

Abstract Multiple mechanisms mediate the risk of allergic diseases, including altered innate and adaptive immune responses, gene-environment interactions, epigenetic regulation, and possibly gene-environment-epigene interactions. The gene-environment interaction adds to the complexity of allergy and asthma and is now being successfully explored using epigenetic approaches. Allergic diseases are inheritable, influenced by the environment, and modified by in utero exposures and aging; all of these features are also common to epigenetic regulation. Environmental exposures, including prenatal maternal smoking, have been associated with allergy-related outcomes that could be explained by epigenetic regulation. In addition, several allergy- and asthma-related genes have been found to be susceptible to epigenetic regulation, including genes important to T effector pathways (IFN-γ, IL-4, IL-13, IL-17) and T-regulatory pathways (FoxP3).

C.-H. Kuo
Department of Pediatrics, Kaohsiung Municipal Ta-Tung Hospital, Kaohsiung, Taiwan, People's Republic of China

C.-H. Kuo · C.-H. Hung (✉)
Department of Pediatrics, Kaohsiung Medical University Hospital, Kaohsiung Medical University, #100, Tz-You 1st Road, Kaohsiung, Taiwan 807, People's Republic of China
e-mail: pedhung@gmail.com

C.-H. Hung
Department of Pediatrics, Faculty of Pediatrics, College of Medicine, Kaohsiung Medical University, Kaohsiung, Taiwan, People's Republic of China

C.-H. Hung
Graduate Institute of Medicine, College of Medicine, Kaohsiung Medical University, Kaohsiung, Taiwan, People's Republic of China

C.-H. Hung
Department of Pediatrics, Kaohsiung Municipal Hsiao-Kang Hospital, Kaohsiung Medical University Hospital, Kaohsiung Medical University, Kaohsiung, Taiwan, People's Republic of China

A. Tao and E. Raz (eds.), *Allergy Bioinformatics*, Translational Bioinformatics 8,
DOI 10.1007/978-94-017-7444-4_4

Therefore, investigation into the relationship between epigenetic regulation and allergic disease is critical for understanding the development of allergy and asthma. Previously published experimental work, with few exceptions, has been comprised mostly of small observational studies and models in in vitro systems and animals. Recently, however, and due to new and advanced technologies, novel translational research and exciting and elegant experimental studies have been published. Epigenetic factors (DNA methylation, modifications of histone tails or noncoding RNAs) work with other components of the cellular regulatory machinery to control the levels of expressed genes. Technologies to investigate epigenetic factors on a genomic scale and comprehensive approaches to data analysis have recently emerged and continue to improve. This chapter addresses the fundamental questions of how genetics and epigenetics influence allergic diseases, including asthma. These include how anti-asthmatic agents can induce epigenetic changes, and how epigenetic regulation can influence cytokine and chemokine production in immune cells. The discovery and validation that genetic and epigenetic biomarkers may be linked to exposure, allergies, or both might lead to better genotyping and epigenotyping of the risk, prognosis, treatment prediction, and development of novel therapies for the treatment of allergy.

Keywords Epigenetics · Allergy · Asthma · Methylation

4.1 Genetics in Allergy

Allergy is a complex condition that is influenced by the interaction between genetics and environmental factors. Asthma, allergic rhinitis, allergic conjunctivitis, and atopic dermatitis (AD) are common allergic diseases and have a collective pathophysiology, including elevated IgE levels, inflammatory processes with a predominant T_H2 response, and a hereditary tendency (Gustafsson 2000). The hereditary tendency of allergic disease indicates the involvement of susceptible genes in the pathogenesis of allergy. A number of susceptibility genes, including regulators of immune cells as well as their relevant responses, epithelial cells, as well as epidermal barrier functions, have been identified by family linkage and candidate-gene studies (Vercelli 2008) as being involved in the development of allergy. Recently, the development of genome-wide association studies (GWASs) using common single nucleotide polymorphisms (SNPs) has accelerated and clarified the novel genes in allergic disease, successfully identifying and confirming the loci of susceptibility genes. These susceptibility genes may play key roles in altering immune responses including the function of immune cells such as dendritic cells, T cells, and B cells as well as monocyte/macrophages, the expression of cytokines/chemokines, as well as the structure and functions of the epidermal/epithelial barrier.

4.2 Susceptibility Genes Associated with Allergic Rhinitis

Allergic rhinitis is characterized by sneezing, nasal pruritus, nasal congestion, and rhinorrhea as a result of an inflammatory process mediated by allergen-specific immunoglobulin E, and its global prevalence is increasing. Allergic rhinitis has a great impact on quality of life and causes a great socioeconomic burden. A study from Singapore (Andiappan et al. 2011) of an ethnic Chinese population (4461) with allergic rhinitis used a GWAS approach. Single nucleotide polymorphisms (SNPs) in two novel candidate genes, mitochondrial ribosomal protein L4 (MRPL4) on chromosome 19q13.2 and the B cell adaptor protein for phosphatidylinositol 3-kinase (BCAP) on chromosome 10q24.1, which are key components of the HIF-1α and PI3 K/Akt signaling pathways, respectively, are associated with allergic rhinitis and atopy. A study from an urban city in China of a Han Chinese population (414 patients versus 293 healthy controls) revealed that SNPs in the tumor necrosis factor-α (TNF-α) and MRPL4 genes occur more frequently in AR patients compared to the control group (Wei et al. 2013). Recently, a large-scale study from Europe investigated four large cohorts of adult Europeans for AR (3933 patients versus 8965 control) and grass sensitization (2315 patients versus 10,032 control). They found that a variant rs7775228 in the HLA locus, variants in a locus near chromosome 11 open reading frame 30 (C11orf30) and leucine-rich repeat containing 32 (LRRC32), and variant rs17513503 in a locus near transmembrane protein 232 (TMEM232) and solute carrier family 25, member 46 (SLC25A46) could be involved. This study also found that variants in thymic stromal lymphopoietin (TSLP), Toll-like receptor 6 (TLR6), and nucleotide-binding oligomerization domain containing 1 (NOD1/CARD4) genes could be important. A study of allergic rhinitis that enrolled 5633 ethnically diverse North American subjects investigated gene expression in disease-relevant tissue (peripheral blood CD4 + lymphocytes) (Bunyavanich et al. 2014). They found that a locus on chromosome 7p21.1 near fer3-like bHLH transcription factor (FERD3L) was associated with allergic rhinitis across all ethnic groups. Further functional studies integrating the GWAS and gene expression data with expression single nucleotide (eSNP), co-expression networks, and pathway approaches revealed that mitochondrial pathways are possibly involved in the pathogenesis of allergic rhinitis.

Allergic rhinitis is highly associated with allergic sensitization. It is known that patients with allergic rhinitis have increased levels of total and allergen-specific IgE. Previously, SNPs from a British Birth Cohort (1083 cases versus 2770 control) were tested across the genome for association with increased specific IgE levels to one or more allergens, including house dust mite (Der p 1), mixed grass, or cat fur (Wan et al. 2011). They found that a single SNP on chromosome 13q14 was weakly associated with increased specific IgE levels. However, the results had no consistent association with atopy as defined by increased specific IgE levels, positive SPT responses, or both, in all study cohorts. Studies replicated across several Caucasian populations are limited and these findings were, therefore, inconclusive in other larger independent populations. Recently, other studies addressing allergic sensitization revealed that four loci (TLR6-TLR1, HLA-DQA2-HLA-DQA, IL2-ADAD1, and LRRC32-C11orf301) are likely to have a key role in the development of allergic disease (Bonnelykke et al. 2013; Hinds et al. 2013; Nilsson et al. 2014; Ramasamy et al. 2011a, b).

4.3 Susceptibility Genes Associated with Asthma

Asthma is a syndrome that is characterized by chronic and persistent inflammation of the airway, with clinical manifestations of chronic cough, wheezing, and shortness of breath. In the most severe phenotype, irreversible airway remodeling and intractable airflow limitation may occur in some patients. Asthma has an increasingly high prevalence worldwide. Asthmatic patients frequently have a strong family history, with inheritability estimated at approximately 60 % (Duffy et al. 1990). Genetic studies can help with understanding the causes of asthma and verifying targets that can be used to treat the syndrome.

Compared to gene studies for allergic rhinitis, there have been more investigations addressing the genes associated with asthma. In 2007, the GABRIEL consortium (A Multidisciplinary Study to Identify the Genetic and Environmental Causes of Asthma in the European Community) performed the first GWAS of asthma (Moffatt et al. 2007). They focused on childhood-onset asthma, enrolling 994 patients and 1243 normal subjects. They revealed multiple markers on chromosome 17q21, and this finding was validated in other independent replication studies enrolling 2320 subjects from a cohort of German children and 3301 subjects from the British 1958 Birth Cohort. They further found that SNPs associated with childhood asthma are associated in cis with transcript levels of ORMDL3, which function as encoding transmembrane proteins anchored in the endoplasmic reticulum. This indicates that genetic variants regulating ORMDL3 expression may contribute to the susceptibility for childhood asthma.

In addition to childhood-onset asthma, there are different subtype/phenotypes of asthma, such as allergic asthma, later-onset asthma, severe asthma, and occupational asthma. A large-scale, consortium-based genome-wide association study of asthma that enrolled 10,365 asthmatic patients and 16,110 controls divided the subjects into groups with either childhood-onset asthma (defined as asthma developing before 16 years of age), later-onset asthma, severe asthma, or occupational asthma (Moffatt et al. 2010). They observed several SNPs associated with asthma, including rs3771166 on chromosome 2 (IL-1RL1/IL-18R1), rs9273349 on chromosome 6 (HLA-DQ), rs1342326 on chromosome 9 (IL-33), rs744910 on chromosome 15 (SMAD3) and rs2284033 on chromosome 22 (IL2RB). The ORMDL3/GSDMB locus on chromosome 17q21 (rs2305480) is specific to childhood-onset disease, and HLA-DR is associated with the total serum IgE concentration.

In childhood asthma, exacerbations are the most frequent cause of hospitalization. Very recently, investigators from Denmark addressed a specific asthma phenotype characterized by severe recurrent exacerbations with an onset between 2 and 6 years of age, enrolling 1173 patients and 2522 controls (Bonnelykke et al. 2014a). They confirmed that GSDMB, IL-33, RAD50, and IL-1RL1, which were previously reported as asthma susceptibility loci, were also involved in their cohort and also newly revealed that encoding cadherin-related family member 3 (CDHR3), which is highly expressed in airway epithelium, is a susceptible gene for asthma exacerbation in childhood.

Although severe asthma occurs in only about 5 % of the total asthma population, it causes a much greater socioeconomic burden than mild to moderate asthma. To identify a susceptibility gene for severe asthma, a GWAS from Europe enrolled 933 individuals with severe asthma based on the global initiative for asthma (GINA) criteria 3 or above and 3346 normal controls (Wan et al. 2012). They confirmed an association in subjects with severe asthma with loci previously found to be associated with mild to moderate asthma. Similar to previous studies for mild to moderate asthma in ethnically diverse populations including Mexican, Puerto Rican, and African American (Galanter et al. 2008; Verlaan et al. 2009) and Japanese (Hirota et al. 2008), the ORMDL3/GSDMB locus on chromosome 17q12-21 (rs4794820) and the L1RL1/IL18R1 locus on chromosome 2q12 (rs9807989) were identified. However, no novel loci for susceptibility to severe asthma were found in this study. Interestingly, in a Japanese population, SNPs (rs9303277, rs7216389, rs7224129, rs3744246, and rs4794820) in the asthma susceptible gene ORMDL3 locus on chromosome 17q21 is also associated with allergic rhinitis. This finding provides genetic evidence for the common clinical observation that asthma and allergic rhinitis frequently coexist in the same patient, and that inflammation of the mucosa is very similar in both asthma and allergic rhinitis.

The T-helper type 2 immune response is important for the pathophysiology of allergy, allergic rhinitis, and asthma. Cytokines play key roles in regulating and maintaining the immune response in the persistent inflammatory process of the airway. On chromosome 5q31.1, RAD50-IL-13 contains important regulator genes for T_H2 immune responses, such as IL-3, IL-4, IL-5, IL-13, and GM-CSF. Variants in the loci are confirmed as having significant association with asthma (Li et al. 2010), as are the susceptibility genes to allergic rhinitis in European (Black et al. 2009) and Asian (Kim et al. 2012; Lu et al. 2011) populations. Interleukin (IL)-33 is an important effector cytokine of type 2 T-helper responses, and the IL-33 receptor is a heterodimer of the IL-1 receptor-like 1 (IL-1RL1) and IL-1 receptor accessory protein (IL-1RAcP) (Ohno et al. 2012). Similar to ORMDL3, IL-33 has been reported in several other GWASs of asthma (Bonnelykke et al. 2014b; Ferreira et al. 2014; Torgerson et al. 2011a). Interestingly, like ORMDL3, genetic variants in IL-33 and the levels of serum IL-33 are also associated with allergic rhinitis (Sakashita et al. 2008). TSLP, an important epithelial cell-derived cytokine, is expressed at the exposure site of allergen in the airways. Common triggers of asthma symptoms can induce the release of TSLP and TSLP levels are positively correlated with disease severity. The activation by TSLP drives the development of T_H2 inflammation (Watson and Gauvreau 2014). Genetic variants in TSLP have been shown to be significantly associated with asthma in one large-scale meta-analysis of a GWAS study in ethnically diverse North American populations (Torgerson et al. 2011b) and, interestingly, variants in or near TSLP are also associated with allergic rhinitis in adults (Ramasamy 2011a, b) and children (Birben et al. 2014), and may be associated with lung function in asthmatic children with allergic rhinitis (Birben et al. 2014).

4.4 Susceptibility Genes Associated with Atopic Dermatitis

AD is a chronic, inflammatory skin disease and is the leading skin disease in Westernized countries, with a rapidly increasing prevalence, affecting approximately 20 % of children and 1–3 % of adults. The pathogenesis of AD is a complicated combination of genetic, environmental, skin barrier, and other immunological aberrances. In 2009, the first GWAS of AD enrolled 939 patients and 975 controls as well as 270 complete nuclear families with two affected siblings, and was supplemented by two additional independent replications including 2637 cases and 3957 controls. SNP rs7927894 on chromosome 11q13.5 near C11orf30, as previously mentioned in allergic rhinitis, was found to be associated with AD with 1.47 times greater risk than noncarriers. A primary epithelial barrier defect has been found to participate in the pathogenesis of AD. Filaggrin can facilitate terminal differentiation of the epidermis and formation of the skin barrier and the lack of normally functioning filaggrin has been suggested in the patients with AD. In a European population, two independent loss-of-function genetic variants (R510X and 2282del4) in the gene encoding filaggrin (FLG) located on chromosome 1p21.3 are significantly associated with AD (Palmer et al. 2006), and this association is the strongest in asthmatic patients in the context of AD. These variants are carried by approximately 9 % of people of European origin. These variants also show highly significant association with asthma occurring in the context of AD. A locus at 1q21.3 (FLG) was later validated by another study in AD in a Chinese Han population that enrolled 1012 cases and 1362 controls, and was followed by a replication study in an additional 3624 cases and 12,197 controls of Chinese Han ethnicity, as well as 1806 cases and 3256 controls from Germany. Susceptibility loci at 5q22.1 (TMEM232 and SLC25A46, rs7701890) and 20q13.33 (TNFRSF6B and ZGPAT, rs6010620) were found in association with AD in the Chinese population, and the 20q13.33 locus was also associated with AD in the German population. A meta-analysis of GWAS of 5606 patients and 20,565 controls from 16 population-based discovery cohorts and of an additional 5419 patients and 19,833 controls from 14 replication cohorts revealed three new risk loci for AD: rs479844 (OVOL1) on the chromosome 11q13 and rs2164983 (ACTL9) on the chromosome 19q13, which are both near genes implicated in epidermal proliferation and differentiation, and rs2897442 (KIF3A) within the cytokine cluster at 5q31.1. Most interestingly, the locus 5q31.1 has also been associated with allergic rhinitis and asthma.

4.5 Conclusion

Allergy is a condition complicated by the interaction between genetic backgrounds and environmental stimulation. The number of people worldwide with allergies is rapidly increasing, particularly in children. AD, one of the most common skin disorders seen in infants and children, is the earliest onset allergic disease seen during the first 6 months of life, and this cutaneous manifestations of atopy often indicates the beginning of the atopic march. Based on longitudinal studies, AD patients, particularly those with the severe phenotype, will develop asthma and allergic rhinitis. Recent studies have revealed that allergic diseases and traits share a large number of genetic susceptibility loci, such as IL-33/IL-1RL1, IL-13-RAD50, and C11orf30/LRRC32, and have been found in more than two allergic diseases. The overlapping of these loci may partly explain the progression of allergic disease from childhood AD to allergic rhinitis and asthma (atopic march). Similarly, allergic rhinitis and asthma that share the same genetic traits has led to the concept of co-morbidity. The understanding of susceptibility genes will help in the development of new and individualized interventions for the treatment of allergic diseases.

4.6 Epigenetics in Allergy

Allergic disease is the most common chronic inflammatory disease in children and causes a substantial morbidity burden (Davies et al. 2003). Evidence indicates that the etiology of asthma and allergic diseases is complex and has strong environmental and genetic components, the later accounting for over half of the risk (Godfrey 2001). Studies have previously documented that DNA sequence variants (SNPs) in multiple genes are involved in innate and adaptive immune pathways either independently or dependently (Hong et al. 2015; Savenije et al. 2014). It is deduced that genes determine in the development of sensitization to environmental allergens in early childhood (Pankratz et al. 2010). Epigenetics is the information passed through cell division other than the DNA sequence itself. Allergic diseases are influenced by the environment and can be modified by exposure in utero, and all of these features are also common to epigenetic regulation. Epigenetic mechanisms provide a new understanding of gene versus environment interactions. With T cell activation, commitment toward an allergic phenotype is regulated by DNA methylation and histone modifications at the T_H2 locus control regions. When normal epigenetic controls are disturbed by environmental exposures, T_H1/T_H2 balance will be affected. Epigenetic changes are not only transferred to daughter cells with cell replication but also can be inherited through several generations. In animal models, with constant environmental pressure, epigenetically determined phenotypes are amplified through generations and can last at least two generations after the environmental stimulation is removed.

Modifications to the epigenome mediate endogenous or exogenous environmental exposures on immune development (Esteller 2008). Epigenetic control of gene expression plays an important role in development, differentiation, and immune regulation in the immune system (Schoenborn et al. 2007). These processes provide regulatory control of gene expression independently of the genomic sequence and vary in response to environmental cues. The genetic and epigenetic factors of allergic diseases interact synergistically with prenatal and early-life exposures (e.g., tobacco smoke, endotoxins, and air pollution) to affect allergic disease risk (Eder et al. 2006). While the enthusiasm for and expectations from GWASs have been slowly fading in the scientific community, findings that environmental exposures could affect the epigenetic profile have begun a new era in allergic disease research by examining epigenetic mechanisms as mediators of these exposures in the occurrence and clinical course of allergic diseases. Epigenetic modifications (DNA methylation, histone modification, and miRNA) can affect transcriptional activity in multiple genetic pathways relevant for the development of asthma and allergic diseases. However, limited work has been carried out thus far in examining the role of epigenetic variations on allergic disease development and management.

Epigenetic mechanisms include DNA methylation, histone variation, chromatin remodeling, posttranslational histone modification, and noncoding RNA. DNA methylation is the first recognized epigenetic mechanism and also the most extensively studied. DNA methylation refers to the covalent addition of a methyl group to a cytosine (C) residue, and the methylation of a gene promoter is correlated with silencing of gene expression. The DNA methylation occurs mostly in the context of CG dinucleotide; however, non-CG methylation has been recently described also at CHG and CHH sites where H is A, C, or T (Lister et al. 2009; Zemach et al. 2010). DNA methyltransferase 1 methylates hemimethylated parent–daughter duplexes during DNA replication. Furthermore, methyltransferases DNA methyltransferase 3a and DNA methyltransferase 3b de novo methylate DNA. The mechanism for demethylating DNA remains controversial. Hypermethylation of CpG sites at gene promoters is associated with transcriptional silencing through precluding the binding of transcription factors to the targeted sequences and increasing affinity to methylated DNA-binding proteins that further recruit other epigenetic modifiers and corepressors (Kuroda et al. 2009; Lister et al. 2009). During cellular differentiation, reprogramming, and development, DNA methylation alterations may also happen not only at promoters but also at other regions that are far away from transcription start sites (Doi et al. 2009; Irizarry et al. 2009; Ji et al. 2010).

Chromatin, the complex of DNA and nucleic proteins in the nucleus, is also a central target of epigenetic modifications. The inactive heterochromatin is densely packed whereas the active euchromatin is less condensed. Changes to these targets can influence DNA folding, DNA-transcription factor interaction, transcript stability, and other methods of gene silencing (heterochromatin) or activation (euchromatin) (Ho 2010). Histone modification via methylation occurs posttranslationally, while miRNAs can control expression of other genes posttranscriptionally (Ho 2010).

Exposure to allergens induces an immune response that triggers the differentiation T_H2 cells, expressing the cytokines IL-4, IL-5, and IL-13, which are responsible for allergic diseases. Loss of DNA methylation and increased association with activating histone marks could establish and maintain a euchromatin structure at the T_H2 locus of T_H2 cells, allowing recruitment of the transcriptional machinery to this region for a rapid and coordinated expression of T_H2-related cytokines (Lee et al. 2002; Tykocinski et al. 2005; Wei et al. 2009). The hypermethylation of the first exon is correlated with promoter hypermethylation resulting in transcriptional silencing.

Core histones of chromatin have long *N*-terminal tails protruding from the nucleosome which undergo posttranslational modifications that alter the interaction with DNA and nuclear proteins. The standard way of reporting the modifications is by naming the histone, followed by the amino acid, and the modification. For example, H3K4me1 denotes single methylation (me1) of lysine 4 (k4) on histone 3 (H3). There is a strong relation between covalent histone modifications and gene expressions (Weissmann and Lyko 2003). Histone acetylation or phosphorylation is associated with an active state. Histone methylation appears to have diverse functions in the control of gene activity, depending on the amino acid and number of methyl- groups added. The addition or removal of the chemical elements on histones could be also catalyzed by histone modifying complexes such as histone acetyl transferase (HAT) and histone deactetylase (HDAC), which add and remove acetyl- groups on histone residues, respectively.

DNA methylation has been associated with changes in IL-4 and IFN-γ transcription (Ho 2010). The early allergic response is marked by increases in IL-4 expression because the GATA-3 transcriptional factor binding sites within the first intron of the gene loses CpG methylation and the IL-4 locus gains H3K9 acetylation and also trimethylation of H3K4 (Fields et al. 2002). T_H2 polarization is associated with loss of IFN-γ expression, which is thought to be mediated by methylation of specific CpGs in its promoter region (Jones and Chen 2006; White et al. 2002). In mouse models, increases in IgE levels are associated with hypomethylation at IL-4 promoter CpG sites and hypermethylation of IL-4 and IFN-γ promoter CpG sites (Liu et al. 2010). Histone acetylation is also associated with IL-4, IL-13, IL-5, IFN-γ, CXCL10, and Foxp3+ transcription patterns (Ho 2010). The miRNA-mediated silencing has been found to repress transcripts associated with HLA-G, IL-13, IL-12p35, and TGF-β (Ho 2010). These studies suggest an important role of epigenetic remodeling in allergic disease.

Allergic diseases can be also found in the epigenetic transgenerational model in which persistent exposure leads to inheritance and augmentation of the phenotype. In a mouse model, diet supplementation in utero with methyl donors has been shown to increase the rate of allergic disease. The F1 progeny that was exposed to a methyl donor diet in utero demonstrated enhancement of the cardinal features of allergic airway disease, including airway hyperreactivity, lung lavage eosinophilia and IL-13, and higher concentrations of serum IgE compared to controls on a normal diet. More importantly, these traits were passed down transgenerationally, although the F2 "grand-children" mice did not have methyl donor supplementation in utero. In humans, transgenerational inheritance influenced by the effects of tobacco may last for two generations. Children whose maternal grandmother smoked during pregnancy had double the chance of developing asthma (Li et al. 2005). This risk was even greater if the mother also smoked during pregnancy (OR = 2.6), supporting the epigenetic transgenerational model in which persistent exposure leads to inheritance and augmentation of the phenotype (Li et al. 2005).

4.7 Epigenomic Study

The most common epigenetic mechanisms include DNA methylation, histone modification, and noncoding RNA. All of above can affect gene transcription through effects on DNA structure and induction of the allergic gene silencing. Microarrays could be the tool of choice for profiling epigenetic marks for allergy genes on a genomic scale, with several platforms and protocols available for DNA methylation (Schoenborn et al. 2007). The above technologies have been widely used for the study of histone marks (ChIP-seq) and miRNAs (miRNA-seq) because they provide superb-quality data. The majority of methylation profiling for allergy genes is still done on array platforms because bisulfite-converted DNA sequencing on the genomic scale is more expensive (Harris et al. 2010). However, several techniques that examine only regions of the genome enriched for methylation marks have been developed (Brinkman et al. 2012; Harris et al. 2010). Recent advances in the development of techniques for epigenomic profiling include attempts to define genome-wide patterns of DNA hydroxymethylation and to study DNA methylation and histone modifications (Brinkman et al. 2012; Song et al. 2012; Statham et al. 2012).

4.8 Environmental Allergy Triggers and Epigenetic Regulations

Maternal and perinatal exposure to pollutants is associated with increased development of allergic diseases. The prenatal and neonatal periods are critical to the development of the immune system and allergic diseases of the airway. Aerosolized pollutants influence development and future phenotype of those systems and diseases. The influence of aerosolized pollutants on the development of immune dysfunction in asthmatics could be mediated through epigenetic remodeling, a mechanism compatible with the early-life programming of this disease. The prevalence of allergic diseases has increased rapidly in the world in only the past few decades and huge variations are observed among populations with similar racial/ethnic backgrounds but different environmental exposures (Alati et al. 2006).

Gene expression modification through epigenetic mechanisms has three primary targets: CpG methylation; amino acid tail modification on histones; and aberrant microRNA expression. The histone modification via methylation occurs posttranslationally, while miRNAs can control expression of other genes post-transcriptionally. Exposure to pollutants could influence DNA folding, DNA-transcription factor interaction, transcript stability, and other methods of gene silencing or activation. Important pollutants that have been implicated in the development of allergic disease are the environmental endocrine-disrupting chemicals (EDC). Exposure to environmental EDCs is associated with allergic diseases via immune mechanisms and epigenetic regulation. For example, nonylphenol and 4-octylphenol may have functional effects on the response of myeloid dendritic cells (mDCs) via, in part, the ER, MKK3/6-p38 MAPK signaling pathway, and histone modifications, with subsequent influence on the T cell cytokine responses (Hung et al. 2010). Phthalates, the common environmental hormone used in the plastic industry, may act as adjuvants to disrupt the immune system and enhance allergic responses. Phthalates can interfere with immunity against infection and promote the deviation of T_H2 responses to increase allergy by acting on human plasmacytoid DCs via suppression of IFN-α/IFN-β and modulation of the ability to stimulate T cell responses (Kuo et al. 2013). Aerosolized pollutants polycyclic aromatic hydrocarbons (PAHs) have been found to be critical constituents in tobacco smoke (Tsay et al. 2013). Tobacco smoke has also shown evidence of inducing epigenetic modification through DNA methylation. Kohli et al. (2012) reports that tobacco smoke exposure is associated with hypermethylation of the promoter region for IFN-γ in the T effector cells and Foxp3 in regulatory T cells.

4.9 Asthma Medications and Epigenetic Regulations

In allergic disease and asthma treatment, some medications also influence the epigenetics in immune cells. Prostaglandin I2 (PGI_2) analog has recently been suggested as a candidate for the treatment of asthma (Hung et al. 2009; Idzko et al. 2007). Iloprost, a PGI_2 analog, enhances H3 acetylation in the MDC/CCL22 (T_H2-related chemokine) promoter area and suppressed H3 acetylation, H3K4, and H3K36 trimethylation in the IP-10 (T_H1-related chemokine) promoter area. The PGI_2 analog also enhances MDC expression via the I prostanoid (IP)-receptor-cAMP, PPAR-α and PPAR-γ, NFκB-p65, MAPK-p38-ATF2 pathways and increasing histone acetylation, and suppresses IP-10/CXCL-10 expression via the IP-receptor-cAMP, PPAR-γ, MAPK- ERK-ELK1 pathways and inhibits histone acetylation and trimethylation in LPS-stimulated monocytes (Kuo et al. 2011). In circulating mDCs, PGI_2 analogs enhance IL-10 and suppress TNF-α expression via IP/EP2/EP4 receptors–cAMP and the EP1 receptor–Ca^{2+} pathway. Iloprost also suppressed TNF α expression via the MAPK-p38-ATF2 pathway and epigenetic regulation by the suppression of histone H3K4 trimethylation (Kuo et al. 2012).

4.10 Conclusions

Allergic diseases and asthma are complex diseases characterized by an intricate interplay of both inheritable and environmental factors. Genetic approaches, especially GWASs, could identify novel genetic targets in the pathogenesis of allergy and asthma, but these targets account for only a small proportion of the inheritability of allergic diseases and asthma. Genetics also fail to explain the sudden rise in allergies as even with significant selection pressure, any change in population genetics necessitates several generations to occur. Epigenetic changes, on the other hand, can be induced more rapidly with various environmental exposures and, as with genetics, the changes can be passed down from parents to offspring. Epigenetics can influence phenotype inheritance through gene imprinting, in utero modifications, and transgenerational inheritance.

Pharmacogenetics is an important example of how gene-environment interactions are already being taken into account in the identification of drug responders and nonresponders. This application represents one component of personalized medicine and places the individual at the center of health care. Gradual accumulation of evidence has solidified the implication of epigenetic regulation as a mediator of complex gene-environment interactions relevant to the development of allergic diseases. Several studies have linked environmental hormones, air pollution, and smoking exposure with allergic diseases via epigenetic mechanisms.

Allergy-related medications could also influence the function of immune cells via epigenetics. Despite the acceptance of epigenetic regulation in the pathogenesis of complex diseases, the extent of environmental epigenetics in the pathogenesis of allergic diseases is just being realized. Large cohort studies are needed to examine the time course and time period of susceptibility for epigenetic regulation following environmental exposures and their contribution to allergic disease. Ultimately, an individual's epigenome early in life may be helpful in determining later risk of allergic diseases and initiating an early intervention or treatment. Studying epigenetics for the associations between environmental exposures, medications, and development of allergic disease holds promise in finding novel pathways for study. The potential to modify genetics and epigenetics may identify approaches to decrease the risk of allergic diseases and improve human health in the future.

References

Alati R, Al Mamun A, O'Callaghan M, Najman JM, Williams GM. In utero and postnatal maternal smoking and asthma in adolescence. Epidemiology. 2006;17(2):138–44.

Andiappan AK, de Wang Y, Anantharaman R, Parate PN, Suri BK, Low HQ, Li Y, Zhao W, Castagnoli P, Liu J, et al. Genome-wide association study for atopy and allergic rhinitis in a Singapore Chinese population. PLoS ONE. 2011;6(5):e19719.

Birben E, Sahiner UM, Karaaslan C, Yavuz TS, Cosgun E, Kalayci O, Sackesen C. The genetic variants of thymic stromal lymphopoietin protein in children with asthma and allergic rhinitis. Int Arch Allergy Immunol. 2014;163(3):185–92.

Black S, Teixeira AS, Loh AX, Vinall L, Holloway JW, Hardy R, Swallow DM. Contribution of functional variation in the IL13 gene to allergy, hay fever and asthma in the NSHD longitudinal 1946 birth cohort. Allergy. 2009;64(8):1172–8.

Bonnelykke K, Matheson MC, Pers TH, Granell R, Strachan DP, Alves AC, Linneberg A, Curtin JA, Warrington NM, Standl M, et al. Meta-analysis of genome-wide association studies identifies ten loci influencing allergic sensitization. Nat Genet. 2013;45(8):902–6.

Bonnelykke K, Sleiman P, Nielsen K. A genome-wide association study identifies CDHR3 as a susceptibility locus for early childhood asthma with severe exacerbations. Nat Genet. 2014a;46(1):51–5.

Bonnelykke K, Sleiman P, Nielsen K, Kreiner-Moller E, Mercader JM, Belgrave D, den Dekker HT, Husby A, Sevelsted A, Faura-Tellez G, et al. A genome-wide association study identifies CDHR3 as a susceptibility locus for early childhood asthma with severe exacerbations. Nat Genet. 2014b;46(1):51–5.

Brinkman AB, Gu H, Bartels SJ, Zhang Y, Matarese F, Simmer F, Marks H, Bock C, Gnirke A, Meissner A, et al. Sequential ChIP-bisulfite sequencing enables direct genome-scale investigation of chromatin and DNA methylation cross-talk. Genome Res. 2012;22(6):1128–38.

Bunyavanich S, Schadt EE, Himes BE, Lasky-Su J, Qiu W, Lazarus R, Ziniti JP, Cohain A, Linderman M, Torgerson DG, et al. Integrated genome-wide association, coexpression network, and expression single nucleotide polymorphism analysis identifies novel pathway in allergic rhinitis. BMC Med Genomics. 2014;7:48.

Davies DE, Wicks J, Powell RM, Puddicombe SM, Holgate ST. Airway remodeling in asthma: new insights. J Allergy Clin Immunol. 2003;111(2):215–25 quiz 26.

Doi A, Park IH, Wen B, Murakami P, Aryee MJ, Irizarry R, Herb B, Ladd-Acosta C, Rho J, Loewer S, et al. Differential methylation of tissue- and cancer-specific CpG island shores distinguishes human induced pluripotent stem cells, embryonic stem cells and fibroblasts. Nat Genet. 2009;41(12):1350–3.

Duffy DL, Martin NG, Battistutta D, Hopper JL, Mathews JD. Genetics of asthma and hay fever in Australian twins. Am Rev Respir Dis. 1990;142(6 Pt 1):1351–8.

Eder W, Ege MJ, von Mutius E. The asthma epidemic. N Engl J Med. 2006;355(21):2226–35.

Esteller M. Epigenetics in cancer. N Engl J Med. 2008;358(11):1148–59.

Ferreira MA, Matheson MC, Tang CS, Granell R, Ang W, Hui J, Kiefer AK, Duffy DL, Baltic S, Danoy P, et al. Genome-wide association analysis identifies 11 risk variants associated with the asthma with hay fever phenotype. J Allergy Clin Immunol. 2014;133(6):1564–71.

Fields PE, Kim ST, Flavell RA. Cutting edge: changes in histone acetylation at the IL-4 and IFN-gamma loci accompany Th1/Th2 differentiation. J Immunol. 2002;169(2):647–50.

Galanter J, Choudhry S, Eng C, Nazario S, Rodriguez-Santana JR, Casal J, Torres-Palacios A, Salas J, Chapela R, Watson HG, et al. ORMDL3 gene is associated with asthma in three ethnically diverse populations. Am J Respir Crit Care Med. 2008;177(11):1194–200.

Godfrey S. Independent inheritance of serum immunoglobulin E concentrations and airway responsiveness. Am J Respir Crit Care Med. 2001;163(4):1030.

Gustafsson DSO, Foucard T. Development of allergies and asthma in infants and young children with atopic dermatitis–a prospective follow-up to 7 years of age. Allergy. 2000;55(5):240–5.

Harris RA, Wang T, Coarfa C, Nagarajan RP, Hong C, Downey SL, Johnson BE, Fouse SD, Delaney A, Zhao Y, et al. Comparison of sequencing-based methods to profile DNA methylation and identification of monoallelic epigenetic modifications. Nat Biotechnol. 2010;28(10):1097–105.

Hinds DA, McMahon G, Kiefer AK, Do CB, Eriksson N, Evans DM, St Pourcain B, Ring SM, Mountain JL, Francke U, et al. A genome-wide association meta-analysis of self-reported allergy identifies shared and allergy-specific susceptibility loci. Nat Genet. 2013;45(8):907–11.

Hirota T, Harada M, Sakashita M, Doi S, Miyatake A, Fujita K, Enomoto T, Ebisawa M, Yoshihara S, Noguchi E, et al. Genetic polymorphism regulating ORM1-like 3 (Saccharomyces cerevisiae) expression is associated with childhood atopic asthma in a Japanese population. J Allergy Clin Immunol. 2008;121(3):769–70.

Ho SM. Environmental epigenetics of asthma: an update. J Allergy Clin Immunol. 2010;126(3):453–65.

Hong X, Hao K, Ladd-Acosta C, Hansen KD, Tsai H-J, Liu X, Xu X, Thornton TA, Caruso D, Keet CA, et al. Genome-wide association study identifies peanut allergy-specific loci and evidence of epigenetic mediation in US children. Nat Commun. 2015; 6.

Hung CH, Chu YT, Suen JL, Lee MS, Chang HW, Lo YC, Jong YJ. Regulation of cytokine expression in human plasmacytoid dendritic cells by prostaglandin I2 analogues. Eur Respir J. 2009;33(2):405–10.

Hung CH, Yang SN, Kuo PL, Chu YT, Chang HW, Wei WJ, Huang SK, Jong YJ. Modulation of cytokine expression in human myeloid dendritic cells by environmental endocrine-disrupting chemicals involves epigenetic regulation. Environ Health Perspect. 2010;118(1):67–72.

Idzko M, Hammad H, van Nimwegen M, Kool M, Vos N, Hoogsteden HC, Lambrecht BN. Inhaled iloprost suppresses the cardinal features of asthma via inhibition of airway dendritic cell function. J Clin Invest. 2007;117(2):464–72.

Irizarry RA, Ladd-Acosta C, Wen B, Wu Z, Montano C, Onyango P, Cui H, Gabo K, Rongione M, Webster M, et al. The human colon cancer methylome shows similar hypo- and hypermethylation at conserved tissue-specific CpG island shores. Nat Genet. 2009;41(2):178–86.

Ji H, Ehrlich LI, Seita J, Murakami P, Doi A, Lindau P, Lee H, Aryee MJ, Irizarry RA, Kim K, et al. Comprehensive methylome map of lineage commitment from haematopoietic progenitors. Nature. 2010; 467(7313): 338–42.

Jones B, Chen J. Inhibition of IFN-gamma transcription by site-specific methylation during T helper cell development. EMBO J. 2006;25(11):2443–52.

Kim WK, Kwon JW, Seo JH, Kim HY, Yu J, Kim BJ, Kim HB, Lee SY, Kim KW, Kang MJ, et al. Interaction between IL13 genotype and environmental factors in the risk for allergic rhinitis in Korean children. J Allergy Clin Immunol. 2012; 130(2): 421–6.

Kohli A, Garcia MA, Miller RL, Maher C, Humblet O, Hammond SK, Nadeau K. Secondhand smoke in combination with ambient air pollution exposure is associated with increasedx CpG methylation and decreased expression of IFN-gamma in T effector cells and Foxp3 in T regulatory cells in children. Clin Epigenetics. 2012;4(1):17.

Kuo CH, Ko YC, Yang SN, Chu YT, Wang WL, Huang SK, Chen HN, Wei WJ, Jong YJ, Hung CH. Effects of PGI2 analogues on Th1- and Th2-related chemokines in monocytes via epigenetic regulation. J Mol Med (Berl). 2011;89(1):29–41.

Kuo CH, Lin CH, Yang SN, Huang MY, Chen HL, Kuo PL, Hsu YL, Huang SK, Jong YJ, Wei WJ, et al. Effect of prostaglandin I2 analogs on cytokine expression in human myeloid dendritic cells via epigenetic regulation. Mol Med. 2012;18:433–44.

Kuo CH, Hsieh CC, Kuo HF, Huang MY, Yang SN, Chen LC, Huang SK, Hung CH. Phthalates suppress type I interferon in human plasmacytoid dendritic cells via epigenetic regulation. Allergy. 2013;68(7):870–9.

Kuroda A, Rauch TA, Todorov I, Ku HT, Al-Abdullah IH, Kandeel F, Mullen Y, Pfeifer GP, Ferreri K. Insulin gene expression is regulated by DNA methylation. PLoS ONE. 2009;4(9):e6953.

Lee DU, Agarwal S, Rao A. Th2 lineage commitment and efficient IL-4 production involves extended demethylation of the IL-4 gene. Immunity. 2002;16(5):649–60.

Li X, Howard TD, Zheng SL, Haselkorn T, Peters SP, Meyers DA, Bleecker ER. Genome-wide association study of asthma identifies RAD50-IL13 and HLA-DR/DQ regions. J Allergy Clin Immunol. 2010; 125(2):328–35.

Li YF, Langholz B, Salam MT, Gilliland FD. Maternal and grandmaternal smoking patterns are associated with early childhood asthma. Chest. 2005;127(4):1232–41.

Lister R, Pelizzola M, Dowen RH, Hawkins RD, Hon G, Tonti-Filippini J, Nery JR, Lee L, Ye Z, Ngo QM, et al. Human DNA methylomes at base resolution show widespread epigenomic differences. Nature. 2009;462(7271):315–22.

Liu F, Killian JK, Yang M, Walker RL, Hong JA, Zhang M, Davis S, Zhang Y, Hussain M, Xi S, et al. Epigenomic alterations and gene expression profiles in respiratory epithelia exposed to cigarette smoke condensate. Oncogene. 2010;29(25):3650–64.

Lu MP, Chen RX, Wang ML, Zhu XJ, Zhu LP, Yin M, Zhang ZD, Cheng L. Association study on IL4, IL13 and IL4RA polymorphisms in mite-sensitized persistent allergic rhinitis in a Chinese population. PLoS ONE. 2011;6(11):e27363.

Moffatt MF, Kabesch M, Liang L, Dixon AL, Strachan D, Heath S, Depner M, von Berg A, Bufe A, Rietschel E, et al. Genetic variants regulating ORMDL3 expression contribute to the risk of childhood asthma. Nature. 2007;448(7152):470–3.

Moffatt MF, Gut IG, Demenais F, Strachan DP, Bouzigon E, Heath S, von Mutius E, Farrall M, Lathrop M, Cookson WO, et al. A large-scale, consortium-based genomewide association study of asthma. N Engl J Med. 2010;363(13):1211–21.

Nilsson D, Henmyr V, Hallden C, Sall T, Kull I, Wickman M, Melen E, Cardell LO. Replication of genomewide associations with allergic sensitization and allergic rhinitis. Allergy. 2014;69(11):1506–14.

Ohno T, Morita H, Arae K, Matsumoto K, Nakae S. Interleukin-33 in allergy. Allergy. 2012;67(10):1203–14.

Palmer CN, Irvine AD, Terron-Kwiatkowski A, Zhao Y, Liao H, Lee SP, Goudie DR, Sandilands A, Campbell LE, Smith FJ, et al. Common loss-of-function variants of the epidermal barrier protein filaggrin are a major predisposing factor for atopic dermatitis. Nat Genet. 2006;38(4):441–6.

Pankratz VS, Vierkant RA, O'Byrne MM, Ovsyannikova IG, Poland GA. Associations between SNPs in candidate immune-relevant genes and rubella antibody levels: a multigenic assessment. BMC Immunol. 2010;11:48.

Ramasamy ACI, Coin LJ, Kumar A, McArdle WL, Imboden M, Leynaert B, Kogevinas M, Schmid-Grendelmeier P, Pekkanen J, Wjst M, Bircher AJ, Sovio U, Rochat T, Hartikainen AL, Balding DJ, Jarvelin MR, Probst-Hensch N, Strachan DP, Jarvis DL. A genome-wide meta-analysis of genetic variants associated with allergic rhinitis and grass sensitization and their interaction with birth order. J Allergy Clin Immunol. 2011a;128:9.

Ramasamy A, Curjuric I, Coin LJ, Kumar A, McArdle WL, Imboden M, Leynaert B, Kogevinas M, Schmid-Grendelmeier P, Pekkanen J, et al. A genome-wide meta-analysis of genetic variants associated with allergic rhinitis and grass sensitization and their interaction with birth order. J Allergy Clin Immunol. 2011b;128(5):996–1005.

Sakashita M, Yoshimoto T, Hirota T, Harada M, Okubo K, Osawa Y, Fujieda S, Nakamura Y, Yasuda K, Nakanishi K, et al. Association of serum interleukin-33 level and the interleukin-33 genetic variant with Japanese cedar pollinosis. Clin Exp Allergy. 2008;38(12):1875–81.

Savenije OE, Mahachie John JM, Granell R, Kerkhof M, Dijk FN, de Jongste JC, Smit HA, Brunekreef B, Postma DS, Van Steen K, et al. Association of IL33-IL-1 receptor-like 1 (IL1RL1) pathway polymorphisms with wheezing phenotypes and asthma in childhood. J Allergy Clin Immunol. 2014;134(1):170–7.

Schoenborn JR, Dorschner MO, Sekimata M, Santer DM, Shnyreva M, Fitzpatrick DR, Stamatoyannopoulos JA, Wilson CB. Comprehensive epigenetic profiling identifies multiple distal regulatory elements directing transcription of the gene encoding interferon-gamma. Nat Immunol. 2007;8(7):732–42.

Song CX, Clark TA, Lu XY, Kislyuk A, Dai Q, Turner SW, He C, Korlach J. Sensitive and specific single-molecule sequencing of 5-hydroxymethylcytosine. Nat Methods. 2012;9(1):75–7.

Statham AL, Robinson MD, Song JZ, Coolen MW, Stirzaker C, Clark SJ. Bisulfite sequencing of chromatin immunoprecipitated DNA (BisChIP-seq) directly informs methylation status of histone-modified DNA. Genome Res. 2012;22(6):1120–7.

Torgerson DG, Ampleford EJ, Chiu GY, Gauderman WJ, Gignoux CR, Graves PE, Himes BE, Levin AM, Mathias RA, Hancock DB, et al. Meta-analysis of genome-wide association studies of asthma in ethnically diverse North American populations. Nat Genet. 2011a;43(9):887–92.

Torgerson DG, Ampleford EJ, Chiu GY, Gauderman WJ, Gignoux CR, Graves PE, Himes BE, Levin AM, Mathias RA, Hancock DB, et al. Meta-analysis of genome-wide association studies of asthma in ethnically diverse North American populations. Nat Genet. 2011b;43(9):887–92.

Tsay JJ, Tchou-Wong KM, Greenberg AK, Pass H, Rom WN. Aryl hydrocarbon receptor and lung cancer. Anticancer Res. 2013;33(4):1247–56.

Tykocinski LO, Hajkova P, Chang HD, Stamm T, Sozeri O, Lohning M, Hu-Li J, Niesner U, Kreher S, Friedrich B, et al. A critical control element for interleukin-4 memory expression in T helper lymphocytes. J Biol Chem. 2005;280(31):28177–85.

Vercelli D. Discovering susceptibility genes for asthma and allergy. Nat Rev Immunol. 2008;8(3):169–82.

Verlaan DJ, Berlivet S, Hunninghake GM, Madore AM, Lariviere M, Moussette S, Grundberg E, Kwan T, Ouimet M, Ge B, et al. Allele-specific chromatin remodeling in the ZPBP2/GSDMB/ORMDL3 locus associated with the risk of asthma and autoimmune disease. Am J Hum Genet. 2009;85(3):377–93.

Wan YI, Strachan DP, Evans DM, Henderson J, McKeever T, Holloway JW, Hall IP, Sayers I. A genome-wide association study to identify genetic determinants of atopy in subjects from the United Kingdom. J Allergy Clin Immunol. 2011; 127(1):223–31.

Wan YI, Shrine NR, Soler Artigas M, Wain LV, Blakey JD, Moffatt MF, Bush A, Chung KF, Cookson WO, Strachan DP, et al. Genome-wide association study to identify genetic determinants of severe asthma. Thorax. 2012;67(9):762–8.

Watson B, Gauvreau GM. Thymic stromal lymphopoietin: a central regulator of allergic asthma. Expert Opin Ther Targets. 2014;18(7):771–85.

Wei G, Wei L, Zhu J, Zang C, Hu-Li J, Yao Z, Cui K, Kanno Y, Roh TY, Watford WT, et al. Global mapping of H3K4me3 and H3K27me3 reveals specificity and plasticity in lineage fate determination of differentiating CD4+ T cells. Immunity. 2009;30(1):155–67.

Wei X, Zhang Y, Fu Z, Zhang L. The association between polymorphisms in the MRPL4 and TNF-alpha genes and susceptibility to allergic rhinitis. PLoS ONE. 2013;8(3):e57981.

Weissmann F, Lyko F. Cooperative interactions between epigenetic modifications and their function in the regulation of chromosome architecture. BioEssays. 2003;25(8):792–7.

White GP, Watt PM, Holt BJ, Holt PG. Differential patterns of methylation of the IFN-gamma promoter at CpG and non-CpG sites underlie differences in IFN-gamma gene expression between human neonatal and adult CD45RO- T cells. J Immunol. 2002;168(6):2820–7.

Zemach A, McDaniel IE, Silva P, Zilberman D. Genome-wide evolutionary analysis of eukaryotic DNA methylation. Science. 2010;328(5980):916–9.

Author's Biography

Dr. Chang-Hung Kuo MD PhD received his Medical Degree from Kaohsiung Medical University, Kaohsiung, Taiwan in 2003 and completed his pediatric residency and allergy, immunology, and rheumatology fellowship at Kaohsiung Medical University Hospital. He is a board-certificated pediatrician, clinical allergist/immunologist, pediatric allergist, immunologist, and rheumatologist. Dr. Chang-Hung Kuo received his PhD from the Graduate Institute of Medicine, Kaohsiung Medical University in 2014 with a thesis on the regulatory factors and mechanisms of the function of asthma-related immune cells. As a Principal Investigator of the National Science Council of Taiwan, his main research interests focus on the function of immune cells, genetic/epigenetic regulation, molecular pharmacology, and the role of environmentally disruptive chemicals in allergic and rheumatologic diseases. Dr. Chang-Hung Kuo was an attending physician at Kaohsiung Medical University Hospital from 2010 to 2014 and has been a visiting staff member since 2014. Dr. Chang-Hung Kuo has also been the Superintendent of Ta-Kuo Clinic since 2014. Dr. Chang-Hung Kuo has published more than 35 scientific papers in internationally well-recognized journals including the journal *Allergy*.

Professor Chih-Hsing Hung PhD is Head of the Department of Pediatrics/Allergy and Immunology Division, Kaohsiung Medical University Hospital, Kaohsiung Medical University, Taiwan and a Professor at the Kaohsiung Medical University. His current research is focused on the effect of environmental hormones on epigenetic regulation and immune cells in allergic diseases. He has found that that several environmental hormones can regulate the function of myeloid and plasmacytoid dendritic cells via epigenetic mechanisms, including histone acetylation and methylation, and has shown the adverse effect of environmental hormones on allergic diseases as well as reveal the detailed effects of environmental hormones on immune cells. Additionally, he is the Vice Secretary of the Taiwan Pediatric Allergy Asthma and Rheumatology Association, and is the Associate Editor of the Journal of Microbiology, Immunology and Infection (Impact factor: 2.349). He has more than 100 publications and is a reviewer of 10 international journals.

Chapter 5
Cross-Reactivity

Wen Li and Zehong Zou

Abstract Cross-reactivity refers to an antigen–antibody reaction that occurs when a specific antibody reacts with antigen(s) other than just the primary sensitizing antigen. This phenomenon is common in allergic diseases. Cross-reactivity among allergens motivated us to reduce the number of allergens used in therapy and research to just the major representative allergens, which have subsequently been confirmed by the results of allergen component-resolved diagnosis and immunotherapy. This chapter addresses the history of the development of the theory of allergen cross-reactivity, the clinical manifestations, the biological basis of allergen cross-reactivity, and a bioinformatics analysis of cross-reactivity.

Keywords Cross-reactivity · Allergen · Co-sensitivity · AFFP · Pan-allergen · Epitopes

5.1 Formation and Development of the Theory of Allergen Cross-Reactivity

Allergens, as the core of the allergic reaction, have been a main focus of allergy research. Throughout a hundred years of history of allergy, allergen research can be divided into seven stages that also provide the foundation for the development of the theory of allergen cross-reactivity.

W. Li · Z. Zou (✉)
Guangdong Provincial Key Laboratory of Allergy and Clinical Immunology,
The State Key Clinical Specialty in Allergy, the State Key Laboratory of Respiratory Disease,
The Second Affiliated Hospital of Guangzhou Medical University,
250# Changgang Road East, Guangzhou, China
e-mail: zouxiaohong128@126.com

A. Tao and E. Raz (eds.), *Allergy Bioinformatics*, Translational Bioinformatics 8,
DOI 10.1007/978-94-017-7444-4_5

5.1.1 Protein Extraction from Allergenic Sources

As early as 1819, John Bostock first reported a case of "hay fever". In 1873, Charles Barkley further proved that "hay fever" is caused by pollen antigens, termed "allergens", which led to the research of allergens (Garrelds et al. 1996). Allergen preparations are indispensible for the diagnosis and treatment of allergic diseases, so the first step into the research of allergens is the preparation of an allergen extract. Early studies of allergens included pollen, fungi, mites, food, etc. Allergen extracts can be used not only in clinical diagnosis (e.g., prick skin test) and in vitro testing but also for allergen-specific immunotherapy. However, batch-to-batch differences in allergen preparations can seriously affect the diagnostic accuracy, the efficacy, and safety of allergen immunotherapy. The document providing guidance on allergen-specific immunotherapy, released by WHO (World Health Organization) in 1998, stated that the quality of the allergen vaccine is critical for the diagnosis and treatment of allergic disease, and that standardized allergen vaccines should be encouraged for clinical use (Bousquet et al. 2008). Both the United States and Europe have developed standardized protocols for allergen preparations (Larsen and Dreborg 2008).

5.1.2 Identification of Novel Allergenic Sources

In recent years, with the development of modern life and genetically modified foods, novel allergens continue to be identified. Every hospital or clinic desires to be able to diagnose allergic disease caused by any and all allergens, thus, the greater number of allergens that are identified, the better allergy research and treatment becomes.

5.1.3 Identification, Characterization, and Cloning of Single Allergens

In the past two decades, with the development of PCR technology, IgE-binding detection technology and cloning and expression technology, allergens have been produced from in vitro expression systems and recombinant hypoallergenic proteins can be obtained through allergen site modification (Walgraffe et al. 2009). According to the nomenclature rules of the allergen nomenclature committee, allergens from the same species are considered to be homologous allergens (isoallergen) if their sequence identity is over 67 % (Radauer et al. 2014). In 1985, 12 novel and unique allergen sequences were cloned and there were no isoallergen genes. From 2000 to 2005, 1085 allergen genes were cloned. However, in this group, 900 allergen genes were found to be isoallergen genes and only 5 were novel allergen genes. Our survey, based on data between 2004 and 2012 from an allergen database (http://www.allergen.org/), showed the same

Table 5.1 Allergen number counted at different times

Group	Species				Allergens			
MM/DD/20YY	11/30/04	06/20/06	05/10/07	06/23/12	11/30/04	06/20/06	05/10/07	06/23/12
Weeds	11	11	11	14	24	31	31	38
Grasses	9	10	11	17	30	30	38	57
Trees	20	20	25	25	46	46	59	66
Mites	9	9	9	9	43	47	55	59
Animals	7	8	8	8	27	27	28	27
Fungi	20	21	23	26	82	90	97	104
Insects	32	32	38	47	75	78	98	120
Foods	56	60	61	91	119	131	142	248
Others	6	8	8	4	30	30	28	8
Total	170	179	194	241	476	510	576	727

trend (Table 5.1). Most allergen research focused on mite allergen and the search for new allergens from related species was continued. A lot of similar work has been carried out on other allergens. Different research institutions study the same allergens either simultaneously or at different times resulting in similar allergen research. Just as Astwood pointed out, the golden age of new allergen discovery is over (Astwood et al. 1995).

5.1.4 Identification of Cross-Reactivity Among Different Allergens

Allergists have found that more than 90 % of *Dermatophagoides pteronyssinus* allergic patients are also allergic to *Dermatophagoides farina* and that *D. pteronyssinus* allergen preparations can be used to treat *D. farina* allergic patients (Fernandez-Caldas and Iraola Calvo 2005). When used in vitro, allergen proteins (mainly similar proteins) isolated from different sources can cross-react with each other. In addition, a comparative analysis of existing allergen sequences showed that the structures of many allergens are similar. Therefore, an interest in the exploration of allergen cross-reactivity began. Cross-reactivity refers to when an antibody to antigen A can react with antigen B and vice versa due to the presence of the same or similar sequences/epitopes/structures (Fig. 5.1).

Barnett defined the concept of allergen cross-reactivity as an allergic reaction caused by an allergen and a specific IgE antibody that was produced to another allergen resulting in a cross-reactivity between these two allergens (Barnett et al. 1987). However, cross-reaction needs to be distinguished from co-sensitization and multiple sensitizations. Co-sensitization or multiple sensitizations are generally related to mixed antigens or allergens, it is a cautious pre-judgment of "cross-reactivity" that a single serum positively reacts to different antigens/allergens. Co-sensitization or multiple sensitizations may be due to the presence of different antibodies reacting with different allergens, or may be caused by the presence of cross-reactive antigens. In general, when allergen cross-reactivity is suspected, co-sensitivity is the preferred term (Aalberse et al. 2001; Sidenius et al. 2001). However, in practice, true cross-reactivity of allergens and co-sensitivity may be difficult to differentiate based on the clinical presentation alone, especially in cases of cross-reactivity between nonhomologous allergens.

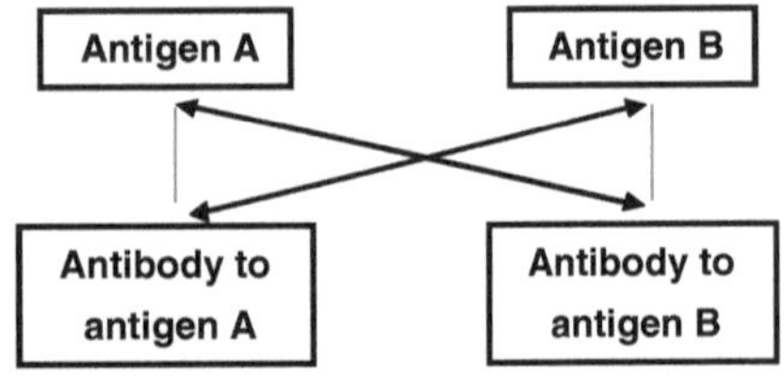

Fig. 5.1 Cross-reactivity of antigen A and antigen B

The cross-reactivity among different allergens occurs on several levels: First, different allergens from the pool of diverse species can have high similarities in their amino acid sequences, resulting in similar primary structures that induce the cross-reaction among different allergens. Analysis of the 280 major allergen entries retrieved from the public database resulted in them being able to be condensed into 21 major representative ones, thus allowing for the use of a smaller number of allergens when conducting comprehensive allergen testing and immunotherapy treatments (He et al. 2014). Second, pan-allergens such as profilin (Tao and He 2004), lipid transfer protein, tropomyosin, etc. are evolutionarily conserved in plants and humans and play an important role in cross-reactivity (Asero et al. 2001; Reese et al. 1999). Third, having highly similar 3D structures confer cross-reactions among different allergens, including those with remote homology. Last, but not least, allergen epitopes [antigenic determinants (AD)] play a key role in the cross-reactivity of allergens (Levin et al. 2014).

5.1.5 Major Representative Allergens

Major allergens are proteins that substantially bind to IgE from more than 50 percent of the patients with that specific allergy (Nordlee et al. 1996). Major allergens have usually been employed as internal standards in order to standardize allergen vaccines (Seppala et al. 2011). Detailed characterization of the important major allergens and their biological sources would allow for improved allergen standardization and, thus, help to obtain more effective and safer regimes for the diagnosis and therapy of many allergic diseases. The growing number of available allergen sequences and major allergens, in addition to the advancements of bioinformatics tools and methods, have made it possible for scientists to shed light on the evolutionary and structural relationships between allergens from different sources. In 2004, Tao and He analyzed the phylogenetic relationships between the newly obtained allergen profilin D106 (GenBank accession No AY268426) from short ragweed (Rg, *Ambrosia artemisiifolia* L.) and its cognates from other species and found that profilin D106 shared a similarity as highly as 54–89 % with different pollen allergens (from mugwort CAD12862 to apple Q9XF41) and as 79–89 % with different food allergens (from peanut Q9SQI9 to peach Q8GT39) (Tao and He 2004). This result made them contemplate the cross-reactivity between different allergens and the selectivity of allergen cloning. In 2005, Breiteneder discovered that most plant food allergens belong to only four structural families, including the cereal prolamin superfamily, the cupins, Bet v 1 (homologues of the major birch pollen allergen) and profilins, indicating that conserved structures and biological activities may play an important role in cross-reactivity (Jenkins et al. 2005). One year later, they found that pollen allergens belong to few protein families and show distinct patterns of species distribution (Radauer and Breiteneder 2006). In 2007,

Tao et al. gradually reduced the number of major allergens according to sequence similarity into fewer major representative ones (Tao et al. 2007), which facilitated recombinant allergen work.

5.1.6 Component-Resolved Diagnosis and Immunotherapy

Individual or few major allergens can be used not only to diagnose the genuine sensitization of patients to a given allergen or to cross-reactive molecules, but also for allergen-specific immunotherapy to yield the same effects as whole allergen extracts in allergic patients (Jahn-Schmid et al. 2003; Pittner et al. 2004). Intensive clinical evidence has proven that specific immunotherapy can prevent both the progression of allergies and the acquisition of new allergies. Immunotherapy with peptides containing T cell epitopes may provide an efficacious and safe alternative to conventional subcutaneous and/or sublingual immunotherapy using native full-length allergen preparations (Linhart et al. 2014).

Allergen is the term for the antigen that induces allergy and the accurate identification of allergens is vital for the treatment of allergic diseases. Allergen extracts are traditionally used in the diagnosis of allergic disease. But the allergen extracts include allergic and nonallergic substances, and natural extracts may contain pan-allergens, which can seriously affect diagnostic accuracy and safety. For example, house dust mite extract contains Der p 10, which belongs to the pan-allergen tropomyosin. If the patient is allergic to tropomyosin from one allergenic source but tolerant to house dust mite, testing using house dust mite extract could result in a misdiagnosis and affect subsequent desensitization therapy. A diagnosis of allergy using a single allergen or a few allergens can more accurately reflect true allergen sensitization or a cross-reactivity condition. It has been reported that treatment with a single allergen can produce the same curative effects as treatment with the allergen extract. Therefore, successful immunotherapy with one allergen can potentially prevent the development of allergy to other allergens.

The clinical use of allergen components can improve the diagnostic accuracy of allergic diseases and efficacy. However, some publications about allergen diagnosis have some general shortcomings: (1) the number of cases is small; (2) the geographical area of research is narrow (mostly concentrated in Europe); (3) the related race is limited; and (4) experiments using the same allergen were not repeated often (Incorvaia et al. 2014; Mohamad Yadzir et al. 2014). Nevertheless, it has been proven that treatment with a single allergen is safe and effective and that use of a single allergen or a few allergens not only identifies the true allergen sensitization or cross-reactive condition but also can be used for immunotherapy. Furthermore, successful immunotherapy with one allergen can prevent the development of allergy to other allergen sources (Ciprandi et al. 2011; Durham 2011; Malling 2003).

5.1.7 Progressive Clustering of Allergen Family Featured Peptides

Some protein families contain both allergens and nonallergens. Some nonallergens have high sequence identity with allergens, which suggests the presence of nonallergen components in allergens. In 2012, we developed the software SORTALLER that helped us reduce published allergens to 72 allergens with 534 characteristic peptides [Allergen Family Featured Peptides (AFFPs)] and can successfully predict more than 2290 nonredundant allergens (Zhang et al. 2012).

Cluster analysis was conducted on allergen sequences in each known allergen family. The allergen sequence was divided into peptides of 6–32 amino acids in length using a 1–10 amino acid sliding window, and then followed by a BLAST (Basic Local Alignment Search Tool, a local alignment search tool basic sequence) analysis with nonallergen sequences. Those peptides that were similar to nonallergen fragments were excluded. If the peptide did not match with nonallergen sequences and the E value of BLAST was lower than 10^{-5}, the peptide was retained and referred to as allergen featured peptide (AFP). AFFPs are each composed of 2–30 small adjacent peptides in the same allergen. IgE cross-reactivity may also result from a single conserved region despite a low overall level of sequence identity (Jenkins et al. 2005). In 2013, Breiteneder identified that only five consecutive amino acid residues are cross-reactive among Ara h 1, Ara h 2, and Ara h 3 (Bublin et al. 2013).

Previously, we retrieved online 478 allergen sequences and clustered them into eight groups by sequence similarity (Tao and He 2005). Subsequently, we focused on the major allergens, retrieved 280 entries, and retained 59 major allergens after initial reduction, which were initially classified into seven clusters. Two or more neighboring clusters were combined to form a new data source for further clustering. This procedure was iterated until the last clustering exhibited 21 allergens that were distantly related to each other (Fig. 5.2). Further alignment showed that several pairs of allergens could be, respectively, grouped together, however, less than 15 % share alignment of their amino acid sequences and most of them exhibited different tertiary structures (Fig. 5.3).

Continuing with the analysis of AFFPs using a progressive clustering procedure similar to the one mentioned above, groups with remote homology (<20–35 % in local region) were represented by single entries and 534 AFFPs were reduced into 20–22 allergens (Fig. 5.4) through five cycles of "cluster-selection-alignment." All the core peptides contain 3–5 matching residues with adjacent mismatches.

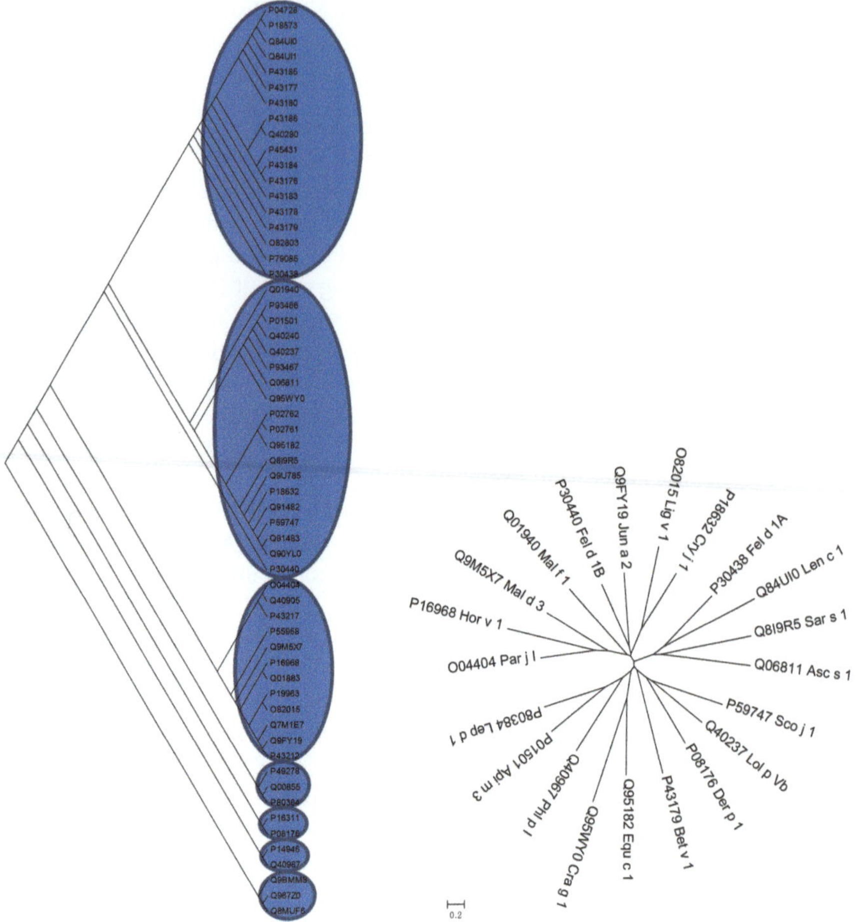

Fig. 5.2 Progressive clustering of the major allergens acquired in UniProtKB/Swiss-Prot database. **a** Phylogeny tree constructed and tested by Maximum Likelihood Estimation for 59 of 280 major allergens. **b** Bootstrap consensus tree of 21 major representative allergens reduced from 59 major allergens (He et al. 2014)

5.2 Cross-Reactivity Among Different Kinds of Allergens

Cross-reactions usually occur between the same species. For example, most dust mite-allergic patients are allergic to both *D. pteronyssinus* and *D. farina* (Fernandez-Caldas and Iraola Calvo 2005). But cross-reactions can exist between different species. For example, birch pollen-sensitive patients can have allergic symptoms after eating celery (Vieths et al. 2002). Cross-reactions between allergens are widespread in allergic disease. Thus, analyzing and understanding the clinical manifestations and molecular mechanisms of cross-reactivity among

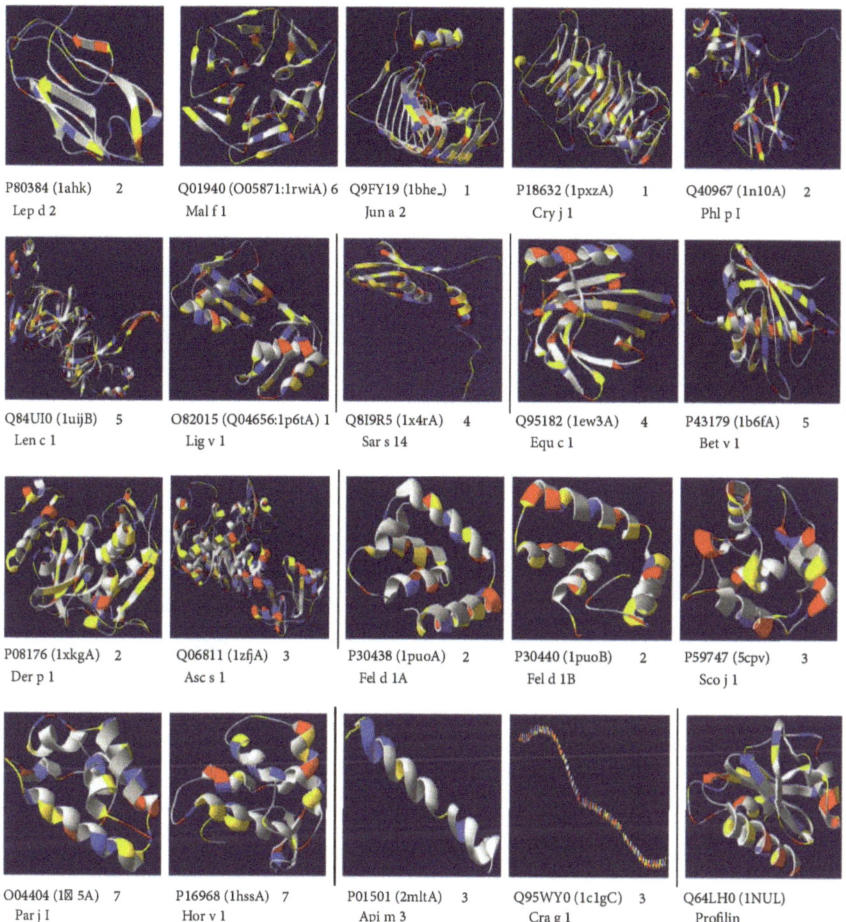

Fig. 5.3 Ribbon diagrams of profilin (Q64LH0) and 19 out of the 21 major representative allergens retained. The seven structural classes are separated by vertical lines. PDB codes are shown in brackets; the following number indicates the initial sequence clustering groups. Q64LH0 was not retrieved as a major allergen, thus is without an initial cluster number. Q9M5X7 and Q40237 were omitted for space consideration (He et al. 2014)

allergens has important significance in guiding the diagnosis and treatment of allergic disease and the development of allergen vaccines.

Allergens are the antigens that can induce the generation of specific IgE antibodies and react with them. They have a wide range of sources, including animals, plants, and microorganisms. Most allergens are proteins and a few of them are glycans, such as *Candida albicans* allergens (Savolainen et al. 1990).

Most allergens have potential biological functions. According to their biological activity or their similarity with proteins of known function, allergens can be

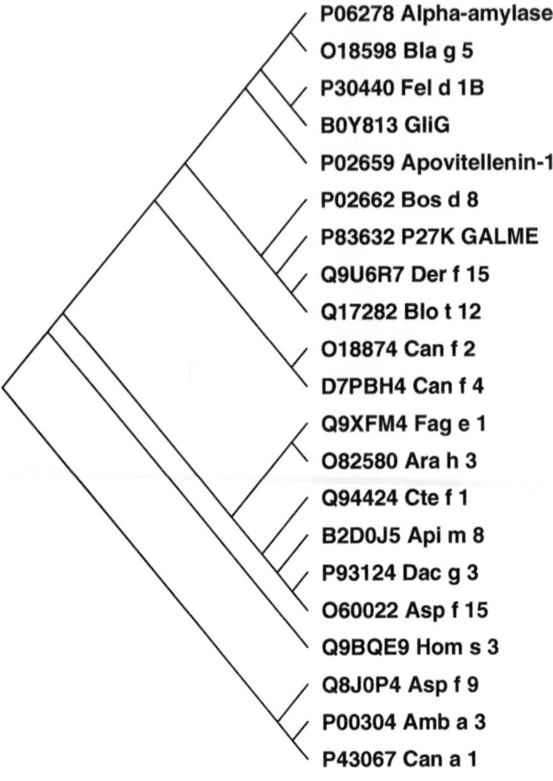

Fig. 5.4 Maximum Parsimony Tree deduced by MEGA5.10 for 21 representative AFFPs (Adapted from the reference He et al. 2014)

divided into several categories, such as enzymes, enzyme inhibitors, transport proteins, and regulatory proteins (Stewart and Thompson 1996). Common allergens include inhalant allergens and food allergens. Cross-reactivity between different inhaled allergens, food allergens, or the reaction between allergens of these two groups is discussed below.

5.2.1 Cross-Reactivity Among Inhalant Allergens

Inhalant allergens are the common causes of allergic rhinitis and asthma. They are usually divided into indoor allergens, outdoor allergens, and occupational allergens. Generally speaking, outdoor allergens are likely to cause seasonal rhinitis and indoor allergens are prone to cause principle perennial rhinitis and asthma (Braun-Fahrlander et al. 1997; Gergen and Turkeltaub 1992). Some of the more common inhalant allergens are dust mites, pollen, and cockroaches.

Most dust mite-sensitive patients are allergic to *D. farinae* and *D. pteronyssinus* (Fernandez-Caldas and Iraola Calvo 2005). In Sweden, *D. siboney* has strong cross-reactivity with the other three kinds of dust mites (*D. farinae, D. pteronyssinus, D. mirocroceras*) (Sidenius et al. 2001). Dust mites and *Euroglyphus maynei* have significant IgE cross-reactivity. Two hundred fifty suspicious allergic patients who had positive prick test results to *E. maynei* also reacted to dust mite. In these two studies, dust mite extract inhibition of reactivity to *E. maynei* was more obvious than the inhibitory effects of *E. maynei* extract on dust mite reactivity (Thomas et al. 2004). In addition, most studies found that storage mite allergy and dust mite allergy were positively correlated, with 60–88 % of the dust mite-sensitive patients also being allergic to storage mite. Cross-reactions between dust mites and *Acarus siro* or *Tyrophagus putrescentiae* are stronger than the cross-reactions between dust mites and *Lepidoglyphus destructor*. The inhibition of dust mite extract to IgE-binding reactivity of *A. siro* or *T. putrescentiae* is more remarkable than that of *A. siro* or *T. putrescentiae* extract to dust mite (Sidenius et al. 2001; Thomas et al. 2004).

Extensive cross-reactions exist among pollen allergens and can even sometimes occur between the pollens from different families. Wahl et al. reported that there was cross-reactivity between birch and ash tree pollen (Wahl et al. 1996). Castro et al. reported that plantain pollen allergens and olive pollen allergen cross-react (Castro et al. 2007). In India, it was reported that cross-reactivity occurs between proteins of the latex from *Hevea brasiliensis*, seeds and pollen of *Ricinus communis*, and pollen of *Mercurialis annua*, members of the *Euphorbiaceae* family (Palosuo et al. 2002). During the process of desensitization therapy, in order to control the disease by avoiding exposure, not only should real allergens be avoided, but also cross-reactive allergens. It is very important for the diagnosis and prevention of hay fever to understand cross-reactivity among pollen allergens.

Most cockroach-sensitive patients are allergic to both American cockroach and German cockroach (Bassirpour and Zoratti 2014). Moreover, cockroach allergy and dust mite allergy are positively correlated. In high-load areas of cockroach, the number of cockroach-allergic patients is higher than in the low-load regions. The positive correlation between cockroach allergy and dust mite allergy is partly because cockroach and dust mite allergens coexist in the same environment. Cross-inhibition between dust mite and cockroach occurs in most sera assays. Furthermore, inhibition of mite on cockroach reactivity is greater than that of cockroach on mite reactivity (Jeong et al. 2004).

5.2.2 Cross-Reactivity Among Food Allergens

It is very common to see cross-reactivity between cow and goat milk. 92 % of cow milk-sensitive patients are also allergic to goat milk (Bellioni-Businco et al. 1999). But only 4 % of the cow milk-sensitive children are allergic to mare milk (Businco et al. 2000). 10–20 % of the cow milk-sensitive children are allergic to beef, while 93 % of the beef-sensitive children are allergic to milk (Martelli et al. 2002; Werfel et al. 1997). In addition, cross-reactivity is widespread among different types of poultry, mammalian meat, eggs (chicken, duck and goose), beans and nuts (Bernhisel-Broadbent and Sampson 1989; Ewan 1996; Langeland 1983; Restani et al. 2009). Mollusks cross-react with crustaceans and other invertebrates (arthropods, sponges, nematomorpha). Thus, avoidance of all crustaceans, mollusks, and other shellfish is generally recommended for people who are allergic to seafood (Leung et al. 1996; Reese et al. 1999).

5.2.3 Cross-Reactivity Between Inhalant Allergens and Food Allergens

Cross-reactions exist between pollen allergens and food allergens, especially vegetables and fruits. In 1942, it was reported that some fresh fruits and vegetables caused contact urticaria in patients with birch hay fever. This type of disease was termed the fruit-vegetable-pollen cross-reactive allergy syndrome or, more recently, pollen allergy syndrome (Caballero and Martin-Esteban 1998). Such cross-reactivity is caused by an actin-binding protein, which is widespread in eukaryotic cells, and is also known as profilin. Dust mite-snail allergy may be a very important issue in those countries that use snail as a food ingredient. Cross-reactions occur between dust mite allergens and other invertebrate allergens and can cause or aggravate allergies to foods such as snails, shrimp, etc. For example, we often find that dust mite-allergic patients who had never eaten snails had an allergic reaction the first time it was ingested and developed severe symptoms such as asthma, anaphylactic shock, generalized urticaria, and/or facial edema. Inhibition tests showed that specific IgE reactivity to snail can be inhibited by dust mite extract. However, snail extract did not significantly inhibit the binding between dust mite extract and dust mite-specific IgE. These results demonstrated the existence of cross-reactivity between dust mite and snail. It also showed that dust mites are a common allergen and that snail allergy was often caused by a cross-reaction with dust mites (Fernandez-Caldas and Iraola Calvo 2005; Sidenius et al. 2001).

5.3 The Biological Basis of Allergen Cross-Reactivity

Cross-reactivity is an immunological phenomenon that is influenced by at least three factors: (1) primary and tertiary structure of the allergen, (2) protein family (Pan-allergen), and (3) allergen epitopes. This section focuses on these three aspects.

5.3.1 Primary and Tertiary Structures of Allergens

The similarity between primary and tertiary structures determines the likelihood of cross-reactivity between allergens. The FAO/WHO predicts the chance of cross-reactivity based on the identity of six or more contiguous amino acids or a minimum 35 % sequence identity over a window of 80 amino acids (Umeda 2006). In particular, point mutations are able to inhibit IgE-binding capacity. The allergenicity of a mutant recombinant allergen was significantly reduced compared to wild type, but the immunogenicity was maintained, a phenomena that can be used to reduce immunotherapy adverse reactions (Vrtala et al. 2007).

Based on the original data released by the International Union of Immunological Societies, Clustal W 1.83, MEGA5.0 and other bioinformatics software that were used and then enhanced by manual selection, 59 major allergens from public databases were progressively analyzed for clustering. Seven clusters (Fig. 5.2a) were found that were then gradually reduced to 21 representative amino acid sequences without any relationship to each other (Fig. 5.2b). The 3D structures of the 21 major representative allergens are depicted in Fig. 5.3. When inspected from their spatial structural orientations and surface exposure of the allergens, all of the 21 allergens were shuffled against the initial clustering and, interestingly, fell into these seven structural classes (Fig. 5.3). However, this classification is complicated by the existence of similar structural scenarios in different structural classes (He et al. 2014).

- **Up-and-down β-barrel**: includes P80384, the major allergen Lep d 2 from Fodder mite (*L. destructor*), as well as other group 2 allergens from dust mite species, for example, Der f 2 (P49278) from *D. farinae*.
- **β-meander and/or ψ-loop constituted calyx**: shortened as (α) calyx, such as hexagon, cradle, and globose twins: includes allergens Q01940, Q9FY19, P18632, Q40967, Q84UI0, and O82015.
- **α-β structured crane:** includes Q8I9R5, an allergen from *Sarcoptes scabiei* type *hominis*.
- **α-β arranged banana string**: includes Q95182 and P43179.
- **(β) formed complex**: includes P08176 and Q06811.
- **α-helix built clips:** a large group including P30438, P30440, P59747, Q40237, O04404, Q9M5X7, and P16968. Interestingly, P30438 and P30440, chain 1 and chain 2 of cat allergen Fel d1, configure a pair of chiral molecules on the 3D level. O04404, Q9M5X7, and P16968, originating from different taxonomic species but classified into the same initial Cluster 7, here exhibited structures similar to each other.
- **α-helix spiral cord**: includes P01501 and Q95WY0, both of which came from the same initial cluster III. P01501 is an allergen Api m 3 from honeybee (*Apis mellifera*), also a main toxin of bee venom with strong hemolytic activity. Q95WY0 is the major oyster allergen and tropomyosin from the Pacific oyster *Crassostrea gigas*. These two allergens exhibited similar structures but displayed different lengths of their spiral cord.

Comparisons of 3D structures suggest that the properties conferred by protein folds in the allergens may potentiate protein immunogenicity, as well as allergenicity. Altogether, combining structural information with analysis of the conservation of primary structure indicates that the conservation of surface features (particularly main chain conformations that cannot be assessed by using simple amino acid similarities) would offer a sounder basis for assessing potential IgE cross-reactivity in novel proteins (Jenkins et al. 2005).

A comparison of sequence conservation within allergen family members, taxonomic distribution of identified allergens, and cross-reactivity data from the literature shows a high degree of correspondence. Omnipresent, highly cross-reactive allergens such as profilins and calcium-binding proteins show sequence similarities of more than 60 % between members of different plant families. In contrast, β-expansins from grasses share less than 50 % of their sequences with nonallergic homologues from other plant families, which do not cross-react with grass pollen expansins (Radauer and Breiteneder 2006).

A review of sequence similarities among allergenic and nonallergenic homologues of pollen allergens shows that the prerequisite for allergenic cross-reactivity is sequence identity of at least 50 %. The Bet v 1 family is the only known exception from this rule. Cross-reactive Bet v 1 homologues share only 37–67 % of their sequences with Bet v 1. This may be explained by the fact that surface exposed residues of Bet v 1 homologues are more conserved than buried ones, resulting in higher levels of similarity between surfaces compared with whole sequences (Radauer and Breiteneder 2006).

5.3.2 *Protein Family (Pan-Allergen)*

In general, for species, the closer they are in classification, the greater the likelihood of a cross-reaction. However, cross-reactivity between allergens from distantly related species has also been found. Allergen components are divided into different protein families according to their function and structure, and IgE cross-reactivity occurs in antibodies to the same protein family. Therefore, classifying allergen components helps to explain the problem of cross-reactivity. A quantitative comparison of plant allergens indicates that, in spite of the number and diversity of proteins encoded in plant genomes, the universe of plant allergens is rather small, with about 65 % of plant food allergens coming from only four protein families. Such information suggests that membership in one of these families is likely to result in a greater potential for a protein to become an allergen, providing that additional factors such as presence in plant foods are taken into account (Jenkins et al. 2005). Below are the four common families of proteins associated with cross-reactivity:

- **Pathogenesis-related protein family (PR)**

Represented by Bet v 1 (PR-10), the pathogenesis-related protein family is commonly found in defense proteins of higher plants. Its expression is induced by pathogens or other external stimulating factors and plays an important role in plant defense against pathogens and external pressure and adaption to adverse environmental processes. Many food allergens have been reported to be members of protein families, such as hazelnut (Cor a 1), celery (Api g 1) and soybean (Gly m 4). Highly similar amino acid sequences in these families lead to cross-reactivity; however, heating or enzymatic hydrolysis easily destroys the binding ability of allergens to IgE. 50–90 % of birch pollen allergic patients develop allergic symptoms when ingesting apple, carrot, celery and hazelnut, which is the result of cross-reactivity with Bet v 1 (Vieths et al. 2002). Mal d 1 is the major apple allergen and has 64 % sequence similarity with Bet v 1, but there are still some differences in serum cross-reactivity and clinical symptoms. The IgE of the vast majority birch allergic patients is capable of binding to Mal d 1, but not all patients exhibit clinical symptoms. Further studies have shown that the cause of this phenomenon is that Bet v 1 has different B cell epitopes that bind IgE (Klinglmayr et al. 2009). The remarkable conservation of both surface residues and main chain conformations in the Bet v 1 family is of great importance in the conservation of IgE-binding epitopes and underlies the fruit-vegetable-pollen cross-reactive allergy syndrome (Jenkins et al. 2005).

- **Lipid transfer protein (LTP)**

Lipid transfer protein is the main allergen in *Rosaceae* fruit and *Amygdaloideae* fruit, especially in pericarp. Enzymatic digestion or heating does not destroy its antigenicity. Allergy symptoms, such as oral allergy syndrome and severe systemic allergic reactions, occur in the absence of pollen allergens. Therefore, LTP is often considered as a nonpollen-related allergen. However, evidence shows that allergy symptoms in some patients with hay fever are caused by LTP. In a study of 24 patients with Artemisia pollen allergy, more than 70 % of these patients had a positive LTP skin test. Mugwort LTP binding to IgE is not inhibited by pear LTP in inhibition ELISA experiments, while 50 % of patients tested for IgE binding to pear LTP showed inhibited IgE binding by mugwort LTP, indicating cross-reactivity with mugwort pollen (Lombardero et al. 2004).

- **Profilins**

As plant allergens, profilins are present in almost all eukaryotic cells. Initially reported as actin-binding proteins in 1977, profilin is the key factor in rapid actin reorganization (Ramachandran et al. 2000). Profilin-sensitized patients are allergic to a large number of pollens and foods. Patients with birch pollen allergy caused by Bet v 2 often develop oral allergy syndrome after ingesting pears, apples, carrots, and celery. In addition, grass allergic patients may react to carrots and celery due to cross-reactivity caused by profilins (Canis et al. 2011).

- **Tropomyosin**

Tropomyosin is a major allergen of invertebrates. This allergen exists in crustaceans (shrimp, lobster and crabs, etc.), mollusks (oyster, snail, abalone, scallop and turbo snail, etc.), and invertebrates (mite, cockroach, worms, etc.) (Table 5.2).

Table 5.2 Identification of the major invertebrate tropomyosin allergens

Species	Allergen name	UniProt number	Type of allergen
Anisakis simplex	Ani s 3	Q9NAS5	Food
Blattella germanica	Bla g 7	Q9NG56	Inhalant
Charybdis feriatus	Cha f 1	Q9N2R3	Food
Chlamys nipponensis	Chl n 1	002389	Food
Crassostrea gigas	Cra g 1	Q95WY0	Food
Dermatophagoides farinae	Der f 10	Q23939	Inhalant
Dermatophagoides pteronyssinus	Der p 10	018416	Inhalant
Haliotis diversicolor	Hal d 1	Q9GZ71	Food
Helix aspersa	Hel as 1	097192	Food
Homarus americanus	Hom a 1	044119	Food
Metapenaeus ensis	Met e 1	Q25456	Food
Periplaneta americana	Per a 7	Q9UB83	Inhalant
Turbo cornutus	Tur c 1	B7XC63	Food

According to the BLAST search results in UniPort, the oyster (*C. gigas*) allergen Cra g 1 shares sequence identity of 51 %, 50 %, 51 %, 52 % and 51 % with worm allergens (Ani s 3), German cockroach (Bla g 7), America cockroach (Per a 7), dust mite (Der f 10), and house dust mite (Der p 10), respectively. The homology of Cra g 1 is 47 % with crabs (Cha f 1), 54 % with Metapenaeus (Met e 1), 53 % with lobster (Hom a 1), and has similarity with other mollusks, such as scallops (Chl n 1) 58 %, abalone (Hal d 1) 66 %, snails (Hel as 1) 65 %, and *Turbo cornutus* (Tur c 1) 62 %. These similarities are near or exceed 50 %, reflecting the high similarity of tropomyosin between different species. Phylogenetic analysis of the main invertebrate allergen tropomyosin conducted by Mega 4.1 show that crustaceans and mollusks are in different clusters (Fig. 5.5). The similarity between mites, cockroaches, and crustaceans is higher than that of mollusks, which suggests that stronger cross-reactivity may exist between mites, cockroaches, worms, and other invertebrates. Mite-allergic patients may need to avoid eating aquatic foods such as crustaceans. But, in fact, cross-reactivity between crustaceans and mollusks is very common, and cross-reactivity between food allergens has been extensively studied. In contrast, cross-reactions between food allergens and inhalant allergens have been less studied, indicating a greater need for research and investigation in this area. In general, cross-reactive species have similar structures between allergens. The main allergic protein of the species above is tropomyosin, thus, the chance of cross-reactivity is very high. However, allergists have gradually reached a consensus that the main cause of cross-reactivity among different foods is due to the high similarity between allergen epitope regions, thus, common allergenic epitopes play an important role in cross-reactions between mites, cockroaches, other invertebrates, crustaceans, and mollusks (Reese et al. 1999).

The AllFam allergen database is allergen family database that elucidates the structure and biochemical properties of protein allergens, allowing for the

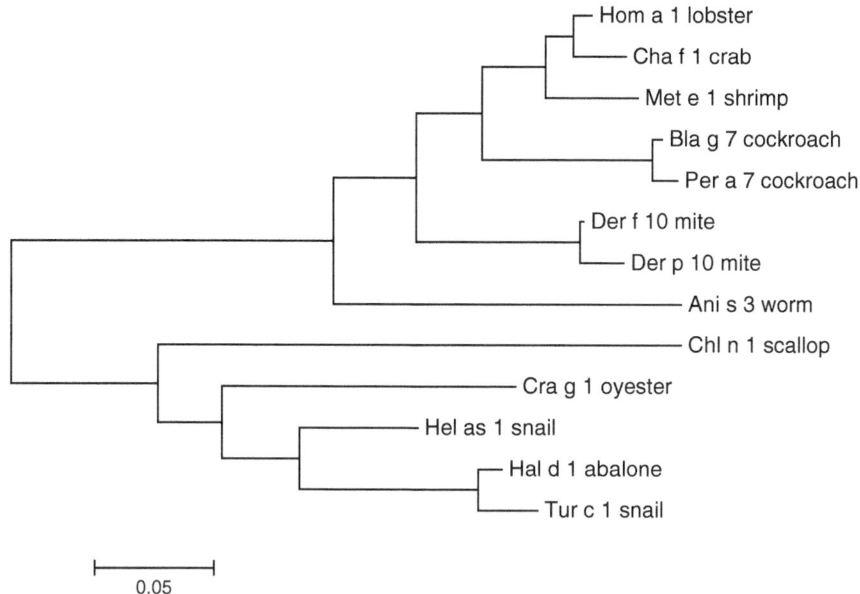

Fig. 5.5 Phylogenetic tree of the main invertebrate allergen tropomyosin (NJ method)

understanding of an allergen from a single allergen or allergen family as well as the systematic analysis of all allergens. This database includes 847 allergens, with 707 allergens classified into 134 AllFam families containing 184 different Pfam domains. Thus, allergen domains are only 2 % of the Pfam protein family database. Thirty-eight allergen families are grouped into 12 superfamilies or clans according to structure similarity or sequence motifs. The most important cluster is the triose phosphate isomerase (TIM) barrel glycosyl hydrolase superfamily containing seven allergens in the glycosyl family, and includes the major allergen families from mites, plants, fungi, and insect venom (Radauer et al. 2008).

As far as their biochemical function is concerned, allergens show a bias toward some activities such as hydrolysis of proteins, polysaccharides, and lipids, binding of metal ions and lipids, transport, storage, and cytoskeleton association. The cysteine protease Der p 1 allergen is a typical example of biochemical function correlated with sensitization (Radauer et al. 2008).

5.3.3 Allergen Epitopes

Allergic epitopes, known as AD, are specific chemical groups that determine antigenic specificity. Antigens bind to a corresponding lymphocyte surface antigen receptor at the AD, leading to activation of lymphocytes, which elicits an immune response. Antigens can also bind specifically to corresponding antibodies or sensitized lymphocytes by the immune epitopes. Antigenic specificity is determined by the features, number, and spatial configuration of the epitopes. Allergens are the antigens that cause allergic reactions. Once an allergen enters the body, B cells are induced to produce IgE antibodies, which bind to surface FcεRI of target cells (mast cells or basophils). After two or more IgE molecules on the mast cell or basophil surface bind to the same bivalent or polyvalent allergen molecule, a cross-link is formed with FcεRI-IgE resulting in the activation of mast cells and basophils that then induce an allergic reactions (Gould and Sutton 2008). Therefore, cross-reactivity usually occurs between allergens that share the same IgE-binding epitope. There may also be the same or similar epitopes between the cross-reactive antigens. B cell epitopes (IgE-binding epitope) include both linear and 3D epitopes with most allergenic epitopes being three-dimensional. Five IgE-binding epitopes of a pan-allergen in invertebrates, tropomyosin, have been identified (Subba Rao et al. 1998). The amino acid sequence of tropomyosin in vertebrates and invertebrates is highly similar, but there are large differences in the amino acid sequence of the corresponding B cell epitope. This suggests that, for the cross-reactive allergens, the B cell epitope sequence is more important than the sequence of the protein (Klinglmayr et al. 2009).

Besides proteins, carbohydrates are another common cause of cross-reactivity, including cross-reactive carbohydrate determinants (CCDs), which generally do not cause clinical symptoms, and saccharides, which can cause other clinical symptoms. CCDs commonly include *N*-terminal glycoprotein oligosaccharides of plants and animals, wherein its two typical structures are α (1, 3) trehalose the *N*-acetyl glucosamine-glucosamine and β (1, 2) xylose connected to the mannose core. These two epitopes, and other parts of sugars, synergistically induce CCD-IgE. Therefore, CCDs are the structural basis for a variety of cross-reactions (Altmann 2007; Vieths et al. 2002). Whether CCD-IgE can induce type I hypersensitivity allergy symptoms remains controversial. CCDs contain monovalent antigen epitopes and can only cause a positive result serological test in vitro. While the occurrence of type I allergy requires basophils or mast cells and other effector cell surface receptors to form dimeric IgE structures, an allergen must have at least two or more binding sites to cause IgE crosslinking. Pineapple enzyme protein only has one sugar chain, so the prick tests are negative. However, some glycoproteins may contain multiple *N*-glycosylation chains such as HRP but cannot cause basophil degranulation, which strongly argues that form-inactive CCD-IgE structures are due to monovalent antigen epitopes. There is also evidence showing that CCD-IgE can stimulate basophils, which can only occur if the IgE antibodies have biological activity, but it does not explain its clinical relevance (Eberlein et al. 2012; Mertens et al. 2010). Furthermore, o-glycans also exist in some plant proteins and have IgE-binding capacity and clinical relevance with the different saccharides mentioned above (Leonard et al. 2005).

Recent studies suggest that specific IgE against galactose-α-1, 3-galactose (α-gal) can cause serious reactions and even fatalities after exposure to cetuximab, and unusual delayed allergic reactions to beef, pork or lamb. This α-gal epitope is highly expressed in nonprimate mammalian tissue. It has been reported that a tick bite causes not only the production of a tick-specific protein but also an IgE response to this carbohydrate (Hamsten et al. 2013).

Because the role of T cells in cross-reactivity is much more complex than that of B cells, it is more difficult to predict cross-reactive T cells, and relevant studies are very few. However, if T cell cross-reactivity exists, it can affect not only the outcome of the immune response, but also the results of the effector phase of allergic reactions (Aalberse 2005).

In-depth study of allergen cross-reactivity is very important for the development of component-resolved diagnosis and immunotherapy. Allergen cross-reactivity indicates that individual or few major allergens can be used not only to diagnose true sensitization of patients to a given allergen or to cross-reactive molecules, but also for allergen-specific immunotherapy to yield the same effects as whole allergen extracts in allergic patients (Blank et al. 2011; Bouaziz et al. 2015). Intensive clinical evidence has proven that specific immunotherapy can prevent both the progression of allergies and the acquisition of new allergic sensitizations. Immunotherapy with peptides containing T cell epitopes may provide an efficacious and safe alternative to conventional subcutaneous and/or sublingual immunotherapy using native full-length allergen preparations (Schein et al. 2007).

5.4 Allergen Family Featured Peptides (AFFP)

Cluster analysis was conducted for allergen sequences in each identified allergen family. The allergen sequence was divided into peptides of 6–32 amino acids in length according to a sliding window of 1–10 amino acids, followed by BLAST (Basic Local Alignment Search Tool, a local alignment search tool basic sequence) using the resulting peptide and nonallergen sequences. Those peptides similar to nonallergen fragments were excluded. If the peptide did not match with nonallergen sequences and the E value of the alignment was lower than 10^{-5}, it was characterized as an AFP. AFFPs are composed of 2–10 adjacent AFPs in the same allergen.

We developed an allergen prediction method, SORTALLER, which predicts allergens by using a novel algorithm on AFFPs that were substantially optimized on most of the SVM-based classifier parameters. SORTALLER outperformed other methods and achieved a perfect balance between high sensitivity and high specificity for discriminating allergenic proteins from diverse sources. This method could be used as the first step in the assessment of programmed allergenicity and a SORTALLER web server has been developed to allow for the simultaneous and rapid prediction of a set of amino acid sequences (Zhang et al. 2012).

A total of 2,359 allergenic protein sequences were first obtained from the Allergome database, the Swiss-Prot Allergen Index (http://www.uniprot.org/docs/allergen.txt), the Food Allergy Research and Resource Program allergen protein database (http://www.allergenonline.org/) and the Allergen Nomenclature database of the International Union of Immunological Societies (http://www.allergen.org/).

After filtering out negative IgE-binding results, 2,290 allergenic protein sequences were retained. The nonallergenic protein sequences were excluded from commonly consumed commodities by humans by searching in the UniProt/Swiss-Prot protein database. After a filtering process, 234,760 nonallergenic protein sequences were retained.

After analyzing the AFFP, we have selected 444–556 AFFPs from 211 known families and ungrouped allergenic proteins by adopting a specific screening procedure with different sliding window sizes. Of all the allergen families, more than half contained only one AFFP, and most AFFPs were shorter than 200 amino acids.

An independent dataset of 1000 sequences (including 500 allergens and 500 nonallergenic proteins) and 14 equivocal proteins (including at least four allergens referenced by IgE experiments) were compared using SORTALLER or with the following other methods: (i) the FAO/WHO evaluation scheme, based on the identity of six or more contiguous amino acids or a minimum 35 % sequence identity over a window of 80 amino acids; (ii) EVALLER; (iii) three different prediction methods of AlgPred; (iv) AllerHunter; (v) APPEL, and (vi) Allermatch. As shown in Table 5.3, the SORTALLER software significantly outperformed the other methods and achieved a perfect balance between specificity and sensitivity in discriminating allergens from diverse sources.

The following aspects may contribute to the superiority of SORTALLER: (1) using the Matthews correlation coefficient (MCC) as the performance measuring end point, which combines two parameters, specificity and sensitivity, to express an unbiased accuracy indicator; (2) having the appropriate number of AFFPs with sufficient information of the allergens and a synergistically optimized sigmoid function, and (3) BLAST for fast and reliable detection of the similarities of query sequences with AFFPs.

Table 5.3 Comparison of different prediction methods

Methods	SE (%)	SP (%)	ACC (%)	MCC
FAO/WHO[a]	99.2	9.6	54.4	0.198
EVALLER[b]	86.6	99.0	92.8	0.863
AlgPred (amino acid)[c]	92.4	80.2	86.3	0.731
AlgPred (dipeptide)[c]	88.8	88.2	88.5	0.770
AlgPred (ARPs BLAST)[c]	81.8	98.0	89.9	0.809
AllerHunter[d]	82.2	99.2	90.7	0.826
SORTALLER[e]	**98.6**	**98.4**	**98.5**	**0.970**

SE sensitivity, *SP* specificity, *ACC* accuracy, *MCC* Matthews correlation coefficient

[a]Allergen sorting was conducted on a Perl script referenced in the FAO/WHO guidelines: cross-reactivity is predicted with an identity of at least six contiguous amino acids or a minimum of 35 % sequence identity with an allergen over a window of 80 amino acids

[b]http://www.slv.se/en-gb/Group1/Food-Safety/e-Testing-of-protein-allergenicity/e-Test-allergenicity/

[c]http://www.imtech.res.in/raghava/algpred/

[d]http://tiger.dbs.nus.edu.sg/AllerHunter/

[e]http://sortaller.gzhmu.edu.cn/

SORTALLER is unique because it is founded on the preferential panning of specific AFFPs and includes ideal information for sorting allergens based on a thoroughly optimized procedure. This novel classifier outperformed other methods and exhibits highly balanced sensitivity and specificity as well as higher accuracy. Moreover, the AFFP dataset incorporated in SORTALLER is useful for developing regimens for component-resolved diagnosis and peptide immunotherapy (Zhang et al. 2012).

5.5 Methods for Allergenicity Evaluation by Amino Acid Sequences

Cross-reactivity is an important phenomenon in allergic reactions. Information regarding known allergen cross-reactions can be searched in many allergen databases such as the Structural Database of Allergenic Proteins (SDAP, http://fermi.utmb.edu/SDAP/). However, a great number of allergen cross-reactions remain to be identified. Currently, the main serological methods used to study cross-reactions in vitro include allergosorbent test (RAST), enzyme-linked immunosorbent assay (ELISA), and protein immunoblotting (Houba et al. 1996; Kondo et al. 2002; Tee et al. 1987). In addition, bioinformatics technology has been widely used to predict unknown allergen cross-reactions before conducting allergenicity evaluating experiments. The potential for allergen cross-reactions is mainly analyzed by amino acid sequence alignment among different allergens. BLAST can be used to align two or more identical amino acid sequences of allergens (Altschul et al. 1990). The FAO/WHO evaluation scheme is based on the identity of six or more contiguous amino acids or a minimum 35 % sequence identity over a window of 80 amino acids (Umeda 2006). Expectation value (*E*-value) and *Z*-score are commonly used for evaluation of the alignment results of allergen amino acid sequences. The *E*-value measures how many matches with the same sequence similarity one would expect to occur by chance (randomly) in a database of a given size. Thus a low *E*-value (e.g., less than 10^{-5}) indicates a high significance of the sequence match. *Z* scores are indications of the quality of the match relative to the database random distribution. Higher Z-scores show better significance for the match (Schein et al. 2007).

We constructed and optimized the SVM-based classifier SORTALLER in concert with two major parameters, i.e. sliding windows of different peptide lengths for AFFP identification and the constant *C* value in sigmoid function for raw BLAST *E*-value scaling. It was shown that a length of 20 amino acids was the optimal peptide length for AFFP screening and a *C* value of 3 in the sigmoid function contributed to optimal allergen prediction. An MCC value as high as 0.970 showed the best prediction results acquired using this method (Zhang et al. 2012).

An integrative approach adopted by FAO/WHO is that any in silico analysis must include environmental factors, such as allergen exposure, dietary habits, and the effect of food processing, as well as individual factors, including genetic background, which may predispose individuals to becoming allergic. It should be noted that available methods cannot, with 100 % accuracy, discriminate between similar proteins according to their allergenicity. Instead, they provide an indication that certain proteins may be cross-reactive (Schein et al. 2007).

References

Aalberse RC. Assessment of sequence homology and cross-reactivity. Toxicol Appl Pharmacol. 2005;207(2 Suppl):149–51.

Aalberse RC, Akkerdaas J, van Ree R. Cross-reactivity of IgE antibodies to allergens. Allergy. 2001;56(6):478–90.

Altmann F. The role of protein glycosylation in allergy. Int Arch Allergy Immunol. 2007;142(2):99–115.

Altschul SF, Gish W, Miller W, Myers EW, Lipman DJ. Basic local alignment search tool. J Mol Biol. 1990;215(3):403–10.

Asero R, Mistrello G, Roncarolo D, de Vries SC, Gautier MF, Ciurana CL, Verbeek E, Mohammadi T, Knul-Brettlova V, Akkerdaas JH, et al. Lipid transfer protein: a pan-allergen in plant-derived foods that is highly resistant to pepsin digestion. Int Arch Allergy Immunol. 2001;124(1–3):67–9.

Astwood JD, Mohapatra SS, Ni H, Hill RD. Pollen allergen homologues in barley and other crop species. Clin Exp Allergy. 1995;25(1):66–72.

Barnett D, Bonham B, Howden ME. Allergenic cross-reactions among legume foods–an in vitro study. J Allergy Clin Immunol. 1987;79(3):433–8.

Bassirpour G, Zoratti E. Cockroach allergy and allergen-specific immunotherapy in asthma: potential and pitfalls. Curr Opin Allergy Clin Immunol. 2014;14(6):535–41.

Bellioni-Businco B, Paganelli R, Lucenti P, Giampietro PG, Perborn H, Businco L. Allergenicity of goat's milk in children with cow's milk allergy. J Allergy Clin Immunol. 1999;103(6):1191–4.

Bernhisel-Broadbent J, Sampson HA. Cross-allergenicity in the legume botanical family in children with food hypersensitivity. J Allergy Clin Immunol. 1989;83(2 Pt 1):435–40.

Blank S, Seismann H, Michel Y, McIntyre M, Cifuentes L, Braren I, Grunwald T, Darsow U, Ring J, Bredehorst R, et al. Api m 10, a genuine *A. mellifera* venom allergen, is clinically relevant but underrepresented in therapeutic extracts. Allergy. 2011;66(10):1322–9.

Bouaziz A, Walgraffe D, Bouillot C, Herman J, Foguenne J, Gothot A, Louis R, Hentges F, Jacquet A, Mailleux AC, et al. Development of recombinant stable house dust mite allergen Der p 3 molecules for component-resolved diagnosis and specific immunotherapy. Clin Exp Allergy. 2015;45(4):823–34.

Bousquet J, Khaltaev N, Cruz AA, Denburg J, Fokkens WJ, Togias A, Zuberbier T, Baena-Cagnani CE, Canonica GW, van Weel C, et al. Allergic Rhinitis and its Impact on Asthma (ARIA) 2008 update (in collaboration with the World Health Organization, GA(2)LEN and AllerGen). Allergy. 2008;63(Suppl 86):8–160.

Braun-Fahrlander C, Wuthrich B, Gassner M, Grize L, Sennhauser FH, Varonier HS, Vuille JC. Validation of a rhinitis symptom questionnaire (ISAAC core questions) in a population of Swiss school children visiting the school health services. SCARPOL-team. Swiss Study on Childhood Allergy and Respiratory Symptom with respect to Air Pollution and Climate. International Study of Asthma and Allergies in Childhood. Pediatr Allergy Immunol. 1997;8(2):75–82.

Bublin M, Kostadinova M, Radauer C, Hafner C, Szepfalusi Z, Varga EM, Maleki SJ, Hoffmann-Sommergruber K, Breiteneder H. IgE cross-reactivity between the major peanut allergen

Ara h 2 and the nonhomologous allergens Ara h 1 and Ara h 3. J Allergy Clin Immunol. 2013;132(1):118–24.

Businco L, Giampietro PG, Lucenti P, Lucaroni F, Pini C, Di Felice G, Iacovacci P, Curadi C, Orlandi M. Allergenicity of mare's milk in children with cow's milk allergy. J Allergy Clin Immunol. 2000;105(5):1031–4.

Caballero T, Martin-Esteban M. Association between pollen hypersensitivity and edible vegetable allergy: a review. J Investig Allergol Clin Immunol. 1998;8(1):6–16.

Canis M, Groger M, Becker S, Klemens C, Kramer MF. Recombinant marker allergens in diagnosis of patients with allergic rhinoconjunctivitis to tree and grass pollens. Am J Rhinol Allergy. 2011;25(1):36–9.

Castro AJ, Alche JD, Calabozo B, Rodriguez-Garcia MI, Polo F. Pla l 1 and Ole e 1 pollen allergens share common epitopes and similar ultrastructural localization. J Investig Allergol Clin Immunol. 2007;17(Suppl 1):41–7.

Ciprandi G, Incorvaia C, Puccinelli P, Soffia S, Scurati S, Frati F. Polysensitization as a challenge for the allergist: the suggestions provided by the Polysensitization Impact on Allergen Immunotherapy studies. Exp Opin Biol Ther. 2011;11(6):715–22.

Durham SR. Allergen immunotherapy: 100 years on. Clin Exp Allergy. 2011;41(9):1171.

Eberlein B, Krischan L, Darsow U, Ollert M, Ring J. Double positivity to bee and wasp venom: improved diagnostic procedure by recombinant allergen-based IgE testing and basophil activation test including data about cross-reactive carbohydrate determinants. J Allergy Clin Immunol. 2012;130(1):155–61.

Ewan PW. Clinical study of peanut and nut allergy in 62 consecutive patients: new features and associations. BMJ. 1996;312(7038):1074–8.

Fernandez-Caldas E, Iraola Calvo V. Mite allergens. Curr Allergy Asthma Rep. 2005;5(5):402–10.

Garrelds IM, Veld Cde G, Wijk RG, Zijlstra FJ. Nasal hyperreactivity and inflammation in allergic rhinitis. Med. Inflamm. 1996;5(2):79–94.

Gergen PJ, Turkeltaub PC. The association of individual allergen reactivity with respiratory disease in a national sample: data from the second National Health and Nutrition Examination Survey, 1976-80 (NHANES II). J Allergy Clin Immunol. 1992;90(4 Pt 1):579–88.

Gould HJ, Sutton BJ. IgE in allergy and asthma today. Nat Rev Immunol. 2008;8(3):205–17.

Hamsten C, Starkhammar M, Tran TA, Johansson M, Bengtsson U, Ahlen G, Sallberg M, Gronlund H, van Hage M. Identification of galactose-alpha-1,3-galactose in the gastrointestinal tract of the tick Ixodes ricinus; possible relationship with red meat allergy. Allergy. 2013;68(4):549–52.

He Y, Liu X, Huang Y, Zou Z, Chen H, Lai H, Zhang L, Wu Q, Zhang J, Wang S, et al. Reduction of the number of major representative allergens: from clinical testing to 3-dimensional structures. Mediators Inflamm. 2014;2014:291618.

Houba R, Van Run P, Heederik D, Doekes G. Wheat antigen exposure assessment for epidemiological studies in bakeries using personal dust sampling and inhibition ELISA. Clin Exp Allergy. 1996;26(2):154–63.

Incorvaia C, Mauro M, Ridolo E, Makri E, Montagni M, Ciprandi G. A pitfall to avoid when using an allergen microarray: the incidental detection of IgE to unexpected allergens. J Allergy Clin Immunol Pract. 2014;pii: S2213-2198(14)00444-9.

Jahn-Schmid B, Harwanegg C, Hiller R, Bohle B, Ebner C, Scheiner O, Mueller MW. Allergen microarray: comparison of microarray using recombinant allergens with conventional diagnostic methods to detect allergen-specific serum immunoglobulin E. Clin Exp Allergy. 2003;33(10):1443–9.

Jenkins JA, Griffiths-Jones S, Shewry PR, Breiteneder H, Mills EN. Structural relatedness of plant food allergens with specific reference to cross-reactive allergens: an in silico analysis. J Allergy Clin Immunol. 2005;115(1):163–70.

Jeong KY, Hwang H, Lee J, Lee IY, Kim DS, Hong CS, Ree HI, Yong TS. Allergenic characterization of tropomyosin from the dusky brown cockroach, Periplaneta fuliginosa. Clin Diagn Lab Immunol. 2004;11(4):680–5.

Klinglmayr E, Hauser M, Zimmermann F, Dissertori O, Lackner P, Wopfner N, Ferreira F, Wallner M. Identification of B-cell epitopes of Bet v 1 involved in cross-reactivity with food allergens. Allergy. 2009;64(4):647–51.

Kondo Y, Tokuda R, Urisu A, Matsuda T. Assessment of cross-reactivity between Japanese cedar (Cryptomeria japonica) pollen and tomato fruit extracts by RAST inhibition and immunoblot inhibition. Clin Exp Allergy. 2002;32(4):590–4.

Langeland T. A clinical and immunological study of allergy to hen's egg white. IV. Specific IGE-antibodies to individual allergens in hen's egg white related to clinical and immunological parameters in egg-allergic patients. Allergy. 1983;38(7):493–500.

Larsen JN, Dreborg S. Standardization of allergen extracts. Methods Mol Med. 2008;138:133–45.

Leonard R, Petersen BO, Himly M, Kaar W, Wopfner N, Kolarich D, van Ree R, Ebner C, Duus JO, Ferreira F, et al. Two novel types of O-glycans on the mugwort pollen allergen Art v 1 and their role in antibody binding. J Biol Chem. 2005;280(9):7932–40.

Leung PS, Chow WK, Duffey S, Kwan HS, Gershwin ME, Chu KH. IgE reactivity against a cross-reactive allergen in crustacea and mollusca: evidence for tropomyosin as the common allergen. J Allergy Clin Immunol. 1996;98(5 Pt 1):954–61.

Levin M, Davies AM, Liljekvist M, Carlsson F, Gould HJ, Sutton BJ, Ohlin M. Human IgE against the major allergen Bet v 1–defining an epitope with limited cross-reactivity between different PR-10 family proteins. Clin Exp Allergy. 2014;44(2):288–99.

Linhart B, Narayanan M, Focke-Tejkl M, Wrba F, Vrtala S, Valenta R. Prophylactic and therapeutic vaccination with carrier-bound Bet v 1 peptides lacking allergen-specific T cell epitopes reduces Bet v 1-specific T cell responses via blocking antibodies in a murine model for birch pollen allergy. Clin Exp Allergy. 2014;44(2):278–87.

Lombardero M, Garcia-Selles FJ, Polo F, Jimeno L, Chamorro MJ, Garcia-Casado G, Sanchez-Monge R, Diaz-Perales A, Salcedo G, and Barber D. Prevalence of sensitization to Artemisia allergens Art v 1, Art v 3 and Art v 60 kDa. Cross-reactivity among Art v 3 and other relevant lipid-transfer protein allergens. Clin Exp Allergy. 2004;34(9):1415–21.

Malling HJ. Immunotherapy for rhinitis. Curr Allergy Asthma Rep. 2003;3(3):204–9.

Martelli A, De Chiara A, Corvo M, Restani P, Fiocchi A. Beef allergy in children with cow's milk allergy; cow's milk allergy in children with beef allergy. Ann Allergy Asthma Immunol. 2002;89(6 Suppl 1):38–43.

Mertens M, Amler S, Moerschbacher BM, Brehler R. Cross-reactive carbohydrate determinants strongly affect the results of the basophil activation test in hymenoptera-venom allergy. Clin Exp Allergy. 2010;40(9):1333–45.

Mohamad Yadzir ZH, Misnan R, Abdullah N, Bakhtiar F, Leecyous B, Murad S. Component-resolved diagnosis (CRD): Is it worth it? frequency and differentiation in rhinitis patients with mite reactivity. Iranian J Allergy Asthma Immunol. 2014;13(4):240–6.

Nordlee JA, Taylor SL, Townsend JA, Thomas LA, Bush RK. Identification of a Brazil-nut allergen in transgenic soybeans. N Engl J Med. 1996;334(11):688–92.

Palosuo T, Panzani RC, Singh AB, Ariano R, Alenius H, Turjanmaa K. Allergen cross-reactivity between proteins of the latex from Hevea brasiliensis, seeds and pollen of Ricinus communis, and pollen of Mercurialis annua, members of the Euphorbiaceae family. Allergy Asthma Proc. 2002;23(2):141–7.

Pittner G, Vrtala S, Thomas WR, Weghofer M, Kundi M, Horak F, Kraft D, Valenta R. Component-resolved diagnosis of house-dust mite allergy with purified natural and recombinant mite allergens. Clin Exp Allergy. 2004;34(4):597–603.

Radauer C, Breiteneder H. Pollen allergens are restricted to few protein families and show distinct patterns of species distribution. J Allergy Clin Immunol. 2006;117(1):141–7.

Radauer C, Bublin M, Wagner S, Mari A, Breiteneder H. Allergens are distributed into few protein families and possess a restricted number of biochemical functions. J Allergy Clin Immunol. 2008;121(4): 847–52.

Radauer C, Nandy A, Ferreira F, Goodman RE, Larsen JN, Lidholm J, Pomes A, Raulf-Heimsoth M, Rozynek P, Thomas WR, et al. Update of the WHO/IUIS Allergen Nomenclature Database based on analysis of allergen sequences. Allergy. 2014;69(4):413–9.

Ramachandran S, Christensen HE, Ishimaru Y, Dong CH, Chao-Ming W, Cleary AL, Chua NH. Profilin plays a role in cell elongation, cell shape maintenance, and flowering in Arabidopsis. Plant Physiol. 2000;124(4):1637–47.

Reese G, Ayuso R, Lehrer SB. Tropomyosin: an invertebrate pan-allergen. Int Arch Allergy Immunol. 1999;119(4):247–58.

Restani P, Ballabio C, Tripodi S, Fiocchi A. Meat allergy. Curr Opin Allergy Clin Immunol. 2009;9(3):265–9.

Savolainen J, Viander M, Koivikko A. IgE-, IgA- and IgG-antibody responses to carbohydrate and protein antigens of *Candida albicans* in asthmatic children. Allergy. 1990;45(1):54–63.

Schein CH, Ivanciuc O, Braun W. Bioinformatics approaches to classifying allergens and predicting cross-reactivity. Immunol Allergy Clin North Am. 2007;27(1):1–27.

Seppala U, Dauly C, Robinson S, Hornshaw M, Larsen JN, Ipsen H. Absolute quantification of allergens from complex mixtures: a new sensitive tool for standardization of allergen extracts for specific immunotherapy. J Proteome Res. 2011;10(4):2113–22.

Sidenius K, Hallas T, Poulsen L, Mosbech H. Allergen cross-reactivity between house-dust mites and other invertebrates. Allergy. 2001;56(8):723–33.

Stewart GA, Thompson PJ. The biochemistry of common aeroallergens. Clin Exp Allergy. 1996;26(9):1020–44.

Subba Rao PV, Rajagopal D, Ganesh KA. B- and T-cell epitopes of tropomyosin, the major shrimp allergen. Allergy. 1998;53(46 Suppl):44–7.

Tao AL, He SH. Bridging PCR and partially overlapping primers for novel allergen gene cloning and expression insert decoration. World J Gastroenterol. 2004;10(14):2103–8.

Tao AL, He SH. Cloning, expression, and characterization of pollen allergens from *Humulus scandens* (Lour) Merr and *Ambrosia artemisiifolia* L. Acta Pharmacol Sin. 2005;26(10):1225–32.

Tao A, Zou Z, Lai H, Huang B, Zhang J, Li Y. Species distribution and progressive clustering of the major allergens worldwide (in Chinese with English abstract). Chin J Microbiol Immunol. 2007;27(5):451–6.

Tee RD, Gordon DJ, Taylor AJ. Cross-reactivity between antigens of fungal extracts studied by RAST inhibition and immunoblot technique. J Allergy Clin Immunol. 1987;79(4):627–33.

Thomas WR, Smith WA, Hales BJ. The allergenic specificities of the house dust mite. Chang Gung Med J. 2004;27(8):563–9.

Umeda T. [Codex ad hoc Intergovernmental Task Force on Foods Derived from Biotechnology]. Shokuhin eiseigaku zasshi. J Food Hyg Soc Jpn. 2006;47(2):J185–8.

Vieths S, Scheurer S, Ballmer-Weber B. Current understanding of cross-reactivity of food allergens and pollen. Ann N Y Acad Sci. 2002;964:47–68.

Vrtala S, Focke M, Kopec J, Verdino P, Hartl A, Sperr WR, Fedorov AA, Ball T, Almo S, Valent P, et al. Genetic engineering of the major timothy grass pollen allergen, Phl p 6, to reduce allergenic activity and preserve immunogenicity. J Immunol. 2007;179(3):1730–9.

Wahl R, Schmid-Grendelmeier P, Cromwell O, Wuthrich B. In vitro investigation of cross-reactivity between birch and ash pollen allergen extracts. J Allergy Clin Immunol. 1996;98(1):99–106.

Walgraffe D, Matteotti C, El Bakkoury M, Garcia L, Marchand C, Bullens D, Vandenbranden M, Jacquet A. A hypoallergenic variant of Der p 1 as a candidate for mite allergy vaccines. J Allergy Clin Immunol. 2009;123(5):1150–6.

Werfel SJ, Cooke SK, Sampson HA. Clinical reactivity to beef in children allergic to cow's milk. J Allergy Clin Immunol. 1997;99(3):293–300.

Zhang L, Huang Y, Zou Z, He Y, Chen X, Tao A. SORTALLER: predicting allergens using substantially optimized algorithm on allergen family featured peptides. Bioinformatics. 2012;28(16):2178–9.

Author's Biography

Dr. Wen Li is working as an Assistant Researcher at Guangdong Provincial Key Laboratory of Allergy & Clinical Immunology, the State Key Clinical Specialty in Allergy, the State Key Laboratory of Respiratory Disease, the Second Affiliated Hospital of Guangzhou Medical University. Email: zouxiaohong128@126.com.

Dr. Li received his master degree in botany in Sun Yat-sen University in 2004 with his thesis on "The Molecular Characterization of Manganese Superoxide Dismutase in Lotus Seed," after which he joined the Second Affiliated Hospital of Guangzhou Medical University as a researcher and took part in the development of a mite allergen vaccine. In 2014, he received his PhD degree in Molecular Medicine from Sun Yat-sen University. Currently, his main research interest focuses on the molecular mechanism of food allergy.

Zehong Zou is a Professor at Guangdong Provincial Key Laboratory of Allergy & Clinical Immunology, the State Key Clinical Specialty in Allergy, the State Key Laboratory of Respiratory Disease, the Second Affiliated Hospital of Guangzhou Medical University. Email: great.liwen@gmail.com.

She is the principal investigator of two provincial research projects and two Key Programs of the Major Project of National Science and Technology on GMO and has worked for 20 years on allergen research and allergy diagnosis. She has been committed to allergenicity evaluation and modification for nearly 10 years with an additional focus in protein purification and identification, fermentation engineering, bioassay, allergen diagnosis, quantitation of tryptase and cytokines in alveolar lavage fluid of asthma patients, etc. She has participated in the research project "Cockroach triggered asthma in Guangdong Province" for which she received an award from Guangdong Medical Scientific and Technological Progress.

Chapter 6
From Allergen Extracts to Allergen Genes and Allergen Molecules

Jiu-Yao Wang

Abstract An allergen is a type of protein capable of instructing the immune system to start producing IgE antibodies. The main purpose for determining the structure of major allergens is to analyze the exposed surface areas and to map the conformational epitopes. These can be determined by experimental methods including crystallographic and NMR-based approaches or predicted by computational methods. Members of the same protein family may share IgE and T cell epitopes, which can cause allergic reactions by cross-reactivity. In the clinical practice, IgE epitopes shared between inhalant and food allergens can induce an immediate IgE-mediated reaction confined to the oral cavity, known as the oral allergy syndrome. Although several structural and functional properties have been identified that contribute to allergenicity, there is not a single common denominator. Allergic sensitization, a multifactorial process, is influenced by a protein's biological and molecular features and by its interaction pathway/s with the immune system. The innate immune system plays a fundamental role in shaping the response to potentially allergenic proteins. In this review, allergenic components of house dust mite that lead to activation of the innate immune system will be discussed as well as how allergen extracts and genes lead to allergen-induced airway inflammation. Understanding the role of mite allergen-induced innate immunity will facilitate the development of therapeutic strategies that exploit innate immunity receptors and associated signaling pathways for the treatment of allergic asthma.

Keywords Allergen · Allergenicity · Cross-reactivity · House dust mite · Innate immune system

J.-Y. Wang (✉)
Department of Pediatrics, Allergy and Clinical Immunology Research (ACIR) Center, College of Medicine, National Cheng Kung University,
No.1, University Road, 70428 Tainan, Taiwan
e-mail: a122@mail.ncku.edu.tw

A. Tao and E. Raz (eds.), *Allergy Bioinformatics*, Translational Bioinformatics 8,
DOI 10.1007/978-94-017-7444-4_6

6.1 Introduction

An allergen is a type of antigen that produces an abnormally vigorous immune response in which the immune system fights off a perceived threat that would otherwise be seen as harmless to the body. In technical terms, an allergen is an antigen capable of stimulating a type I hypersensitivity reaction in atopic individuals through Immunoglobulin E (IgE) responses (Wasserman 2012). Most humans mount significant IgE responses only as a defense against parasitic infections. However, some individuals may respond to many common environmental antigens. This hereditary predisposition is called atopy. In atopic individuals, non-parasitic antigens stimulate inappropriate IgE production, leading to type I hypersensitivity. Sensitivities vary widely from one person (or other animal) to another and a very broad range of substances can be allergens in sensitive individuals. Recently, the determination of allergenicity, i.e., causing allergic sensitization to an allergen, through investigation of genes and allergen structure, has started to reveal the biological function of allergens, which may provide a tool in treating allergen-induced inflammation in allergic diseases.

6.2 Types of Allergens

Allergens can be found in a variety of sources, such as dust mite excretion, pollen, pet dander, or even royal jelly (Rosmilah et al. 2008). Food allergies are not as common as food sensitivities, but some foods such as peanuts (a legume), nuts, seafood, and shellfish are the causes of serious allergies in many people. Officially, the United States Food and Drug Administration recognizes eight foods as being a common cause of allergic reactions in a large segment of the sensitive population. These include peanuts, tree nuts, eggs, milk, shellfish, fish, wheat and their derivatives, soy and their derivatives, as well as sulfites (chemical based, often found in flavors and colors in foods) at concentrations of 10 ppm and over (http://www.fda.gov. for complete details). It should be noted that in other countries, in view of the differences in the genetic profiles of their citizens and different levels of exposure to specific foods due to different dietary habits, the "official" allergen list changes from country to country. Canada recognizes all eight of the allergens recognized by the US, and, in addition, recognizes sesame seeds and mustard. The European Union additionally recognizes celery. Another type of allergen is urushiol, a resin produced by poison ivy and poison oak, which causes the skin rash condition known as urushiol-induced contact dermatitis by changing the configuration of skin cells so that they are no longer recognized by the immune system as part of the body (Llanchezhian et al. 2012). Various trees and wood products such as paper, cardboard, MDF, etc. can also cause mild to severe allergy symptoms such as asthma and skin rash (Cook and Freeman 1997). An allergic reaction can be caused by any form of direct contact with the allergen–ingestion of food or drink, inhalation of pollen, perfume or pet dander or direct contact with an allergy-causing plant that one is sensitive to. Other common causes of serious allergies are wasp, fire ant and bee stings, penicillin, and latex. A life-threatening form of allergic reaction is called anaphylaxis. One form of treatment for anaphylaxis is the administration of sterile epinephrine that suppresses the body's overreaction to the allergen and allows time for the patient to be transported to a medical facility.

6.3 Allergen Nomenclature

Hundreds of allergen-containing reagents are routinely employed for skin testing and IgE antibody assays. Allergen products have been previously classified by the manufacturer into different allergen groups, designated by a coding system. Different manufacturers may assign a different code for the same allergen or they may use the same code for an allergen product derived from different species of the same genus or from different allergen sources all together. This lack of consistency has caused a lot of confusion and difficulties with the comparision of data. The official nomenclature for allergenic proteins is based on the Linnaean binominal nomenclature, identifying genus and species of all organisms, and was first published in 1986 (Marsh et al. 1986) and revised in 1994 (King et al. 1994a, b, 1995). The committee was founded in 1984 to establish a system for the nomenclature of allergens and is composed of leading experts in allergen characterization, structure, function, molecular biology, and bioinformatics. Allergens for which names have been updated include respiratory allergens from birch and ragweed pollen, midge larvae, and horse dander as well as food allergens from peanut, cow's milk, tomato, and cereal grains. The International Union of Immunological Societies (IUIS) Allergen Nomenclature Sub-Committee encourages researchers to use these updated allergen names in future publications. The allergen nomenclature is overseen by the IUIS Allergen Nomenclature Sub-Committee under the auspices of the World Health Organization (WHO) and the (IUIS). This committee maintains the database of approved allergen names (www.allergen.org), which has developed from a plain text list to a fully functional, searchable database. In order to maintain a consistent allergen nomenclature that complies with the guidelines established by the subcommittee, researchers are required to submit newly described allergens to the Allergen Nomenclature Sub-Committee before submitting their manuscript to a journal for consideration for publication. Submissions are kept confidential by the subcommittee, and no specific information other than the name of the new allergen will be disclosed on the Web site before publication. The submission form is available at www.allergen.org (Radauer et al. 2014).

6.4 The Structure and Function of Allergens

Elucidating the three-dimensional structure of clinically relevant allergens is of central importance for the following reasons: (1) It allows the visualization and analysis of surface-exposed residues and, in combination with experimental or computational methods, the actual or putative B cell epitopes can be elucidated. (2) Structure can yield information about bound ligands (proteins and/or small molecules), which may modulate the protein's allergenicity. (3) The allergen structure forms the basis for the rational design of hypoallergenic derivatives, which may be generated through various methods (point mutations, truncations, mosaic proteins, fusion with carrier proteins, etc.) (Dall'Antonia et al. 2014).

Most allergens are relatively small, stable, and well-structured proteins. Therefore, they are perfectly suited for structural studies by both X-ray crystallography and NMR spectroscopy. Patients with type I allergy make IgE antibodies against some, but not all, environmental or dietary proteins they are exposed to. In fact, most allergens belong to a rather limited number of protein families. An absolute prerequisite for a molecule to be designated an allergen is that it binds to specific IgE antibodies. The minimum requirement for a molecule to be designated as an allergen is that it binds IgE antibodies and it can itself act as primary sensitizer, i.e., be able to instruct the immune system to start producing these IgE antibodies. Clear examples of allergens that are not able to do so are those in fruits, nuts, and vegetables that are cross-reactive with the major birch pollen Bet v1. Their allergenicity is dependent on their structural (and functional) similarity to the "parent" molecule Bet v 1, the primary sensitizer (Thomas 2013).

The more intriguing question is, however, what determines whether a protein is capable of being a primary sensitizer. Sensitization is a complex interplay of the susceptibility of the exposed individual (inherited risk of becoming allergic), the timing of exposure (the immune system is more susceptible to sensitization but also to the induction of tolerance earlier in life), the dose (high exposure early in life may skew toward tolerance), the context of exposure (environmental exposures such as pollution, microbes, parasites, diet, and lifestyle can exacerbate the development of atopy), and endogenous properties of the protein.

Glycosylation per se has often been mentioned as marker for allergenicity, but convincing evidence for such a general claim cannot be found. Some properties of proteins, including specific types of glycosylation and binding of lipids, seem to determine their role as allergens via interaction with the innate immune system. Many known allergens are indeed lipid-binding proteins (e.g., Bet v 1 and homologues, house dust mite (HDM) group 2 allergens, lipocalins of pets, plant lipid transfer proteins) and some are glycoproteins (e.g., peanut Ara h 1 and grass pollen Phl p 1) (Vieths et al. 2002). Their lipid ligands and conjugated glycans have been shown to interact with pathogen-recognition receptors such as Toll-like receptors (TLRs) and C-type lectins on antigen-presenting cells, thereby skewing the immune systems toward T_H2-type responses and IgE production. In addition, protease activity, such as that of the cysteine protease Der p 1, has been shown to drive T_H2 inflammation. It is important to note that all these innate T_H2-skewing properties may also turn other proteins without these proallergenic properties into allergens during simultaneous exposure (Chapman et al. 2007).

In summary, several structural and functional properties have been identified that contribute to allergenicity, but there is not a single common denominator.

6.5 Allergen Cross-Reactivity

Allergens belong to a relatively low number of different protein families according to their intrinsic features, e.g., similar amino acid sequences and/or three-dimensional folding (Radauer et al. 2008). Members of the same protein family may share IgE and T cell epitopes, which can cause allergic reactions by cross-reactivity. In this context, birch pollen-related food allergy has been well studied (Bohle et al. 2006). This special form of food allergy affects more than 70 % of birch pollen-allergic patients and is one of the most frequent food allergies in adults. Bet v 1, the single major birch pollen allergen belongs to the Pathogenesis-related Protein Family 10 and homologous molecules are present in various foods, e.g., Mal d 1 in apple, Pru av 1 in cherry, Gly m 4 in soy, and Ara h 8 in peanut. Although these proteins derive from plant species nonrelated to birch trees, their primary and tertiary structures are highly homologous with Bet v 1. Bet v 1 contains mainly conformational IgE epitopes, as destruction of its three-dimensional structure leads to a dramatic reduction of its IgE-binding capacity (Dall'Antonia et al. 2011). Due to similar protein folding, Bet v 1 homologs contain surface patches forming epitopes that may be recognized by Bet v 1-specific IgE antibodies. Although not all IgE epitopes are shared, Bet v 1-related food allergens contain sufficient epitopes to achieve cross-linkage of IgE bound to the surface of mast cells and basophils. In most cases, this induces the oral allergy syndrome, an immediate IgE-mediated reaction confined to the oral cavity (Jahn-Schmid et al. 2005). Destruction of the three-dimensional protein structure, e.g., by gastrointestinal degradation or heat processing, reduces IgE cross-reactivity, which explains why cooked foods containing Bet v 1-related proteins usually are tolerated by birch pollen-allergic patients.

Cross-reactivity at the T cell level depends on amino acid sequence homologies. After uptake by antigen-presenting cells, allergens are degraded into short linear peptides, which are then loaded onto MHC class II molecules to be presented to T cells. Proteins with homologous amino acid sequences are processed in analog fashion resulting in similar peptides. These activate cross-reactive T cells to proliferate and produce cytokines. Clinically, T cell activation by Bet v 1-related food allergens may result in a worsening of atopic eczema in birch pollen-allergic patients.

The analysis of the immune mechanisms underlying birch pollen-related food allergy has markedly contributed to our understanding of how immunological cross-reactivity can induce allergy. Birch pollen-related food allergy is now currently being investigated as a disease model to elucidate whether immunological cross-reactivity can cure allergy, e.g., by cross-reactive regulatory T cells and/or cross-reactive blocking IgG4 antibodies.

In summary, members of the same protein family may share IgE and T cell epitopes, which can cause allergic reactions by cross-reactivity. In the clinical practice, shared IgE epitopes between inhalant allergens and food allergens can induce an immediate IgE-mediated reaction confined to the oral cavity, known as the oral allergy syndrome. Therefore, cross-reactivity at the T cell level represents one of the mechanisms responsible for the worsening of atopic eczema in birch pollen-allergic patients. In the future, immunological cross-reactivity should be explored as a possible cure for allergy by inducing cross-reactive regulatory T cells and/or cross-reactive blocking IgG4 antibodies.

6.6 Mechanisms of Protein Antigens Becoming Allergens

Allergens interact with various parts of the innate immune system, which plays a fundamental role in shaping adaptive immune responses (Thomas 2013): C-type lectin receptors (CTRs) interact with carbohydrate moieties present on allergens; protease-activated receptor (PAR) 2 acts via the allergens' proteolytical activity; and inflammasome complexes are stimulated (Dai et al. 2011; Dombrowski et al. 2012; Varga et al. 2013). In addition, the presentation of lipids from the allergen source by CD1d to invariant natural killer T cells (iNKTs) (Brennan et al. 2013), which enhances sensitization or in some cases even drives it, has been demonstrated. The innate immune system comprises several cell types that express pattern recognition receptors (PRRs). PRRs recognize pathogen- or damage-associated molecular patterns (PAMPs, DAMPs), which are frequently present on allergens.

TLRs are a conserved family of PRRs. The allergens Der p 2 (house dust mite), Fel d 1 (cat), and Can f 6 (dog) bind lipopolysaccharide (LPS) and interact with TLR4, shifting the LPS-response curve to a T_H2-inducing range. Bacterial contaminants present on pollen, as shown for ryegrass and Parietaria, are responsible for triggering TLR2, TLR4, and TLR9 signaling. CTRs contain carbohydrate recognition domains that bind glycosylated allergens and trigger pathways that determine T cell polarization. Ara h 1 (peanut), Der p 1 (HDM), and Can f 1 (dog) interact with the CTRs DC-SIGN. Ara h 1, Der p 1 and 2, Fel d 1, Can f 1, and Bla g 2 (cockroach) interact with the mannose receptor. PARs signal in response to extracellular proteases. Allergens of HDM (Der p 1, 3, 9) and mold (Penc 13) activate PAR-2 and induce IL-25 and thymic stromal lymphopoietin (TSLP). NOD-like receptors (NLRs) sense cytoplasmic PAMPs and DAMPs. Some NLRs are core components of inflammasomes, protein complexes involved in generating the proinflammatory cytokines IL-1β and IL-18. Inflammasomes are triggered by Der p 1, Api m 4 (bee), and Ambrosia artemisiifolia pollen extracts. Surfactant associated proteins (SP), found in the alveoli of the lungs, bind inhaled glycosylated allergens via a carbohydrate recognition domain. Der p 1 and Der f 1 degrade SP-A resulting in increased degranulation of mast cells and basophils triggered by these allergens.

In addition, other cells of the innate immune system, such as epithelial cells, function as a physical barrier whose tight junction proteins are degraded by proteases including Der p 1 and Act d 1 (kiwi). Following allergen contact, epithelial cells produce TSLP, IL-25, and IL-33 and instruct dendritic cells to induce T_H2 responses. Dendritic cells bridge innate and adaptive immunity and polarize the T helper cell response (Bublin et al. 2014; Gomez-Casado et al. 2013; Kitzmuller et al. 2015). Polarization toward a T_H2 response in dendritic cell-T cell cocultures has been shown for Bet v 1 (birch pollen) and Pru p 3 (peach) when cells were derived from allergic donors. iNKTs recognize lipids presented by CD1d (Salio et al. 2013) and secrete IL-4, -5 and -13 when presented with lipids from Brazil nut or sphingomyelin from milk. Codelivery of certain lipids and potentially allergenic proteins can determine the outcome of the sensitization process.

In summary, allergic sensitization is a multifactorial process that is influenced by a protein's biological and molecular features and by the interaction pathway/s with the immune system. The innate immune system plays a fundamental role in shaping the response to potentially allergenic proteins. Proteins can manifest their allergenicity through interaction with Toll-like-, C-type lectin-, NOD-like-, and PARs (present on epithelial cells and dendritic cells) or with surfactant proteins (present in soluble form). Moreover, lipids or glycans (directly bound by allergens, present in the allergen source, or originating from microbial contaminations) can also modulate the immune response of predisposed individuals by interacting with the innate immune system.

6.7 Representative Allergen: House Dust Mite

Hypersensitivity to (HDM; *Dermatophagoides* sp.) allergens is one of the most common allergic responses, affecting up to 85 % of asthmatics (Thomas et al. 2010). HDMs are of paramount importance in all but a few regions of the world, where they do not survive due to aridness, extreme cold, or high altitude (Thomas 2010). Sensitization to indoor allergens is the strongest independent risk factor associated with asthma. Additionally, >50 % of children and adolescents with asthma are sensitized to HDM. IgE-binding studies have found that while a wide range of mite proteins can elicit antibodies most induce low or sporadically detectable titers. The main species that cause allergic sensitization are *Dermatophagoides pteronyssinus* and *Dermatophagoides farinae*. The most abundant, *D. pteronyssinus*, is essentially the only species found in Australia and the United Kingdom, whereas mixed species are found elsewhere except for the *D. farinae*-rich regions of central and northern Korea, northern Italy, and the high-latitude areas of eastern USA, where mites are found in low numbers. The allergens from these species cross-react extensively so species specificity cannot be determined by skin test.

Only three allergens of *D. pteronyssinus* bind IgE from most people at high titers, Der p 1 and 2 and the recently recognized Der p 23 (Weghofer et al. 2013). The allergens Der p 4, 5, 7, and 21 each elicit IgE antibodies in 30–50 % of mite-allergic subjects and collectively, and sometimes individually, induce titers of a magnitude considered important for the induction of disease. The group 1 and 2 allergens can be readily detected in dust and proprietary HDM extracts made from mites cultured in optimized allergen-producing conditions. The distribution of other allergens in the environment is largely unknown and frequently cannot be detected in proprietary extracts. The evolutionary conservation of the amino acid sequence of tropomyosin Group 10 makes them a potential source of cross-reactivity with allergens from a wide range of species. In most regions of the world, however, they only induce IgE in about 10 % of mite-allergic subjects. The possibility that the biochemical functions of the allergens might help promote their allergenicity has been discussed, especially the cysteine protease activity of Der p 1, the lipopolysaccharide binding activity of Der p 2, and the chitin-binding of Der p 23, but this remains unproven. Blomia tropicalis, a glycyphagoide mite found with *D. pteronyssinus* in some tropical and subtropical environments, provides another source of allergens, where Blo t 5 and Blo t 21 are the major allergens and Blo t 2 is a minor one (Thomas 2010).

Although allergen-specific $CD4^+$ T_H2 cells orchestrate the HDM allergic response through induction of IgE directed toward mite allergens, activation of innate immunity also plays a critical role in HDM-induced allergic inflammation (Wang 2013). Previously, we have found HDM components that lead to activation of the innate immune response (Wang et al. 1996, 1998; Chun-Keung et al. 1996; Chen et al. 2003; Yu and Chen 2003; Liu et al. 2005; Ye et al. 2011; Huang et al. 2011). This activation could be due to HDM proteases. Proteases may be recognized by protease-activated receptors (PARs), TLRs, or CTRs, or act as a molecular mimic for PAMP activation signaling pathways. Understanding the role of mite allergen-induced innate immunity will facilitate the development of therapeutic strategies that exploit innate immunity receptors and associated signaling pathways for the treatment of allergic asthma.

References

Bohle B, Zwolfer B, Heratizadeh A, Jahn-Schmid B, Antonia YD, Alter M, et al. Cooking birch pollen-related food: divergent consequences for IgE- and T cell-mediated reactivity in vitro and in vivo. J Allergy Clin Immunol. 2006;118:242–9.

Brennan PJ, Brigl M, Brenner MB. Invariant natural killer T cells: an innate activation scheme linked to diverse effector functions. Nat Rev Immunol. 2013;13:101–17.

Bublin M, Eiwegger T, Breiteneder H. Do lipids influence the allergic sensitization process? J Allergy Clin Immunol. 2014;134:521–9.

Chapman MD, Wünschmann S, Pomés A. Proteases as Th2 adjuvants. Curr Allergy Asthma Rep. 2007;7:363–7.

Chen CL, Lee CT, Liu YC, Wang JY, Lei HY, Yu CK. House dust mite *Dermatophagoides farinae* augments proinflammatory mediator productions and accessory function of alveolar macrophages: implications for allergic sensitization and inflammation. J Immunol. 2003;170:528–36.

Chun-Keung Yu, Lee Shiour-Ching, Wang Jiu-Yao, Hsiue Tzuen-Ren, Lei Huan-Yao. Early-type hypersensitivity-associated airway inflammation and eosinophilia induced by *Dermatophagoides farinae* in sensitized mice. J Immunol. 1996;156:1923–30.

Cook DK, Freeman S. Allergic contact dermatitis to multiple sawdust allergens. Australas J Dermatol. 1997;38(2):77–9.

Dai X, Sayama K, Tohyama M, Shirakata Y, Hanakawa Y, Tokumaru S, et al. Mite allergen is a danger signal for the skin via activation of inflammasome in keratinocytes. J Allergy Clin Immunol. 2011;127:806–14.

Dall'Antonia F, Gieras A, Devanaboyina SC, Valenta R, Keller W. Prediction of IgE-binding epitopes by means of allergen surface comparison and correlation to cross-reactivity. J Allergy Clin Immunol. 2011;128:872–879.

Dall'Antonia F, Pavkov-Keller T, Zangger K, Kellerc W. Structure of allergens and structure based epitope predictions. Methods. 2014;66(1):3–21.

Dombrowski Y, Peric M, Koglin S, Kaymakanov N, Schmezer V, Reinholz M, et al. Honey bee (Apis mellifera) venom induces AIM2 inflammasome activation in human keratinocytes. Allergy. 2012;67:1400–7.

Gomez-Casado C, Garrido-Arandia M, Gamboa P, Blanca-Lopez N, Canto G, Varela J, et al. Allergenic characterization of new mutant forms of Pru p 3 as new immunotherapy vaccines. Clin Dev Immunol. 2013;2013:385615.

Hammad H, Lambrecht BN. Dendritic cells and airway epithelial cells at the interface between innate and adaptive immune responses. Allergy. 2011;66:579–87.

Huang HJ, Lin YL, Liu CF, Kao HF, Wang JY. Mite allergen decreases DC-SIGN expression and modulates human dendritic cell differentiation and function in allergic asthma. Mucosal Immunol. 2011;4:519–27.

Jahn-Schmid B, Radakovics A, Luttkopf D, Scheurer S, Vieths S, Ebner C, et al. Bet v 1142-156 is the dominant T-cell epitope of the major birch pollen allergen and important for cross-reactivity with Bet v 1-related food allergens. J Allergy Clin Immunol. 2005;116:213–9.

King TP, Hoffman D, Lowenstein H, Marsh DG, Platts-Mills TA, Thomas W. Allergen nomenclature. Int Arch Allergy Immunol. 1994a;105:224–33.

King TP, Hoffman D, Lowenstein H, Marsh DG, Platts-Mills TA, Thomas W. Allergen nomenclature. Bull World Health Organ. 1994b;72:797–806.

King TP, Hoffman D, Lowenstein H, Marsh DG, Platts-Mills TA, Thomas W. Allergen Nomenclature. J Allergy Clin Immunol. 1995;96:5–14.

Kitzmuller C, Zulehner N, Roulias A, Briza P, Ferreira F, Fae I, et al. Correlation of sensitizing capacity and T-cell recognition within the Bet v 1 family. J Allergy Clin Immunol. 2015;136:151–8.

Liu CF, Chen YL, Chang WT, Shieh CC, Yu CK, Reid KB, Wang JY. Mite allergen induces nitric oxide production in alveolar macrophage cell lines via CD14/toll-like receptor 4, and is inhibited by surfactant protein D. Clin Exp Allergy. 2005;35:1615–24.

Llanchezhian R, Joseph CR, Rabinarayan A. Urushiol-induced contact dermatitis caused during Shodhana (purificatory measures) of Bhallataka (Semecarpus anacardium Linn) fruit. Ayu. 2012;33(2):270–3.

Marsh DG, Goodfriend L, King TP, Lowenstein H, Platts-Mills TA. Allergen nomenclature. Bull World Health Organ. 1986;64:767–74.

Radauer C, Bublin M, Wagner S, Mari A, Breiteneder H. Allergens are distributed into few protein families and possess a restricted number of biochemical functions. J Allergy Clin Immunol. 2008;121:847–52.

Radauer C, Nandy A, Ferreira F, Goodman RE, Larsen JN, Lidholm J, Pomés A, Raulf-Heimsoth M, Rozynck P, Thomas WR, Breiteneder H. Update of the WHO/IUIS allergen nomenclature database based on analysis of allergen sequences. Allergy. 2014;69:413–9.

Rosmilah M, Shahnaz M, Patel G, Lock J, Rahman D, Masita A, Noormalin A. Characterization of major allergens of royal jelly Apis mellifera. Trop Biomed. 2008;25(3):243–51.

Salio M, Ghadbane H, Dushek O, Shepherd D, Cypen J, Gileadi U, et al. Saposins modulate human invariant Natural Killer T cells self-reactivity and facilitate lipid exchange with CD1d molecules during antigen presentation. Proc Natl Acad Sci USA. 2013;110(49):E4753–61.

Thomas WR. Geography of house dust mite allergens. Asian Pac J Allergy Immunol. 2010;28:211–24.

Thomas WR. Innate affairs of allergens. Clin Exp Allergy. 2013;43:152–63.

Thomas WR, Hales BJ, Smith WA. House dust mite allergens in asthma and allergy. Trends Mol Med. 2010;16:321–8.

Varga A, Budai MM, Milesz S, Bácsi A, Tőzsér J, Benkő S. Ragweed pollen extract intensifies lipopolysaccharide-induced priming of NLRP3 inflammasome in human macrophages. Immunology. 2013;138:392–401.

Vieths S, Scheurer S, Ballmer-Weber B. Current understanding of cross-reactivity of food allergens and pollen. Ann N Y Acad Sci. 2002;964:47–68.

Wang Jiu-Yao. The innate immune response in house dust mite-induced allergic inflammation. Allergy Asthma Immunol Res. 2013;5:68–74.

Wang Jiu-Yao, Kishore Uday, Lim Boon-Leong, Strong Peter, Reid Kenneth BM. Interaction of human surfactant proteins A and D with mite (Dermatophagoid ptyrosynissus) allergens. Clin Exp Immunol. 1996;106:367–73.

Wang Jiu-Yao, Shieh CC, You PF, Lei HY, Reid KBM. The inhibitory effect of pulmonary surfactant protein A and D on allergen-induced lymphocyte proliferation and histamine released in asthmatic children. Am J Respir Crit Care Med. 1998;158:510–8.

Wasserman SI. Approach to the patient with allergic or immunologic disease. In: Goldman L, Schafer AI, editors. Cecil Medicine. 2012.

Weghofer M, Grote M, Resch Y, Casset A, Kneidinger M, Kopec J, et al. Identification of Der p 23, a peritrophin-like protein, as a new major *Dermatophagoides pteronyssinus* allergen associated with the peritrophic matrix of mite fecal pellets. J Immunol. 2013;190:3059–67.

Ye YL, Wu HT, Lin CF, Hsieh CY, Wang JY, Liu FH, Ma CT, Bei CH, Cheng YL, Chen CC, Chiang BL, Tsao CW. *Dermatophagoides pteronyssinus* 2 regulates nerve growth factor release to induce airway inflammation via a reactive oxygen species-dependent pathway. Am J Physiol Lung Cell Mol Physiol. 2011;300:L216–24.

Yu CK, Chen CL. Activation of mast cells is essential for development of house dust mite *Dermatophagoides farinae*-induced allergic airway inflammation in mice. J Immunol. 2003;171:3808–15.

Author Biography

Dr. Jiu-Yao Wang, Ph.D. is now a Distinguished Professor of Pediatrics and Director of the Allergy and Clinical Immunology Research (ACIR) center at National Cheng Kung University Medical Center, Tainan, Taiwan. After clinical residency and subspecialty training in pediatric allergy, clinical immunology and rheumatology at the National Taiwan University Hospital, Taipei from 1984 to 1987, he was appointed as Lecturer (1989), Associate Professor (1992), and Professor (2000) at the College of Medicine, National Cheng Kung University, Tainan, Taiwan. He obtained his Doctor of Philosophy (D.Phil.) degree from the MRC Immunochemistry Unit, University of Oxford, United Kingdom in 1996. For more than 25 years, Dr. Wang's lab has been focused on the study of the genetic, environmental, and immunological factors involved in the innate immunity of allergic and asthmatic children in Taiwan. These studies have produced more than 110 SCI journal papers that have been cited more than 2000 times. He has received several academic awards, including the 37th Ten Outstanding Young Persons in Medical Research Award (1999) sponsored by Taiwan Junior Commercial Chamber, Taiwan, and Gold Medal of K.T. Li Foundation for the Development of Science and Technology (2014). He has been the president of the Taiwan Academy of Pediatric Allergy, Asthma, and Clinical Immunology (TAPAACI), Board Member and Congress Chair of the 2013 Asian Pacific Academy of Allergy, Asthma and Clinical Immunology (APAAACI), and currently serves as a National Delegate and Board Member of the Educational Council of the World Allergy Organization (WAO).

Chapter 7
Allergen Gene Cloning

Xue-Ting Liu and Ailin Tao

Abstract For gene cloning of known allergens, primers are designed according to individual allergen genes and selective amplification by PCR, however, for cloning of an unknown or unstudied species, the design of the degenerate primers is crucial to its successful amplification by PCR. In order to achieve high expression efficiency of the desired allergen gene in a prokaryotic system or other protein expression system, it is necessary to optimize the allergen gene codon. After codon optimization of the allergen gene sequences, protecting bases and an affinity purification tag are added to the 3′ and 5′ end restriction loci and then sent out for sequence synthesis. The prokaryotic protein expression system is the most commonly used expression system and is also the most affordable. During the prokaryotic protein expression process, high yields of the target protein can be obtained by optimizing the host strain and induction conditions including inducing temperature, inducing agent, inducing time, and IPTG concentration. In the *Eschericha coli* system, most recombinant proteins are expressed in inclusion bodies. The denaturation and refolding of inclusion bodies are very important to obtain allergens that have biological activities. Finally, recombinant allergens can be purified using chromatography, isoelectric point precipitation, and salt fractionation.

Keywords Degenerate primer · Codon optimization · Prokaryotic expression · Inclusion body · Refolding · Purification of allergen

X.-T. Liu · A. Tao (✉)
Guangdong Provincial Key Laboratory of Allergy and Clinical Immunology, The State Key Clinical Specialty in Allergy, The State Key Laboratory of Respiratory Disease, The Second Affiliated Hospital of Guangzhou Medical University, 250# Changgang Road East, Guangzhou, China
e-mail: Aerobiologiatao@163.com

X.-T. Liu
e-mail: jane102514@163.com

A. Tao and E. Raz (eds.), *Allergy Bioinformatics*, Translational Bioinformatics 8,
DOI 10.1007/978-94-017-7444-4_7

7.1 Introduction

Allergen gene cloning is the premise and foundation for the creation of recombinant allergens (Tao and He 2005). Primers are designed according to known individual allergen genes and selective amplification by PCR is the method most frequently used for allergen gene cloning. However, due to the differences between species DNA sequences and in those species less studied or with lower homologies, the PCR method of gene cloning is not ideal and is therefore limited in its application. Sequence homology exists in different species of allergens and lays a solid foundation for the design of common primers and subsequent gene clones. Cloning of an allergen for a species that has not yet been studied can be accomplished by designing degenerate primers (Linhart and Shamir 2005).

7.2 Cloning of Allergen Genes and Designing of Degenerate Primers

Due to differences in nucleotide sequences among species, the gene sequence cannot always be directly applied when cloning a homologous gene, especially in an unknown or unstudied species. In this case, the amino acid sequences that are conserved between the different species can be used. On the basis of sequence homology between a large number of allergens in a biological information database and by designing degenerate primers, we can perform selective PCR amplification on the cDNA pool using the Touchdown program. At the same time, the creation of degenerate primers can be further intensified and homologous gene allergens can be cloned by utilizing RACE technology to obtain full-length cDNA. The design of the degenerate primer is crucial to its successful amplification by PCR.

7.2.1 The Search for Homologous Amino Acid Sequences

The search begins with entering the common name of the gene to be studied into the NCBI database search text box and then searching the “Protein” database to obtain the amino acid sequence of the desired gene, which is not limited to any particular species. If the nucleotide sequence of the target gene is already known, another method can be employed, namely, the translation of the nucleotide sequence into an amino acid sequence that is then used to run a BLAST program in the NCBI database to obtain the homologous amino acid sequences.

7.2.2 Sequence Alignment to Determine Representative Sequences of a Homologous Gene

The Clustal W multiple sequence alignment program and the Mega3.0 software program can be used to bring the searched amino acid sequence into multiple sequence alignment (Chenna et al. 2003; Kumar et al. 2004). When multiple genes are involved, there may be more differences in the amino acid sequences and more clusters may be found; in such cases, a plurality of the representative sequences should be selected to carry out the next step.

7.2.3 Search for a Homologous Nucleotide Sequence

Use a representative nucleotide sequence to search for a homologous nucleotide sequence(s). This search process is the same as that using a homologous amino acid sequence and uses the BLAST program in the NCBI database to obtain a homologous nucleotide sequence (She et al. 2009).

7.2.4 Alignment of a Homologous Nucleotide Sequence and Identification of Conserved Regions

The Clustal W multiple sequence alignment procedures and the Mega3.0 software program, as well as other programs, can be used to align the obtained homologous nucleotide sequences and identify a relatively conserved nucleotide region to design the primers. If the conserved region is weak, primers can be created by designing degenerate bases at the changeable sites. When cloned by reverse transcription PCR, the desired target gene fragment can be obtained through combining universal primers with the upstream primer created from the conserved region. The full-length gene can be obtained through designing primers with specific sequences from gene fragments found using the 5′-RACE technique.

7.2.5 Degenerate Primer PCR

Although degenerate primers can be designed, they do not always work well, so it is very important to optimally adjust the reaction conditions. For example, try varying possible annealing temperatures or using the Gradient PCR or Touchdown PCR protocols, which should generally be successful in obtaining the desired target gene (Linhart and Shamir 2007; Souvenir et al. 2007).

(1) **Touchdown PCR**

The so-called annealing temperature is the temperature when half of the double-stranded DNA denatures. That is, at this temperature, half of the DNA may still exist in a double-stranded state. For a primer, the annealing temperature refers to the temperature at which the primer can still be combined with the template. Under these denaturing conditions, only the primer that is perfectly matched with the template can amplify, and, in spite of low amplification efficiency, the amount of primer is limited so as to avoid nonspecific amplification. This method is quite effective when cloning genes in a DNA or cDNA pool. Touchdown PCR is specifically designed to use this principle to enhance PCR amplification.

For the touchdown protocol, if the annealing temperature is normally Tm °C, then the annealing temperature should be set to Tm + 5 °C for a few cycles to specifically amplify the desired gene, with the main loop amplification finally set at Tm °C. Notably, the above-mentioned annealing temperatures are obtained using the OMEGA software. If you obtain annealing temperatures using a different software program, then the above-recommended annealing temperature adjustment may not be optimal.

Generally, the shorter the annealing time is, the stronger the specificity of the amplification. In order to enhance the specificity of amplification, the Touchdown program gradually reduces the annealing time so as to improve the specificity of primers. For example, the beginning annealing time is set to 45 s, then the annealing time is shortened by 1–2 s after every cycle, with the main cycling set for 30 s. This procedure is more efficient for the amplification of degenerate primers.

(2) **Gradient PCR**

When using degenerate primers for amplification, there can be uncertainty as to whether the primer is fully matched with the template. Under normal circumstances, the optimal annealing temperature of the reaction can be determined through trial and error. If using a gradient PCR system, you can accomplish this step all at once in one reaction plate. When setting the annealing temperature, select the "Gradient" program and, depending on the range of the various primer annealing temperatures to be tested, different rows (or columns) in the reaction plate can be programmed to achieve different annealing temperatures. The products of the different amplification reactions can be evaluated on an agarose gel. Based on these results, the optimum annealing temperature of the reaction can be determined, as well as whether the degenerate primers accept the original degenerate bases and whether the original gene primer sequences (using the RACE method) need to be confirmed after the gene is cloned.

(3) **RACE**

Rapid amplification of cDNA ends (RACE) is a PCR-based method for the rapid amplification of cDNA with a low number of transcripts at the 5′ and 3′ ends (Frohman 1993). Obtaining a complete cDNA sequence is essential for gene studies, protein expression, and gene function. Complete cDNA sequences can be acquired by library screening and end cloning techniques. The classic RACE technology was developed by Frohman et al. (Frohman et al. 1988), mainly using RT-PCR technology and including unilateral PCR and anchored PCR in order to get the complete cDNA 5′ and 3′ terminal using a known cDNA sequence. Some improvements to the traditional RACE technology are mainly in primer design and improvements in RT-PCR technology are as follows: (a) the use of locking primers to construct the first-strand cDNA, in which in the oligo (dT) primers of the 3′ end introduce two degenerate nucleotides (5′-Oligo(dT)$_{16\text{-}30}$MN-3′,M=A/G/C;N=A/G/C/T), making the primer starting point at the poly(A) tail; (b) tailing at the 5′ end with poly(C) rather than poly(A); (c) using the RNase H-Mononey Murine Leukemia Virus (MMLV) reverse transcriptase and thermophilic DNA polymerase, which can effectively reversely transcribe mRNA and amplify DNA at high temperatures (60–70 °C) to eliminate the influence of the mRNA secondary structure acting on reverse transcription due to the high CC content at the 5′ end; and (d) the adoption of the hot-start PCR technique and the Touchdown PCR protocol to improve the specificity of the PCR reactions.

Compared with library screening, RACE has several advantages: (a) the methods are implemented using PCR technology, do not need the establishment of a cDNA library, and useful information can be obtained in a very short period of time; (b) the cost and experimental time are reduced; and (c) as long as the primer is designed correctly, a full-length coding region of the interested gene(s) can be obtained.

7.3 Codon Optimization

Whether the genes themselves match with the carrier and the host system is a very critical problem in gene protein expression. Codons are different in eukaryotic and prokaryotic cells, therefore, when we express eukaryotic genes in a prokaryotic system, some of the resulting eukaryotic genes may contain a rare codon for prokaryotic cells, resulting in low expression levels and low expression efficiency (Elena et al. 2014).

In order to achieve high expression efficiency of the desired allergen gene in a prokaryotic system or other protein expression system, it is necessary to optimize the allergen gene codon. This includes eliminating rare codons and using optimized codons adjusting GC content as well as the secondary structure of the RNA nucleotide sequence so that the formation of a complex stem-loop structure is avoided in order to have the most efficient mRNA translation.

This method was developed from many attempts by groups of graduate students in the Guangdong Provincial Key Laboratory of Allergy and Clinical Immunology, by the analysis of both positive and negative examples, and finalized by Professor Ailin Tao.

7.3.1 The Overall Principles of Codon Optimization

Codon optimization should be conducted according to the following principles:

(1) The initiation codon AUG should not be wrapped in the stem-loop structure and terminator structures at the start site are prohibited. In order to improve efficiency, translation should be initiated while in a single-stranded state or on completely open linear structures.
(2) Based on codon degeneracy, replace the rare codons in *Eschericha coli* with the optimal or suboptimal codons but strictly keep the target amino acid sequence unchanged.
(3) Due to limitations on the amount of tRNAs, a single optimal codon cannot always be reused. Consider alternatively using the suboptimal codon instead.
(4) Adjust the AT/GC content by alternative use of degenerate codons. GC and AT contents in the whole genome should be balanced and localized accumulation of GC or AT should be prevented. In order to avoid localized accumulation of CG or AT, the optimal codon(s) could be replaced by suboptimal codon(s).
(5) The use of five or more consecutive identical nucleotides/bases should be prevented.

7.3.2 Specific Steps for Codon Optimization

The following steps are usually used for codon optimization:

(1) Use the online software provided by the website (http://www.kazusa.or.jp/codon/) to initially optimize the codons of the allergen coding sequence.
(2) Determine the rarely used codons. When the protein-expressing host is determined, codon usage is hence defined. For example, the codon usage of the commonly used prokaryote *E. coli* (strain k12) is shown in Table 7.1, which comes from the website (http://www.sci.sdsu.edu/~smaloy/MicrobialGenetics/topics/invitro-genetics/codon-usage.html) and originally adapted from a book (Maloy et al. 1995), and can be utilized to make further codon optimization.
(3) Optimize the RNA secondary structure using the DNAStar software. After initial codon optimization, the RNA secondary structure should be taken into consideration because the RNA secondary structure directly impedes protein translation; the more complex the RNA structure, the harder transcription and translation becomes. Therefore, it should be ensured that there has been no change to the amino acid sequence while the simplest RNA secondary structure is formed by applying abundant or sub-abundant RNA codons, when appropriate. The RNA secondary structure should not form complex winding or long hairpin structures. A large circular structure is optimal and the formation of six or more RNA stem-loop hydrogen bonds should be prevented. The RNA secondary structures of the rice allergen RA17, deduced before and after codon optimization, is exemplified in Fig. 7.1.

Table 7.1 Codon usage in *E. coli* K12 genes[a]

	Codon	Amino acid[b]	%[c]	Ratio[d]	Codon	Amino acid	%	Ratio	Codon	Amino acid	%	Ratio	Codon	Amino acid	%	Ratio	
U	UUU	Phe(F)	1.9	0.51	UCU	Ser(S)	1.1	0.19	UAU	Tyr(Y)	1.6	0.53	UGU	Cys(C)	0.4	0.43	U
	UUC	Phe(F)	1.8	0.49	UCC	Ser(S)	1.0	0.17	UAC	Tyr(Y)	1.4	0.47	UGC	Cys(C)	0.6	0.57	C
	UUA	Leu(L)	1.0	0.11	UCA	Ser(S)	0.7	0.12	UAA	STOP	0.2	0.62	UGA	STOP	0.1	0.30	A
	UUG	Leu(L)	1.1	0.11	UCG	Ser(S)	0.8	0.13	UAG	STOP	0.03	0.09	UGG	Trp(V)	1.4	1.00	G
C	CUU	Leu(L)	1.0	0.10	CCU	Pro(P)	0.7	0.16	CAU	His(H)	1.2	0.52	CGU	Arg(R)	2.4	0.42	U
	CUC	Leu(L)	0.9	0.10	CCC	Pro(P)	0.4	0.10	CAC	His(H)	1.1	0.48	CGC	Arg(R)	2.2	0.37	C
	CUA	Leu(L)	0.3	0.03	CCA	Pro(P)	0.8	0.20	CAA	Gln(Q)	1.3	0.31	CGA	Arg(R)	0.3	0.05	A
	CUG	Leu(L)	5.2	0.55	CCG	Pro(P)	2.4	0.55	CAG	Gln(Q)	2.9	0.69	CGG	Arg(R)	0.5	0.08	G
A	AUU	Ile(I)	2.7	0.47	ACU	Thr(T)	1.2	0.21	AAU	Asn(N)	1.6	0.39	AGU	Ser(S)	0.7	0.13	U
	AUC	Ile(I)	2.7	0.46	ACC	Thr(T)	2.4	0.43	AAC	Asn(N)	2.6	0.61	AGC	Ser(S)	1.5	0.27	C
	AUA	Ile(I)	0.4	0.07	ACA	Thr(T)	0.1	0.30	AAA	Lys(K)	3.8	0.76	AGA	Arg(R)	0.2	0.04	A
	AUG	Met(M)	2.6	1.00	ACG	Thr(T)	1.3	0.23	AAG	Lys(K)	1.2	0.24	AGG	Arg(R)	0.2	0.03	G
G	GUU	Val(V)	2.0	0.29	GCU	Ala(A)	1.8	0.19	GAU	Asp(D)	3.3	0.59	GGU	Gly(G)	2.8	0.38	U
	GUC	Val(V)	1.4	0.20	GCC	Ala(A)	2.3	0.25	GAC	Asp(D)	2.3	0.41	GGC	Gly(G)	3.0	0.40	C
	GUA	Val(V)	1.2	0.17	GCA	Ala(A)	2.1	0.22	GAA	Glu(E)	4.4	0.70	GGA	Gly(G)	0.7	0.09	A
	GUG	Val(V)	2.4	0.34	GCG	Ala(A)	3.2	0.34	GAG	Glu(E)	1.9	0.30	GGG	Gly(G)	0.9	0.13	G
	U				C				A				G				

aThe data shown in this table is from the Arabidopsis Research Companion on the World Wide Web (//weeds/mgh.harvard.edu). Condon frequencies for many other bacteria can be found at http://morgan.angis.su.oz.au/Angis/Tables.html

[b]The letter in parenthesis represents the one letter code for the amino acid

[c]% represents the average frequency this codon is used per 100 codons

[d]Ratio represents the abundance of that codon relative to all of the codons for that particular amino acid

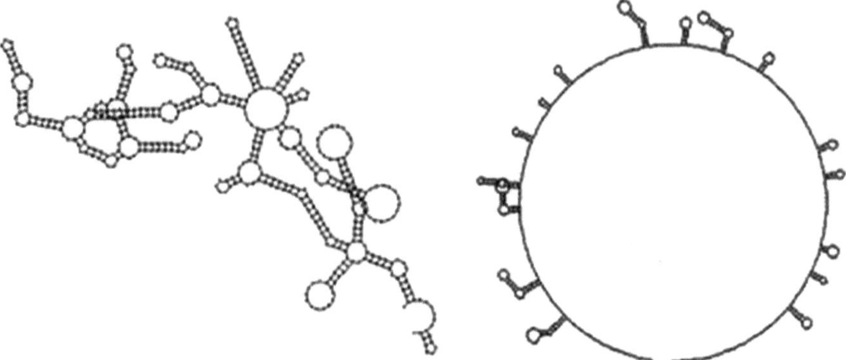

Fig. 7.1 Comparison of RNA secondary structures of the main rice allergen RA17 before and after codon optimization

7.3.3 Verification of Codon Optimization

In order to prevent an error in the optimization process, the optimized sequence should be validated. This can be accomplished using translation software (http://expasy.org/tools/dna.html etc.) to translate optimized nucleotide sequences into amino acid sequences and then see whether there is 100 % identity between the original target amino acid sequence and the optimized nucleotide sequence by applying a special BLAST program "bl2seq" (http://blast.ncbi.nlm.nih.gov/Blast.cgi?PAGE_TYPE=BlastSearch&PROG_DEF=blastn&BLAST_PROG_DEF=megaBlast&BLAST_SPEC=blast2seq), selecting "blastp" for amino acid sequence comparison. Gene synthesis should be considered only when there are no amino acid errors.

7.3.4 Synthesis of the Allergen Gene

After codon optimization of the allergen gene sequences, restriction sites should be added to the 3′ and 5′ ends of the target gene as well as affinity tags (such as Strep II, 6*His, etc.) in order to facilitate construction, digestion, and purification, after which the gene-coding sequence can be synthesized. There are some general considerations for choosing restriction sites:

(1) The inside of target gene fragment should not contain selected restriction sites, otherwise, it will cleave off the target gene when it is identified as a positive recombinant with double enzyme digestion.
(2) The multiple cloning sites in the vectors should contain the selected restriction sites to ensure that the direction of the target gene to be inserted into the

vector is correct (directional cloning). Otherwise, when changing the vector to express the allergen proteins, primers must be redesigned to introduce the new restriction sites.

(3) As much as possible, choose frequently used restriction sites that have high efficiency and activity in commonly used buffers.

7.4 Prokaryotic Expression of Allergens

7.4.1 Vector Selection

A vector, by definition, is a medium carrying a substance. Cloning vectors or gene vectors can carry an endogenous gene of interest into the receptor cells, achieve asexual reproduction, or express a protein of interest. There are a great variety of vectors and a comprehensive analysis of them should be done when designing experiments. A literature search can be used to find the classic vectors, along with existing laboratory techniques and goals (such as eukaryotic or prokaryotic expression, inclusion body expression or soluble expression, resistance, etc.) that should then be taken into consideration when selecting the proper ones. Vector selection should be based on the purpose of gene construction and with consideration of the appropriate restriction sites. If the construction is for the expression of a specific gene, a suitable expression vector should be chosen.

(1) According to their functions, vectors can be divided into cloning or expression vectors. Cloning vectors have a relaxed replicon, can help a foreign gene amplify in the host cell, and are mainly used for cloning and amplification of DNA fragments (gene). Expression vectors are capable of importing exogenous genes, transcribing them, and then translating them into proteins.
(2) According to the type of receptor cell, vectors can be divided into prokaryotic vectors (such as plasmids in bacteria),eukaryotic vectors (such as plasmids in yeast), and shuttle vectors. Shuttle vectors are carrier molecules that can replicate and carry target genes back and forth between the two entities. Shuttle plasmids contain both prokaryotic and eukaryotic replicons to ensure amplification in two types of cells.
(3) Attention should be centered on prokaryotic expression vectors. Select an appropriate promoter and the corresponding recipient strain with attention to reading frame dislocation when expressing eukaryotic proteins. Express the native protein or a fusion protein as a reference.
(4) The restriction enzyme sites and the composing direction of the multiple cloning sites vary among different vectors; attention must be given to link the target gene with the vector without reading frame dislocation.

7.4.2 Plasmid Reconstruction

Bacterial strains containing the target plasmid are grown and the plasmid DNAs are extracted followed by a double restriction digestion to obtain the desired sticky gene fragments. In the meantime, the expression vector is double digested with the same restriction enzymes and the digested vector and the target gene are ligated by T4 DNA ligase, which uses ATP as energy source cofactor.

7.4.3 Allergen Protein Expression

After treatment with Ca^{2+}, original *E. coli* cells are made competent and able to uptake the constructed plasmid DNAs. The positive clones are then selected by restriction and colony PCR identification, or even by sequencing. The verified plasmids from the positive clones are then transformed into an expression host, such as *E. coli* Rosetta and alike, which contain several rare codons to improve protein expression.

During the expression process, high yields of the target protein can be obtained by optimizing the induction conditions, such as inducing temperature and time duration, and inducer concentration.

The protein of interest is inducibly expressed in the transformed host strain and is purified from the precipitate (inclusion bodies) and/or the supernatant (cytoplasm). In the *E. coli* system, many proteins are expressed in the form of inclusion bodies.

7.4.4 Formation of Inclusion Bodies

There is a synergistic effect of many factors that influence the formation of inclusion bodies (Strandberg and Enfors 1991). High expression rates and incorrect disulfide bond pairing may trigger the formation of inclusion bodies because the recombinant proteins cannot form the correct secondary bonds due to the lack of the cofactors. Culture conditions such as temperature, pH, medium composition, and other factors can also affect the formation of inclusion bodies. Moreover, in prokaryotic expression systems that lack a modification process for the newly expressed protein, inclusion bodies are easily formed (Ramon et al. 2014).

7.4.5 Denaturation of Inclusion Bodies

The denaturation of inclusion bodies usually includes the following processes:

(1) **Cell lysis**

To harvest inclusion bodies from genetically engineered bacteria, cell lysis can be accomplished either by mechanical methods, namely high-pressure homogenizer and ultrasound or, when the sample is relatively small, by nonmechanical methods (Rodriguez-Carmona et al. 2010). The main nonmechanical methods include chemical disruption with sodium dodecyl sulfate (SDS) containing detergents, osmotic shock, freeze-thaw, enzyme lysis, etc. (Scotto-Lavino et al. 2006).

(2) **Inclusion body washing**

Crude inclusion bodies can be harvested by centrifugation of the cell lysate, which contains not only the recombinant protein but also impurities such as lipids, hybrid proteins, nucleic acids, lipopolysaccharides, etc. These impurities will seriously affect the refolding and yield of the target protein, as well as the biological activity of the protein and the downstream purification process. It is difficult to remove the membrane proteins and nucleic acids in the crude inclusion bodies, hence, some types of detergents, such as Triton X-100, NP40, or Tween 20, can be used for inclusion body washing. Sodium chloride is often added when washing the inclusion bodies because the washing power of detergents is enhanced with increased ionic strength. Usually, a high concentration of denaturant (e.g., 6 mol/L ~ 8 mol/L urea or guanidine hydrochloride, etc.) is used for dissolving and a low concentration of denaturant is used for washing. In addition, EDTA is also added into the wash solution; it chelates metal ions and prevents the target protein from degradation by any proteases.

(3) **Protein dissolving**

To obtain active protein, the first step is to dissolve the inclusion bodies using denaturing agents, which are generally either surfactants and/or chaotropic agents.

Surfactants: Surfactants, namely detergents, are an economical agent that can be used to dissolve the inclusion body protein. SDS and N-dodecyl sarcosine are the commonly used surfactants. The major advantage of surfactants is that the biological activity of the dissolved protein can be maintained and the aggregation of the dissolved protein is less than with the other solvents. The main drawback of surfactants is that they influence the subsequent protein refolding and purification. They strongly bind with proteins and are difficult to remove, thus, they impede ion exchange and hydrophobic interaction chromatography because of their dissociation function and the hydrophobic action of the ionic surfactants. Additionally, detergents can solubilize and activate the proteases in the membrane proteins, which may reduce the efficiency of the dissolving and refolding processes.

Chaotropic agents: The chaotropes guanidine hydrochloride and urea are commonly used during inclusion body protein treatment as a denaturant/solubilizing agent at a general concentration of 6–8 mol/L with the protein concentration being 1–10 mg/mL. Guanidine hydrochloride is a relatively expensive and stronger denaturant compared to urea. It can dissolve the inclusion bodies that urea cannot. In addition, the adventitious accumulation of cyanate in urea solutions during product manufacturing can cause unwanted carbamylation of proteins, leading to alterations in protein structure, stability, and function. These effects can be prevented by the use of an anionic buffer (such as glycinamide pH8, ethylenediamine pH7, citrate pH5, taurine pH9, etc.) and/or ion exchange chromatography prior to urea preparation (Lin et al. 2004). The dissolving buffers can be held at ambient temperature for 14 days (Lin et al. 2004).

Extreme pH: This method dissolves certain specific inclusion body proteins. Organic acids, low in cost and high in efficiency, are generally used only for the dissolving of some inclusion bodies since some irreversible unwanted modifications or acid degradation may occur under the harsh conditions of extreme pH.

(4) **Refolding of inclusion bodies**

Heterologous proteins overexpressed in *Escherichia coli* often result in a significant proportion of the target protein accumulating in dense insoluble aggregates known as inclusion bodies, which are still unstructured even after being solubilized according to the above steps. The denatured unstructured proteins do not have biological activity, thus, it is necessary to refold these proteins to maximize the yield of soluble functional protein (Hwang et al. 2014).

An efficient refolding process transforms proteins from a denatured heat-unstable state to a thermodynamically stable state with a native structure and biological activity. Hence, the denaturing agents, along with fusion partners that have been specifically designed to accumulate in insoluble inclusion bodies, should first be removed. However, this process often leads to intermolecular misfolding and aggregation, and results in low refolding efficiency. The refolding efficiency actually depends on the balance between correct folding and reaggregation, the latter of which should be avoided.

Generally, when using urea or guanidine hydrochloride as the denaturants at a concentration of 6–8 mol/L, refolding begins when the concentration is reduced to 4–6 mol/L and ends after the concentration is gradually dropped to about 1–1.5 mol/L. There are many methods for the refolding of inclusion body proteins (Yamaguchi and Miyazaki 2014), therefore, it is necessary to comprehensively analyze the relevant literature and the appropriate refolding methods that are available, together with other information regarding the protein specificities such as disulfide bonds, the molecular isoelectric points, etc. (Burgess 2009) before commencing refolding. In any case, refolding efficiency is affected by several factors including the initial concentration of the refolding protein, the refolding buffer, pH, temperature, the interval time of refolding, the refolding sample methods, etc., all of which should be optimized accordingly (Porowinska et al. 2012; Su et al. 2011).

7.5 The Purification of Allergens

Similar to the usual methods, purification of recombinant proteins should take advantage of their physical and chemical properties, i.e., molecular size, shape, solubility, isoelectric point, hydrophilicity and affinity with other molecules, etc. There are many methods for protein purification. At present, the main purification methods include concentrated precipitation, chromatography, and electrophoresis.

Comprehensive analysis of the relevant literature should be done before designing experiments; common protein purification methods should be analyzed and summarized and then combined with the specific properties of the protein (isoelectric point, molecular weight, hydrophobic and hydrophilic properties, etc.) in order to choose the appropriate method of purification (Structural Genomics Consortium et al. 2008). Then, if problems are encountered, the purification scheme should be adjusted.

References

Burgess RR. Refolding solubilized inclusion body proteins. Methods Enzymol. 2009;463:259–82.

Chenna R, Sugawara H, Koike T, Lopez R, Gibson TJ, Higgins DG, Thompson JD. Multiple sequence alignment with the clustal series of programs. Nucleic Acids Res. 2003;31(13):3497–500.

Elena C, Ravasi P, Castelli ME, Peiru S, Menzella HG. Expression of codon optimized genes in microbial systems: current industrial applications and perspectives. Front Microbiol. 2014;5:21.

Frohman MA. Rapid amplification of complementary DNA ends for generation of full-length complementary DNAs: thermal RACE. Methods Enzymol. 1993;218:340–56.

Frohman MA, Dush MK, Martin GR. Rapid production of full-length cDNAs from rare transcripts: amplification using a single gene-specific oligonucleotide primer. Proc Natl Acad Sci U S A. 1988;85(23):8998–9002.

Hwang PM, Pan JS, Sykes BD. Targeted expression, purification, and cleavage of fusion proteins from inclusion bodies in *Escherichia coli*. FEBS Lett. 2014;588(2):247–52.

Kumar S, Tamura K, Nei M. MEGA3: integrated software for molecular evolutionary genetics analysis and sequence alignment. Brief Bioinform. 2004;5(2):150–63.

Lin MF, Williams C, Murray MV, Conn G, Ropp PA. Ion chromatographic quantification of cyanate in urea solutions: estimation of the efficiency of cyanate scavengers for use in recombinant protein manufacturing. J Chromatogr B Anal Technol Biomed Life Sci. 2004;803(2):353–62.

Linhart C, Shamir R. The degenerate primer design problem: theory and applications. J Comput Biol. 2005;12(4):431–56.

Linhart C, Shamir R. Degenerate primer design: theoretical analysis and the HYDEN program. Methods Mol Biol. 2007;402:221–44.

Maloy S, Stewart VJ, Taylor RK. Genetic analysis of pathogenic bacteria: a laboratory manual. New York: Cold Spring Harbor Laboratory Press; 1995.

Porowinska D, Marszalek E, Wardecka P, Komoszynski M. In vitro renaturation of proteins from inclusion bodies. Postepy Hig Med Dosw (Online). 2012;66:322–9.

Ramon A, Senorale-Pose M, Marin M. Inclusion bodies: not that bad. Front Microbiol. 2014;5:56.

Rodríguez-Carmona El, Cano-Garrido O, Seras-Franzoso J, Villaverde A, García-Fruitós E. Isolation of cell-free bacterial inclusion bodies. Microb Cell Fact. 2010;9:71.

Scotto-Lavino E, Du G, Frohman MA. 3′ end cDNA amplification using classic RACE. Nat Protoc. 2006;1(6):2742–5.

She R, Chu JS, Wang K, Pei J, Chen N. GenBlastA: enabling BLAST to identify homologous gene sequences. Genome Res. 2009;19(1):143–9.

Souvenir R, Buhler J, Stormo G, Zhang W. An iterative method for selecting degenerate multiplex PCR primers. Methods Mol Biol. 2007;402:245–68.

Strandberg L, Enfors SO. Factors influencing inclusion body formation in the production of a fused protein in *Escherichia coli*. Appl Environ Microbiol. 1991;57(6):1669–74.

Structural Genomics Consortium, China Structural Genomics Consortiumm Northeast Structural Genomics Consortium, Gräslund S, Nordlund P, Weigelt J, Hallberg BM, Bray J, Gileadi O, Knapp S, et al. Protein production and purification. Nat Methods. 2008;5(2):135–46.

Su Z, Lu D, Liu Z. Refolding of inclusion body proteins from *E. coli*. Methods Biochem Anal. 2011;54:319–38.

Tao AL, He SH. Cloning, expression, and characterization of pollen allergens from *Humulus scandens* (Lour) Merr and *Ambrosia artemisiifolia* L. Acta Pharmacol Sin. 2005;26(10):1225–32.

Yamaguchi H, Miyazaki M. Refolding techniques for recovering biologically active recombinant proteins from inclusion bodies. Biomolecules. 2014;4(1):235–51.

Author's Biography

Dr. Xue-Ting Liu is a Docent and Junior Research Scientist at Guangdong Provincial Key Laboratory of Allergy and Clinical Immunology, the State Key Clinical Specialty in Allergy, the State Key Laboratory of Respiratory Disease, the Second Affiliated Hospital of Guangzhou Medical University. Email: jane102514@163.com.

Dr. Xue-Ting Liu received her Ph.D. from Jinan University majoring in Biochemistry and Molecular Biology. From 2005 to 2011, she worked on the antitumor effects of the protein drug sFGFR2 and how it exerts its antitumor activity. Since 2011, she has been working at the Guangdong Provincial Key Laboratory of Allergy and Clinical Immunology. Her main research interests focus on the mechanisms responsible for chronic urticaria caused by Staphylococcal Enterotoxin A (SEA) and reducing the antigenicity of protein drugs, such as SEA. Dr. Liu has published several scientific papers within internationally well-recognized journals. She has participated in several projects for the Science Foundation of National and Guangdong Provincial and presided over a project from the National Natural Science Foundation.

Professor Ailin Tao is a Professor at Guangzhou Medical University, Director of Guangdong Provincial Key Laboratory of Allergy and Clinical Immunology, Principal Investigator of the State Key Laboratory of Respiratory Disease, Deputy Director of the State Key Clinical Specialty in Allergy of the Second Affiliated Hospital of Guangzhou Medical University, Member of the State Committee for Transgenic Safety Assessment, Standing Committee Member of Allergy Branch of Guangdong Medical Association, Member of Guangdong Provincial Committee for Transgenic Safety Assessment and Master Tutor.

Prof. Ailin TAO earned his doctorate degree from the State Key Laboratory of Crop Genetic Improvement of Huazhong Agricultural University in 2002, followed by a postdoctoral training at Postdoctoral Station of Basic Medicine in Shantou University Medical College, majoring in allergen proteins. His most recent research has been on allergy bioinformatics, allergy and clinical immunology and disease models such as allergic asthma, allergic rhinitis, infection and inflammation induced by allergy, inflammatory and protracted diseases caused by antigens or superantigens. He has gained experience in the field of allergology including the mechanisms of immune tolerance, allergy triggering factors and chronic inflammation pathways, and allergenicity evaluation and modification for food and drugs. He proposed some new concepts including “Representative Major Allergens”, “Allergenicity Attenuation” of immunotoxin and allergens, “broad-spectrum immunomodulator” as well as the theoretical hypothesis of “Balanced Stimulation by Whole Antigens”. Prof. TAO’s laboratory focuses on the diagnosis of allergic disease and the medical evaluation of food and drug allergenicity and its modification. Prof. TAO has now constructed a system for the prediction, quantitative assessment, and simultaneous modification of epitope allergenicity, which has been applied to more than 20 allergens, and he also developed a bioinformatics software program for allergen epitope prediction, SORTALLER (ht tp://sortaller.gzhmu.edu.cn), which performed significantly better than other existing softwares, reaching a perfect balance of high specificity (98.4 %) and sensitivity (98.6 %) for discriminating allergenic proteins from several independent datasets of protein sequences of diverse sources. Furthermore, this program has a Matthews correlation coefficient as high as 0.970, a fast running speed, and can rapidly predict a set of amino acid sequences with a single click. The software has been frequently used by researchers from many institutions in China and over 30 countries worldwide, thus becoming the number one allergen epitope prediction software program. Prof. TAO has set up an allergen database ALLERGENIA (http://ALLERGENIA.gzhmu.edu.cn) that has several advantages over other databases such as a wide selection of nonredundant allergens, excellent astringency and accuracy, and friendly and usable analytical functions.

Chapter 8
High Throughput Screening of Allergy

Junyan Zhang and Ailin Tao

Abstract Allergen avoidance and specific immunotherapy that are based on an accurate diagnosis of allergy have proven to be helpful for the prevention and treatment of allergic diseases. Detection of specific IgE against different allergens in serum or plasma can be used as an important criterion for allergy diagnosis. Conventional methods for the detection of IgE in the clinic are mainly enzyme-linked immunosorbent analysis (ELISA), fluorescence enzyme immunoassays, and radioimmunoassays (RIAs). However, it is difficult to achieve high throughput screening by these conventional methods. Microarrays can test more than 1000 known allergens and are able to resolve reactivities to various allergen components, which is necessary for clinical diagnosis and treatment. Thus, allergen microarrays provide a versatile platform for detecting specific IgE or other types of antibodies against thousands of allergens in a parallel and high throughput way and are currently used in diagnosis of allergy. In this chapter we will provide an overview of allergen microarrays.

Keywords Allergen microarrays · High throughput screening · Allergy diagnosis

J. Zhang · A. Tao (✉)
Guangdong Provincial Key Laboratory of Allergy and Clinical Immunology,
The State Key Clinical Specialty in Allergy, The State Key Laboratory of Respiratory Disease,
The Second Affiliated Hospital of Guangzhou Medical University,
250# Changgang Road East, Guangzhou, China
e-mail: Aerobiologiatao@163.com

J. Zhang
e-mail: kuhn2000@163.com

A. Tao and E. Raz (eds.), *Allergy Bioinformatics*, Translational Bioinformatics 8,
DOI 10.1007/978-94-017-7444-4_8

8.1 Introduction: Allergy Diagnosis

Allergen avoidance can effectively reduce the risk of allergies while specific immunotherapy is known to be an effective therapy for allergic diseases. When developing a treatment plan for the allergic patient, typically allergen avoidance or specific immunotherapy, it is necessary to clearly identify what allergens the patient is sensitive to. Therefore, accurate determination of specific allergens is very important for the prevention and treatment of allergic diseases.

At present, the diagnosis of allergy relies mainly on clinical history, in vivo tests, and other laboratory experiments (Hamilton 2010; Maloney et al. 2008; Soares-Weiser et al. 2014). Clinical history is helpful, but is sometimes not definitive or accurate. In vivo tests include provocation tests and skin tests. Provocation tests are the gold standard for allergy diagnosis, but they risk the induction of a severe allergic reaction. Provocation tests must be performed in a controlled situation with good medical first aid, and only implemented after careful and comprehensive assessment of risk and benefit. Skin tests are not always reliable due to a certain number of false positive and negative results and, therefore, must be combined with the clinical history or other tests in order to make an accurate judgment. Allergen-specific IgE testing is the main laboratory test used for allergy diagnosis, and is frequently used in the clinic, but the rate of allergen-specific positivity is not high. Testing of allergen-specific IgE in 200,000 samples of patients with suspected atopic diseases by Peking Union Medical College Hospital in China indicated that the rate of detection of specific IgE for many allergens is low (Wang and Zhang 2012) (see Tables 8.1 and 8.2). Other methods, such as the basophil activation test in vitro, have a relatively high accuracy rate (Sturm et al. 2009), but are infrequently used in clinical practice due to their complexity and the lack of uniform laboratory practices for the detection of histamine or other inflammatory mediators.

Table 8.1 Rate of positive-specific IgE detection to suspected allergens

Suspect allergen	Number tested	Number positive	% positive
d1, *Dermatophagoides pteronyssinus*	23,662	9078	38.4
d2, *Dermatophagoides farinae*	20,196	8218	40.7
f24, shrimp	2233	511	22.9
f4, wheat	1809	483	26.7
f1, egg white	1299	382	29.4
f2, milk	1516	381	25.1
f23, crab	1508	213	14.1
f91, mango	492	84	17.1
f3, fish	1024	36	3.5

Referenced from: Wang and Zhang (2012)

Table 8.2 Rate of positive-specific IgE detection to suspected major mixed allergens

Mix inhaled allergen				Mix food allergen			
Suspect allergen	Number tested	Number positive	% positive	Suspect allergen	Number tested	Number positive	% positive
phad	26,728	10,556	39.5	fx5	10,578	1300	12.3
mx2	9527	1537	16.1	fx2	3417	349	10.2
tx6	2983	1209	40.5	fx1	570	86	15.1
tx5	2910	850	29.2	fx3	638	83	13.0
wx7	1198	502	41.9	fx10	1704	76	4.5
tx7	464	126	27.2	fx21	234	41	17.5
gx2	536	118	22.0	fx22	349	26	7.5

Referenced to: Wang and Zhang (2012)

Detection of specific IgE against different allergens in serum or plasma can be used as an important criterion for the clinical diagnosis of allergy (Canonica et al. 2013). The conventional methods for IgE detection are mainly enzyme-linked immunosorbent analysis (ELISA), fluorescence enzyme immunoassays (FEIA), and radioimmunoassays (RIAs). However, these methods have several disadvantages: (1) The execution of these methods is relatively complex; (2) Precious amounts of serum are consumed in one test; (3) The cost can be very high when several allergens require testing; and (4) The detected IgE is usually against whole allergen extracts, not individual allergenic components. Currently, there are more than 1000 known allergens and component-resolved diagnosis based on these allergens is becoming increasingly necessary for improving the accuracy of allergy diagnosis (Treudler and Simon 2013). For all these reasons, it has been difficult to achieve high throughput screening by conventional methods.

Protein microarrays have advantages over conventional-specific IgE testing methods because they can provide high specificity, high sensitivity, high throughput, good repeatability, and require only a small serum sample. Currently, microarrays are being widely used in drug screening (Kumble 2007), proteome research (Templin et al. 2003), and clinical diagnosis (Smith et al. 2004; Spisak and Guttman 2009) as well as for high throughput screening of allergy (Harwanegg and Hiller 2004).

8.2 Allergen Microarrays

Allergen microarrays provide a versatile platform for the detection specific IgE or other types of antibodies against thousands of allergens in a parallel and high throughput way. Currently, they are being used in the diagnosis of aeroallergens, food allergens, and contact allergens such as pollen (Cabauatan et al. 2014), peanut, cow's milk (CM), and egg (Ott et al. 2008), as well as latex (Ebo et al. 2010). See the data collected by Wayne G. Shreffler in Table 8.3 (Shreffler 2011).

Table 8.3 Microarray-based molecular allergen assays

	Allergy type	Allergens	Allergen source	Array used	Patients	Reference
Aeroallergens	Tree/grass pollen	Phl p 1, 2, 5, and 6 and Bet v 1 and 2	Recombinant	In house	51 allergic (history and testing), 11 nonatopic control subjects	Jahn-Schmid et al.
	Ragweed/mugwort	Amb a 1, 5, 6, 8, 9; Art v 1, 3, 4, 5, 6	Natural and recombinant	In house	10 ragweed allergic (history and SPT); 9 mugwort allergic (history and SPT)	Gadermaier et al.
Food allergens	Peanut allergy/ tolerance	Ara h 1, 2, 3, and 8; Pru p 3; Bet v 1; Phl p 1, 4, 5b, 7, and 12; CCD	Natural and recombinant	In house	108 peanut-sensitized subjects identified from a healthy cohort of 1085, of which 17 were proved to be allergic and 52 were proved to be tolerant; an additional 12 with peanut allergy were specifically recruited	Nicolau et al.
	Baker's asthma	Six *Escherichia coli*—expressed wheat proteins selected with sera from patients with baker's asthma; Phl p 1, 5, 7, and 12	Recombinant	In house	22 patients with baker's asthma (history and provocation testing); 32 with wheat allergy (history and provocation)	Constantin et al.

(continued)

Table 8.3 (continued)

	Allergy type	Allergens	Allergen source	Array used	Patients	Reference
	Cow's milk	Bos d 4, 5.0101, 5.0102, 8αs1, 8αs2, 8β, and 8κ	Natural and recombinant	In house	78 milk-sensitized subjects of varying ages and clinical features	Hochwallner et al.
	Oral allergy syndrome	Mal d 1	Recombinant	Commercial		Ebo et al.
Contact allergens	Latex	Hev b 1, 3, 5, 6, and 8; CCD	Recombinant except nAna c 2 as CCD	Commercial	22 allergic (history and testing); 20 asymptomatic with positive serum IgE levels; 26 healthy control subjects	Ebo et al.
	Latex	Hev b 1, 3, 5, 6, 8, 9, 10, and 11	Recombinant	Commercial, customized	52 adult patients with latex allergy; 50 control subjects with venom allergy	Ott et al.

CCD Cross-reactive carbohydrate determinants
Referenced from: Shreffler (2011)

Ott et al. (2008) investigated 130 infants and children with suspected allergy to CM and hen's egg (HE) using allergen microarrays. They compared the detection result of the allergen microarray method with that obtained by extract-based FEIA and skin prick testing (SPT). Their research found that allergen microarrays show performance characteristics comparable to the current diagnostic tests in the diagnosis of symptomatic CM and HE allergies. The combination of allergen microarray results of α-casein, β-casein, κ-casein, Bosd4, and Bosd5 for CM allergy and Gal d1, Gal d2, and Gal d4 for HE allergy both generated the same or highly similar AUC values as compared to FEIA testing in the diagnosis of both HE and CM allergies. See Tables 8.4 and 8.5.

Hiroshi Kido et al. detect specific IgE against more than 20 natural allergen extracts or purified allergens including milk, egg, wheat, pollen, and house dust using a diamond-like, carbon-coated DLC allergen microarray (Suzuki et al. 2011). They found that there were high correlations (0.822–0.966) between allergen-specific IgE values for these 20 allergens determined by the DLC allergen microarray and the UniCAP system. The DLC allergen microarray showed higher sensitivity than the UniCAP system for the detection of allergen-specific IgE. Lowering the limit of dilution rate in UniCAP system to further dilution at 4–8 folds also could be detected by DLC allergen microarray. When used to detect IgE

Table 8.4 Comparison of assay performance in the diagnosis of CM allergy

Allergen microarray							FEIA	SPT
Allergen	α-casein	β-casein	κ-casein	Bos d 4	Bos d 5	Combination[a]	CM extract	Native CM
AUC	0.6	0.6	0.6	0.7	0.6	0.7	0.7	0.7
Sensitivity (%)	26.2	26.2	38.1	50	23.9	59.5	71.1	60.7
Specificity (%)	97.7	93	88.4	93	95.3	83.7	81.4	48.2
P-value[b]	0.29	0.27	0.43	0.61	0.18	0.5	–	–

[a]Combination of all fluorescence intensity values of single CM allergen components
[b]*P*-values of nonparametric paired Wilcoxon tests of FEIA and allergen microarray results with respect to AUC values

Table 8.5 Comparison of assay performance in the diagnosis of HE allergy

Allergen microarray					FEIA	SPT
Allergen	Gal d 1	Gal d2	Gal d4	Combination[a]	HE extract	Native HE
AUC	0.8	0.77	0.6	0.8	0.8	0.8
Sensitivity (%)	57.8	57.8	17.8	53.3	71.1	60.7
Specificity (%)	80.7	80	100	100	86.7	100
P-value[b]	0.61	0.2	0.044	0.98	–	–

[a]Combination of all fluorescence intensity values of single HE allergen components
[b]*P*-values of nonparametric-paired Wilcoxon test FEIA and allergen microarray with respect to AUC values

Table 8.6 Comparison of assay sensitivity in detecting allergen-specific IgE in CB and MB

	CB (1:1 dilution)		MB (1:1 dilution)	
Allergen	DLC chip	UniCAP	DLC chip	UniCAP
	(Bue/mL)	(Ua/mL)	(Bue/mL)	(Ua/mL)
Food				
Egg white	30.35	ND[a]	77.41	0.545
	11.02	ND	23.65	ND
Ovomucoid	180	ND	134.8	0.96
	84.89	ND	64.66	ND
	13.3	ND	23.15	ND
Milk	221.4	ND	182.7	1.095
	30.9	ND	61.71	0.54
	18.05	ND	23.15	ND
Inhalant				
Cedar pollen	55.55	ND	90.98	0.96
	21.78	ND	32.2	ND
Dermatophagoides farinae	54.01	ND	80.76	1.275
	47.38	ND	25.53	ND
Dermatophagoides pteronyssinus	63.04	ND	215.6	2.95
	26.48	ND	60.7	ND

CB and MB serums (1:1 dilution) were used for the measurement of allergen-specific IgE levels using the UniCAP system and the DLC chip. Detection limit on the DLC chip: 10 Bue/mL. [a]ND, Not detectable by UniCAP assay: <0.35 Ua, arbitrary unit

in cord blood (CB) and maternal blood (MB), they found that IgE was detected by the DLC allergen microarray but not by the UniCAP system (Kamemura et al. 2012) (see Table 8.6). Their work laid a foundation for further research into the relationship between IgE levels in CB and MB with infant allergy.

Lin et al. (2012) successfully improved the accuracy rate in the diagnosis of symptomatic peanut allergy using allergen peptide microarray immunoassays and bioinformatic methods. They found that receiver operating characteristic (ROC) curves for the detection of specific IgE against Ara h 1, Ara h 2, and Ara h 3 alone by allergen peptide microarray were better than the detection of total IgE by the UniCAP method. The ROC curve of the machine learning (SVM) method of analysis, which is based on the combination of IgE results for Ara h 1, Ara h 2, and Ara h 3, was best at predicting peanut allergy (Fig. 8.1). Models created using machine learning methods can predict the outcome of double-blind, placebo-controlled food challenges with high accuracy. Jing Li et al. also used an allergen peptide microarray immunoassay to analyze the epitopes of peanut allergen. They develop a peptide microarray immunoassay using 419 overlapping peptides (15 mers, 3 offset) covering the amino acid sequences of Ara h 1, Ara h 2, and Arah 3. Specific IgE and IgG4 from patients with symptomatic peanut allergy and from patients who were sensitized but clinically tolerant to peanut were tested and

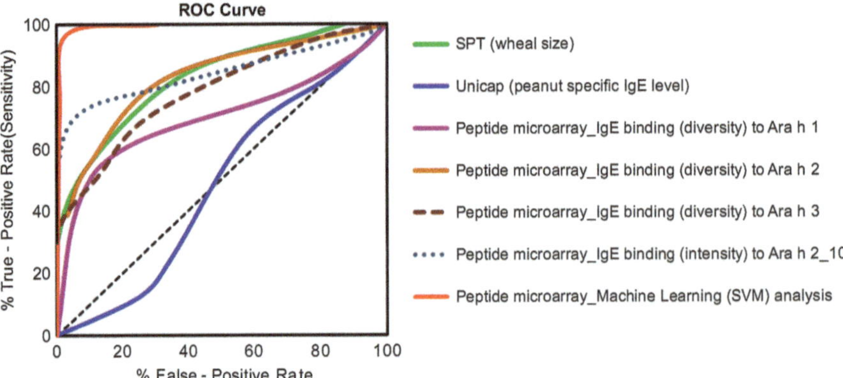

Fig. 8.1 Comparison of the diagnostic performance of different allergy tests and analytical methods in predicting the outcome of double-blind, placebo-controlled food challenges. The area under the *ROC curve* indicates how well a test method can distinguish between two diagnostic groups (peanut allergic vs. peanut tolerant). The *diagonal line* indicates a completely random guess. Both IgE-binding diversity (expressed as the number of positive peptides) and intensity (express as *Z* score) were measured using peptide microarray (Lin et al. 2012)

compared for binding to these overlapping peptides. Consequently, four peptide biomarkers were found that could better predict peanut allergy.

Allergen microarrays based on detection of specific IgE are not the only tests using patient serum used in allergy diagnosis, but can also be coupled with the basophil activation test. Lin et al. (2007) were the first to couple the allergen microarray method with the basophil activation test in 2007. In their work, the basophilic cell lines KU-812 and RBL-703/21 and human basophils purified from the peripheral blood of healthy donors were resensitized with the serum or IgE preparation to be tested and incubated with manually spotted allergen array chips. Basophil activation, which indicated the effective cross-linking of IgE by allergens, was monitored by upregulation of a basophil activation surface marker (CD63). Their research indicated that coupling the allergen microarray method with the potential functionality and biological activity of a cell-based test is feasible and could result in a new system to detect allergic sensitization. However, the basophil activation test adds more complexity to the diagnosis of allergic disease than using just microarrays based on specific IgE detection, since it is difficult to control the state and response of human basophils.

In summary, with the growing development of allergen microarray technology and its wide application, this method has gradually become an effective tool in high throughput screening for the diagnosis of allergy. It is particularly suitable for patients who require allergen avoidance or specific immunotherapy but have not identified the allergens they are sensitive to and which one is the main cause of their symptoms using traditional diagnostic methods and clinical history. Therefore, allergen microarray technology is a helpful tool to gain information for the prevention and treatment of allergic diseases.

8.3 Preparation of Allergen Microarrays

In order to detect the major allergens using a microarray method, individual allergens are first coated on a solid support. After blocking and washing, the microarray is incubated with the sera of allergic patients followed by the addition of anti-IgE antibodies labeled with either immunofluorescence or enzyme. After washing away unbound anti-IgE antibodies, fluorescence signals are read by a fluorescence scanner (or by confocal microscopy) or photos are taken by charge-coupled device (CCD) camera after chemical color development. By comparing the results with cutoff values and the reference serums, it can be determined which allergens the patients are sensitive to. In order to further improve the sensitivity of the detection system, biotin–avidin or a dendrimer system can also be used to amplify the detection signal. For example, anti-IgE antibodies could be labeled with biotin while streptavidin labeled with fluorescent substances.

The following sections describe the key points of allergen preparation for a microarray system.

8.3.1 Selecting the Substrates

Selecting a substrate that can efficiently immobilize proteins on its surface and keep their biological activity is one of the key steps in the development of a protein microarray. A good substrate must meet the following requirements: (1) The substrate should have the reactive groups on its surface to allow interaction with the biomolecules; (2) Maximum binding capacity with biomolecule should be achieved; (3) The substrate should have sufficient stability, including mechanical, physical, and chemical stabilities; and (4) The substrate should have good biological compatibility.

There are two kinds of commonly used substrates:

1. **Filter membrane and gel film**

   The filter membrane and gel film are soft substrates that easily adsorb proteins. They can maintain an aqueous environment to preserve protein activity. The biological and chemical properties of the filter membrane are relatively stable. It is helpful for protein immobilization, but has higher background noise and a lower signal-to-noise ratio, and the spotting density is limited. The optical property of the gel film is better than the filter membrane and gel films are good at maintaining protein activity because they have a hydrophilic porous structure. However, the deficiencies of these two types of substrates are that they have high nonspecific adsorption and the possibility of interference between proteins after diffusion.
2. **Enhanced glass slides**
   Glass slides have an advantage in that they are inexpensive, have a smooth surface, low detection costs, and stable properties. Modifying the glass surface

is very important for making a protein microarray (Kusnezow et al. 2003). Commonly, the surface of glass slides have been modified with either an amino group, aldehyde group, epoxy group, Mercaptopropyltrimethoxysilane group, polystyrene-based modified surface, and so on (Angenendt 2005). Such substrates immobilize the proteins mainly by covalent bonds or by high-affinity binding between special molecules, which results in tight binding and fewer conformational changes. They have an advantage in maintaining the natural activity of the proteins, as demonstrated by the carboxylated DLC film-coated glass slide that Hiroshi Kido et al. utilized to demonstrate the features of improved glass slides (see Fig. 8.2).

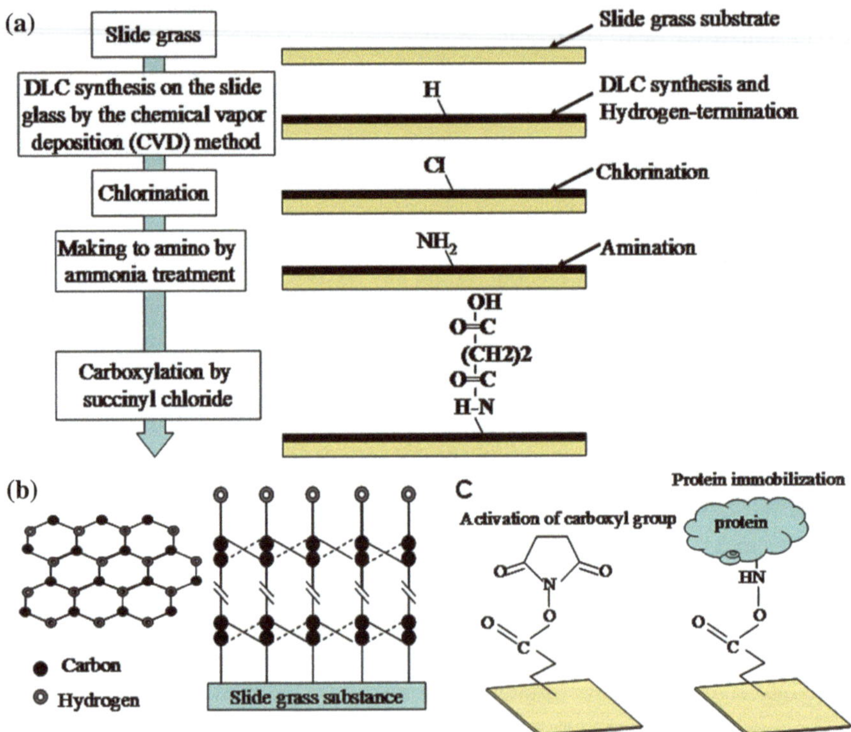

Fig. 8.2 Schematic illustration of the DLC chip, carboxylation, and antigen immobilization. **a** Chemical treatment processes for DLC surface carboxylation. **b** Schematic arrangement of carbon and hydrogen on the DLC chip. **c** Antigen immobilization processes on the DLC chip

8.3.2 Obtaining the Allergens

The allergens used for the preparation of allergen microarrays can be obtained as follows:

1. **Allergen extracts**

   Allergen extracts from dust mites, pollen, mold, cat or dog hair, eggs, milk, crabs, and other allergens are commercially available. However, the components of the extracts are complex and unknown. Using these crude extracts for diagnosis can only roughly determine which allergens the patient is sensitive to with a low accuracy. It does not precisely determine which allergenic components in the extract the patient is sensitive to. Thus, it does not allow for high throughput screening and other clinical requirements necessary for the diagnosis of allergic disease.
2. **Native allergen components that have been isolated and purified**
   Individual allergen components can be obtained by further separating and purifying them from the crude extracts. For example, Asturias, J.A. et al. purified an 18-kDa protein named Pla a 1 from the *P. acerifolia* pollen extract using ion exchange, gel filtration, and reverse-phase chromatography (Asturias et al. 2002). Riecken et al. (2008) purified a natural Ara h 8 in peanut extract using a unique combination of purification steps, including optimized extraction conditions, size exclusion and ion exchange chromatography, and treatment of the interfering contaminants with iodoacetic acid. Natural allergens that have been separated and purified from extracts have certain advantages in allergy diagnosis, but the purification process is tedious and high volume production is difficult.
3. **Recombinant allergens**
   An in vitro expression system can produce abundant amounts of allergens that have advantages such as known molecular weight, stability, quality, high purity, and high volume. Furthermore, their physical properties, chemical properties, and immunological activities are very similar to the native allergens. Therefore, recombinant allergens are an important source for allergens that are used in component-resolved diagnostics and high throughput screening. However, it is still a challenge to obtain some allergens with important posttranslational modifications or that need a fully native tertiary structure (Shreffler 2011).
4. **Allergens synthesized by a cell-free system in situ**
   In situ cell-free protein synthesis is a new technique that uses an exogenous DNA or specific mRNA sequence fixed on the substrate surface as a template to synthesize target proteins (Jackson et al. 2004; Stoevesandt et al. 2011). Tedious steps, such as protein extraction and purification, are avoided and the problem of protein preservation is also easily resolved. With the continued improvements of the in situ cell-free protein synthesis system, the use of a cell-free expression system for preparing allergen microarrays may be a good choice.

8.3.3 Preparation of Detection Probes

Allergens immobilized on solid phase substrates capture the specific IgE, which is followed by the addition of labeled anti-IgE antibodies that detect the allergen-specific IgE. Anti-IgE antibodies can be labeled with fluorescent substances, enzymes, or isotopes. The fluorescently labeled methods are widely used in protein analysis because they have high sensitivity, good selectivity, and a large dynamic rang. Enzymatic labeling methods with enzymes such as horseradish peroxidase (HRP) and alkaline phosphatase (AP) are also commonly used. Enzymatic labeling methods have high sensitivity, however, significant variations can arise due to different incubation times (Angenendt 2005). Isotope labeling methods are traditionally one of the most sensitive labeling procedures but are gradually being replaced by other detection methods because of the risks of radioactive contamination.

8.3.4 Detection and Analysis of Signals

In fluorescence-labeled allergen microarrays, a laser scanning confocal microscope can be used to scan for signal, and the mean fluorescence intensity of each point is then analyzed by computer. For enzyme-labeled microarrays, pictures are taken by a CCD camera after color development. The signals are processed by computer to obtain a gray scale for each point. After background signals are subtracted, qualitative and quantitative analysis is performed by comparing the relative amount of signal. Positive and negative and reference serums should always be included in each assay and each sample be tested multiple times to assure reproducibility and confidence in the result.

8.4 Challenges in High Throughput Screening Using Allergen Microarrays

8.4.1 Standardization of Allergens

Allergen components, concentration, and biological activity can all affect the detection of specific IgE. In order to improve the accuracy of IgE detection, close attention should be paid to allergen standardization, and particularly for isolated and purified allergen extracts or allergen components. Different extraction and separation methods cause there to be differences in allergen components and quality that lead to varying abilities to bind with specific IgE. Thus, methods are needed for allergen standardization. The most common allergen standardized method is a competitive IgE-binding inhibition test (e.g., ELISA or RAST), and it should always be used in the evaluation of the total allergenic activity of an allergen (Grier 2001). Relative potency can be obtained by comparing the target allergen extract with a reference allergen extract. Other biochemical and immunological methods such as SDS-PAGE, immunoelectrophoresis, circular dichroism spectrum analysis, immunoblotting, mass spectrometric analysis, dot blot analysis, proteomic approaches, etc. can all be used in checking and controlling the consistency of composition and activity of the allergen product (Jeong et al. 2011; Larsen and Dreborg 2008). Recombinant allergen proteins also need to be standardized, but it is easier to achieve standardization with recombinant proteins compared to allergen extracts and purified allergen components.

8.4.2 Cross-Reactivity in Allergy Screening

Cross-reactivity is caused by homology of the antigen sequence and structure with other proteins. It often occurs between allergenic molecules in closely related species (e.g., between grass or between mite allergens) or in evolutionarily preserved molecules with similar functions that are present in widely different species and belong to the same protein family (e.g., members of the tropomyosin protein family, such as Der p 10 in house dust mite and Pen m 1 in black tiger shrimp) (Canonica et al. 2013). Allergens that caused cross-reactivity can heavily influence the diagnosis of allergy. For example, birch pollen has cross-reactivity with peanut (Mittag et al. 2004). A patient with birch pollen allergy had high levels of sIgE against birch pollen as well as increased levels of sIgE against peanut, leading to the question of whether the patient was truly sensitive to both of these allergens. Further in depth examination found that the patient's level of sIgE against the main allergenic components of peanut such as Ara h2, Ara h1, and Ara h3 was very low, while only sIgE against Ara h 8 was significantly increased. Given that Ara h 8 can cross-react with birch pollen, this patient was determined to be sensitive only to birch pollen and not also to peanut. Therefore, in order to make an accurate diagnosis of allergic sensitivity, sIgE testing against the main allergenic components as well as to cross-reactive allergens should be performed at the same time (Predki et al. 2005), rather than testing for sIgE against the whole allergen extracts (Pfiffner et al. 2012).

8.4.3 High Throughput Screening for Drug Allergy

Drug hypersensitivity reactions (DHRs) are common adverse drug reactions that clinically resemble allergic reactions. When DHRs are mediated by a definite immunological mechanism (either drug-specific antibody or T cell), drug allergies is the preferred term (Demoly et al. 2014). Drugs are capable of inducing all types of immunological reactions, described by Gell and Coombs (Pichler 2003), but the most common are IgE and T cell-mediated reactions. It is very hard to distinguish between immune mediated drug allergies and nonimmune mediated DHRs just through clinical manifestations and without further laboratory examination. The diagnosis of drug allergies is the same as for the diagnosis of other allergies, such as food allergy, and includes provocation testing, skin testing, and specific antibody detection, as well as the basophil activation test in vitro (Ebo et al. 2011; Romano and Demoly 2007). Detection of sIgE against a drug is helpful for the diagnosis of drug allergy as was shown by Romano A et al. who tested for sIgEs of 70 patients with sensitivity to cephalosporin (demonstrated mostly by anaphylactic shock) using a Sepharose-radioimmunoassay (RIA) method. sIgEs against cephalosporins were detected in 67.1 % of the patients tested (Romano et al. 2005). Manfredi et al. also measured sIgEs of 55 patients with immediate reactions to quinolones, and found positive sIgEs in 54.5.1 % of the patients tested (Manfredi et al. 2004). Protein microarrays have been used in high throughput allergy screening, however, high throughput microarrays for drug allergies based on detection of specific antibody (especially IgE) have not yet been reported. With the development of chip technology and the deepening of drug allergy research, the use of high throughput screening to identify drug allergies and guide safe drug use in the clinic is feasible in the future.

8.4.4 Allergen Microarray Quality Control

Quality control is an important challenge for allergen microarrays. Without sound quality control, the results produced by experimental microarrays may not be convincing. Several critical steps are involved in the preparation of protein microarrays. The accuracy and repeatability of the results, as well as the quality of the allergen microarrays, depend on many factors such as: (1) Heterogeneity caused by allergen microarray spotting techniques; (2) The quality of the allergen immobilized to the solid phase substrates; (3) Microarray storage conditions-both surface dust aggregation and dewetting impact the quality of the allergen microarray. Also, some allergens are susceptible to degradation, thus, a suitable storage condition is important; (4) Microarray image processing and data acquisition; (5) Other antibody subtypes such as IgG can also affect the detection of allergen-specific IgE. In brief, many factors can interfere with and influence the results of the allergen microarray and cause artificial signals. These interfering signals are not easily identified, therefore, it is necessary to establish criteria that strictly evaluate the results obtained by allergen microarrays.

References

Angenendt P. Progress in protein and antibody microarray technology. Drug Discovery Today. 2005;10(7):503–11.

Asturias JA, Ibarrola I, Bartolome B, Ojeda I, Malet A, Martinez A. Purification and characterization of Pla a 1, a major allergen from *Platanus acerifolia* pollen. Allergy. 2002;57(3):221–7.

Cabauatan CR, Lupinek C, Scheiblhofer S, Weiss R, Focke-Tejkl M, Bhalla PL, Singh MB, Knight PA, van Hage M, Ramos JD, et al. Allergen microarray detects high prevalence of asymptomatic IgE sensitizations to tropical pollen-derived carbohydrates. J Allergy Clin Immunol. 2014;133(3):910–914e5.

Canonica GW, Ansotegui IJ, Pawankar R, Schmid-Grendelmeier P, van Hage M, Baena-Cagnani CE, Melioli G, Nunes C, Passalacqua G, Rosenwasser L, et al. A WAO—ARIA—GA(2)LEN consensus document on molecular-based allergy diagnostics. World Allergy Organ J. 2013;6(1):17.

Demoly P, Adkinson NF, Brockow K, Castells M, Chiriac AM, Greenberger PA, Khan DA, Lang DM, Park HS, Pichler W, et al. International consensus on drug allergy. Allergy. 2014;69(4):420–37.

Ebo DG, Hagendorens MM, De Knop KJ, Verweij MM, Bridts CH, De Clerck LS, Stevens WJ. Component-resolved diagnosis from latex allergy by microarray. Clin Exp Allergy: J Br Soc Allergy Clin Immunol. 2010;40(2):348–58.

Ebo DG, Leysen J, Mayorga C, Rozieres A, Knol EF, Terreehorst I. The in vitro diagnosis of drug allergy: status and perspectives. Allergy. 2011;66(10):1275–86.

Grier TJ. Laboratory methods for allergen extract analysis and quality control. Clin Rev Allergy Immunol. 2001;21(2–3):111–40.

Hamilton RG. Clinical laboratory assessment of immediate-type hypersensitivity. J Allergy Clin Immunol. 2010;125(2):S284–96.

Harwanegg C, Hiller R. Protein microarrays in diagnosing IgE-mediated diseases: spotting allergy at the molecular level. Expert Rev Mol Diagn. 2004;4(4):539–48.

Jackson AM, Boutell J, Cooley N, He M. Cell-free protein synthesis for proteomics. Briefings Funct Genomics Proteomics. 2004;2(4):308–19.

Jeong KY, Hong CS, Lee JS, Park JW. Optimization of allergen standardization. Yonsei Med J. 2011;52(3):393–400.

Kamemura N, Tada H, Shimojo N, Morita Y, Kohno Y, Ichioka T, Suzuki K, Kubota K, Hiyoshi M, Kido H. Intrauterine sensitization of allergen-specific IgE analyzed by a highly sensitive new allergen microarray. J Allergy Clin Immunol. 2012;130(1):113–121e2.

Kumble KD. An update on using protein microarrays in drug discovery. Expert Opin Drug Discov. 2007;2(11):1467–76.

Kusnezow W, Jacob A, Walijew A, Diehl F, Hoheisel JD. Antibody microarrays: an evaluation of production parameters. Proteomics. 2003;3(3):254–64.

Larsen JN, Dreborg S. Standardization of allergen extracts. Methods Mol Med. 2008;138:133–45.

Lin J, Bruni FM, Fu Z, Maloney J, Bardina L, Boner AL, Gimenez G, Sampson HA. A bioinformatics approach to identify patients with symptomatic peanut allergy using peptide microarray immunoassay. J Allergy Clin Immunol. 2012;129(5):1321–1328e5.

Lin J, Renault N, Haas H, Schramm G, Vieths S, Vogel L, Falcone FH, Alcocer MJ. A novel tool for the detection of allergic sensitization combining protein microarrays with human basophils. Clin Exp Allergy: J Br Soc Allergy Clin Immunol. 2007;37(12):1854–62.

Maloney JM, Rudengren M, Ahlstedt S, Bock SA, Sampson HA. The use of serum-specific IgE measurements for the diagnosis of peanut, tree nut, and seed allergy. J Allergy Clin Immunol. 2008;122(1):145–51.

Manfredi M, Severino M, Testi S, Macchia D, Ermini G, Pichler WJ, Campi P. Detection of specific IgE to quinolones. J Allergy Clin Immunol. 2004;113(1):155–60.

Mittag D, Akkerdaas J, Ballmer-Weber BK, Vogel L, Wensing M, Becker WM, Koppelman SJ, Knulst AC, Helbling A, Hefle SL, et al. Ara h 8, a Bet v 1-homologous allergen from peanut, is a major allergen in patients with combined birch pollen and peanut allergy. J Allergy Clin Immunol. 2004;114(6):1410–7.

Ott H, Baron JM, Heise R, Ocklenburg C, Stanzel S, Merk HF, Niggemann B, Beyer K. Clinical usefulness of microarray-based IgE detection in children with suspected food allergy. Allergy. 2008;63(11):1521–8.

Pfiffner P, Stadler BM, Rasi C, Scala E, Mari A. Cross-reactions vs co-sensitization evaluated by in silico motifs and in vitro IgE microarray testing. Allergy. 2012;67(2):210–6.

Pichler WJ. Delayed drug hypersensitivity reactions. Ann Intern Med. 2003;139(8):683–93.

Predki PF, Mattoon D, Bangham R, Schweitzer B, Michaud G. Protein microarrays: a new tool for profiling antibody cross-reactivity. Hum Antibodies. 2005;14(1–2):7–15.

Riecken S, Lindner B, Petersen A, Jappe U, Becker WM. Purification and characterization of natural Ara h 8, the Bet v 1 homologous allergen from peanut, provides a novel isoform. Biol Chem. 2008;389(4):415–23.

Romano A, Demoly P. Recent advances in the diagnosis of drug allergy. Curr Opin Allergy Clin Immunol. 2007;7(4):299–303.

Romano A, Gueant-Rodriguez RM, Viola M, Amoghly F, Gaeta F, Nicolas JP, Gueant JL. Diagnosing immediate reactions to cephalosporins. Clin Exp Allergy: J Br Soc Allergy Clin Immunol. 2005;35(9):1234–42.

Shreffler WG. Microarrayed recombinant allergens for diagnostic testing. J Allergy Clin Immunol. 2011;127(4):843–9 quiz 50-1.

Smith AH, Vrtis JM, Kodadek T. The potential of protein-detecting microarrays for clinical diagnostics. Adv Clin Chem. 2004;38:217–38.

Soares-Weiser K, Takwoingi Y, Panesar SS, Muraro A, Werfel T, Hoffmann-Sommergruber K, Roberts G, Halken S, Poulsen L, van Ree R, et al. The diagnosis of food allergy: a systematic review and meta-analysis. Allergy. 2014;69(1):76–86.

Spisak S, Guttman A. Biomedical applications of protein microarrays. Curr Med Chem. 2009;16(22):2806–15.

Stoevesandt O, Taussig MJ, He M. Producing protein microarrays from DNA microarrays. Methods Mol Biol. 2011;785:265–76.

Sturm GJ, Kranzelbinder B, Sturm EM, Heinemann A, Groselj-Strele A, Aberer W. The basophil activation test in the diagnosis of allergy: technical issues and critical factors. Allergy. 2009;64(9):1319–26.

Suzuki K, Hiyoshi M, Tada H, Bando M, Ichioka T, Kamemura N, Kido H. Allergen diagnosis microarray with high-density immobilization capacity using diamond-like carbon-coated chips for profiling allergen-specific IgE and other immunoglobulins. Anal Chim Acta. 2011;706(2):321–7.

Templin MF, Stoll D, Schwenk JM, Potz O, Kramer S, Joos TO. Protein microarrays: promising tools for proteomic research. Proteomics. 2003;3(11):2155–66.

Treudler R, Simon JC. Overview of component resolved diagnostics. Curr Allergy Asthma Rep. 2013;13(1):110–7.

Wang RQ, Zhang HY. Two hundred thousands results of allergen specific IgE detection. Chin J Allergy Clin Immunol. 2012;6(1):18–23.

Author's Biography

Dr. Junyan Zhang is a research associate working in Guangdong Provincial Key Laboratory of Allergy & Clinical Immunology, the Second Affiliated hospital of Guangzhou Medical University.

He is a medical doctor graduated from Southern Medical University. Now his main research direction is drug allergy. He is working on evaluating drug allergenicity by several models and establishing laboratory examination method for drug allergy diagnosis as well as investigating the mechanism of drug allergy.

Professor Ailin TAO is a Professor of Guangzhou Medical University, Director of Guangdong Provincial Key Laboratory of Allergy & Clinical Immunology, Principal Investigator of the State Key Laboratory of Respiratory Disease, Deputy Director of the State Key Clinical Specialty in Allergy of the Second Affiliated Hospital of Guangzhou Medical University, Member of the State Committee for Transgenic Safety Assessment, Standing Committee Member of Allergy Branch of Guangdong Medical Association, Member of Guangdong Provincial Committee for Transgenic Safety Assessment, Master Tutor.

Prof. Ailin TAO earned his doctorate degree from the State Key Laboratory of Crop Genetic Improvement of Huazhong Agricultural University in 2002, followed by a postdoctoral training at Postdoctoral Station of Basic Medicine in Shantou University Medical College, majoring in allergen proteins. Recent studies of his lab mainly focus on allergy bioinformatics, allergy and clinical immunology, disease models such as allergic asthma, allergic rhinitis, infection, and inflammation induced by allergy, inflammatory, and protracted diseases caused by antigen or superantigen. He has accumulated some experience in the allergology field including the mechanism on immune tolerance, allergy triggering factors and chronic inflammation pathways, and allergenicity evaluation and modification for food and drugs. He proposed some new concepts such as "Representative Major Allergens", "Allergenicity Attenuation" of immunotoxin and allergens, "broad-spectrum immunomodulator", etc. He put forward the theoretical hypothesis of "Balanced Stimulation by Whole Antigens". He focuses on the diagnosis of allergic disease, and the allergenicity medical evaluation and modification for food and drugs. He constructed the method system for prediction, quantitative assessment, and modification in parallel on allergenicity epitopes, which has been applied for more than 20 cases of allergens. He has developed a bioinformatics software for allergen discriminating, SORTALLER (http://sortaller.gzhmu.edu.cn), which performed significantly better than other existing softwares and reached a perfect balance with high specificity (98.4 %) and sensitivity (98.6 %) for discriminating allergenic proteins from several independent datasets of protein sequences of diverse sources, also highlighting with the Matthews correlation coefficients as high as 0.970, fast running speed, and rapidly predicting a batch of amino acid sequences with a single click. The software has been frequently used by researchers from lots of institutions in more than 20 Chinese cities and over 30 countries worldwide, thus becoming the TOP 1 in this field. He has set up an allergen database ALLERGENIA (http://ALLERGENIA.gzhmu.edu.cn), which has obvious international advantages in five aspects: wide allergen coverage, nonredundancy, astringency, accuracy, and user-friendly analysis functions.

Chapter 9
Surrogate Markers for Allergen-Specific Immunotherapy

Jiu-Yao Wang

Abstract Allergen-specific Immunotherapy (AIT) still remains the only option for the long-term management of allergic rhinitis and allergic asthma, including disease due to house dust mite (HDM) allergy. In vivo provocation tests (e.g., skin testing and conjunctival provocation testing) are useful indicators of clinical improvement, but the identification of in vitro markers for monitoring the effects of AIT has been a long-sought goal. Previous studies on AIT in IgE-mediated allergic diseases found that suppression of allergen-induced late skin responses occurred as early as 2–4 weeks and was accompanied by increases in IL-10, whereas suppression of the immediate skin response to allergen occurred later, at 6–12 weeks, and was accompanied by increases in serum 'blocking' IgG4 antibodies that had the capacity to suppress both allergen-triggered basophil histamine release and the binding of IgE-allergen complexes to B cells. It has been claimed that a flow cytometry-based assay (IgE-FAB) can be used to demonstrate IgE-facilitated antigen presentation and activation of T cells. Measurement of peripheral T cell responses is complex and difficult to standardize for routine clinical use. Whether quantitative measures of IgG-associated inhibition (in contrast to IgG levels) and/or suppression of early and/or late skin responses may predict AIT efficacy and long-term benefits in individual patients remains to be determined. The search for adequate biological markers that reflect the immune status of patients before, during, and after AIT is critical to evaluate its clinical efficacy and the future of personalized medicine in treating allergic diseases.

Keywords Allergen · Immunotherapy · Biomarkers · IgG4 blocking antibody · Basophils activation tests

J.-Y. Wang (✉)
Department of Pediatrics, Allergy and Clinical Immunology Research (ACIR) Center, College of Medicine, National Cheng Kung University, No.1, University Road, Tainan 70428, Taiwan
e-mail: a122@mail.ncku.edu.tw

A. Tao and E. Raz (eds.), *Allergy Bioinformatics*, Translational Bioinformatics 8,
DOI 10.1007/978-94-017-7444-4_9

9.1 Introduction

Allergen immunotherapy is highly effective in selected patients with IgE-mediated allergic rhinitis and mild asthma including those who fail to respond to usual anti-allergic drugs. Clinical efficacy may be accompanied by disease remission following discontinuation of treatment and prevention of new sensitizations and/or progression from rhinitis to asthma (Passalacqua and Durham 2007). These clinical observations imply long-term effects on the immune system with induction of immunological tolerance. Data from studies of peripheral blood and target organs suggest that induction of regulatory T cells and/or deviation from antigen-specific T_H2 in favor of T_H1 responses are relevant. Serum measurements imply additional altered long-term memory B cell responses, including increases in allergen-specific IgG, particularly IgG4 and IgA, especially, the IgA2 subclass. Recent studies of the mechanisms of sublingual immunotherapy have identified more modest but similar changes compared to the subcutaneous route (Lee et al. 2013; Moingeon et al. 2006; Scadding and Durham 2009). A critical question is whether these studies may provide biomarkers that may be indicative and/or predictive of the clinical response to immunotherapy.

9.2 The Need for Biomarkers

Biological markers are objectively measured indicators of normal biological processes, pathogenic processes, or pharmacologic responses to therapeutic intervention (Group 2001). Currently, they are of major interest for patient management in future approaches to treatment in all domains of immunotherapy (e.g., allergy, cancer, autoimmunity). For the diagnosis and treatment for allergic diseases, biomarkers should allow allergists to accomplish two things: (1) to identify, prior to immunotherapy, responder versus nonresponder patients with a high degree of confidence and (2) to follow the immunological status of patients during immunotherapy in order to modulate and adjust treatment schemes. Therefore, biomarkers can be classified into three main types: those able to (1) predict a response to treatment and safety, (2) document short-term efficacy, and (3) monitor long-term efficacy. Each type of biomarker is valuable in the support of patient management, before, during, and after immunotherapy. Whereas significant efforts have been made to identify such surrogates in allergen-specific immunotherapy, documenting the impact on symptom and medication scores in the context of double-blind, placebo-controlled clinical trials remains the only true demonstration of clinical efficacy (Prentice 1989).

9.3 Current Status of Biomarkers in Allergic Diseases

Currently, the diagnosis of allergy relies upon skin prick tests and measurement of allergen-specific IgE reactivity using in vitro diagnostic assays such as ELISA (CAP system, Vidas) or microfluidic assay (BioIC) (Shyur et al. 2010). Although these diagnostic methods can define a pattern of sensitization for any given patient, and in some cases precisely identify the allergens involved, they are not appropriate for following changes in the immune status of the patient during treatment. Several markers have been proposed to monitor allergic inflammation (e.g., eosinophil counts or ECP levels in sputum, tryptase in nasal fluid). Another indirect and noninvasive biological measure of airway hyperreactivity (AHR) is the NO content in exhaled air. All of these markers relate to inflammatory symptoms associated with late immune effector mechanisms and as such are likely to be relevant when following the impact of symptomatic treatments such as corticosteroids or anti-histamine. However, allergen-specific immunotherapy (AIT) is a causal treatment that corrects the immune imbalance in allergic patients (from a T_H2 to T_H1 or Treg response) in an allergen-specific manner (Akdis and Blaser 2000; Akdis et al. 2004). Therefore, a biomarker for AIT would be one relevant to the modification of an immune parameter that is involved in early inflammation. Classically, only a limited number of immunological parameters have been monitored in the course of allergy immunotherapy studies and were essentially restricted to measuring modulation of the ratio between IgE/IgG4 titers in the serum. More recently, a potential blocking activity for allergen-specific antibodies has been assessed (Francis et al. 2008). It was shown that allergen-specific blocking IgG antibodies inhibit allergen-induced mast cell and basophil degranulation as well as IgE-facilitated allergen presentation to T cells and is associated with a reduction of in vivo sensitivity (Dreborg et al. 2012; Niederberger et al. 2004; Pauli et al. 2008; Shamji and Durham 2011). In addition, measurement of IgE-mediated activation of basophils has been performed (looking either at histamine release or at the expression of the activation marker CD203c). Cellular assays (e.g., basophil activation assays and FAB assay) (Lichtenstein et al. 1973; Shamji et al. 2006) may allow for the uncovering and measurement of the effects of allergen-specific blocking antibodies that interfere with allergen-IgE interaction as well as for the correlation of in vitro results with clinical outcomes but they are quite cumbersome. Presently, reduction in allergen-specific IgE binding can be measured by microarray, which may become a possible surrogate marker for the effects of specific immunotherapy (Skrindo et al. 2015). Several studies identified shifts from T_H2 to T_H1 and/or regulatory T cell responses by analyzing PBMC. These latter findings, however, were not reproduced in other studies and, as of today, none of those parameters has been confirmed to correlate with clinical efficacy. They have only demonstrated the establishment of an immune response induced by the vaccine. Thus, the search for adequate biological markers that reflect the immune status of patients and clinical efficacy is critical to support the use of allergen-specific immunotherapy.

9.4 Candidate Biomarkers in Clinical Trials for Allergen-Specific Immunotherapy

Immunotherapy still remains the only option for the long-term management of AR and allergic asthma, including disease due to HDM allergy. AIT is currently administered via the subcutaneous (SCIT) and sublingual (SLIT) routes, which have been shown to be effective and induce long-term clinical and immunologic tolerance. (Durham et al. 1999, 2012; Francis et al. 2008; James et al. 2011). Unlike other pharmacological treatment options, immunotherapy acts on symptoms as well as altering the natural course of the disease. A search for biomarkers can ideally be conducted in allergic patients undergoing immunotherapy. In this context, a rational approach is to collect samples during clinical studies or to set up dedicated clinical studies to assess immunological responses (both humoral and cellular) induced during immunotherapy in association with clinical efficacy. Immunologic changes after immunotherapy include suppression of allergen-specific T_H2 responses, induction of regulatory T cells (IL-10^+ $CD4^+$ CD25hi, and $CD4^+$ CD25hi forkhead box protein 3, Foxp3,—positive cells) and the appearance of "protective" allergen-specific IgG antibodies, particularly of the IgG4 subclass (Akdis et al. 1998; Francis et al. 2003; James et al. 2011; Shamji et al. 2012; van Neerven et al. 1999). Suppression of the cutaneous early allergic response after immunotherapy is temporarily associated with increases in serum IgG-associated inhibitory activity but the detailed mechanism of AIT remains elusive.

9.5 Comparison of Immune Responses to Allergen Between Atopic and Healthy Individuals

An additional approach to identify and validate biomarkers is to analyze potential protective responses observed in healthy individual exposed to allergens. Although it was previously thought that nonallergic patients are anergic to allergens from the environment, there is now evidence that they indeed mount allergen-specific responses preventing allergic symptoms. For example, the IgE and IgG reactivity profiles to HDM allergens, as well as IgE levels to certain allergen components, differed considerably between children with and without asthmatic symptoms caused by HDM allergy (Resch et al. 2015). In fact, asthmatic children were characterized by an expanded IgE repertoire regarding the numbers of recognized allergen components and by increased specific IgE levels. To analyze and compare allergen-specific T cell responses in healthy and allergic patients, allergen-specific HLA class II tetramers have been developed. In this approach, T cell epitopes are identified within the allergen, and the major immunodominant epitope is combined with multimeric recombinant HLA molecules labelled with a fluorescent dye. It has been reported that such a reagent was used as a probe to detect and characterize birch pollen and house dust mite-specific $CD4^+$ T lymphocytes in the blood of healthy people (Van Overtvelt et al. 2008). The detailed characterization (e.g. TCR, V beta usage, cytokine profile, transcriptome) of allergen reactive T cells detected in healthy donors or allergic patients using such tetramers could lead to the identification of immunological biomarkers that correlate with protection.

9.6 Biomarkers for Allergen-Specific Immunotherapy? An Unmet Need

In the future, when patients receive allergen-specific immunotherapy, biomarkers should be quantitative and easily measured in readily accessible body fluids (e.g., blood, saliva, nasal secretions) (Horak et al. 2009). Serologic parameters (e.g., antibody titers) can easily be monitored even in the context of large-scale clinical studies because they usually require small volumes of blood and simple logistics. The functionality of these antibodies can further be assessed by measuring their affinity and their blocking potential. Such analyses can be extended to various isotypes possibly involved in immunotherapy efficacy. Exploratory technologies, like proteomic approaches, can also be used to identify new markers modulated during immunotherapy. In contrast, cellular responses can only be assessed in small-scale clinical studies due to the complexity of T cell analysis using fresh blood. There are several assays that can be used to monitor allergen-specific T cells in peripheral blood (e.g., proliferation of CFSE-labeled cells, intracellular cytokine production, ELISPOT, gene expression by real-time PCR) (Keilholz et al. 2002). These assays allow discriminating between allergen-specific T_H1, T_H2, and Treg responses. In addition, the development of new tools such as tetramers now permit the detection of allergen-specific cells and may enable better identification of changes during treatment.

9.7 Conclusion

The benefits conveyed by the discovery of biomarkers in the field of immunotherapy for common allergies are numerous. Particularly, biomarkers would be extremely valuable in (1) identifying patients likely to benefit from the treatment (i.e., establishing their immune status before the treatment), (2) adapting treatment modalities such as immunization schemes and dosing to the immune status of the patient ("personalized medicine"), (3) providing tools for immunomonitoring during and after the treatment, (4) establishing biological mechanisms of action of successful immunotherapy, and (5) supporting the approval process for a drug or treatment. We currently lack quantitative biomarkers that are able to reveal the clinical efficacy of allergen-specific immunotherapy. The development of methods to monitor allergen-specific immune responses (or biological changes reflecting the early consequences of immune stimulation with the allergen) should provide immunomonitoring tools that would be extremely useful to allergists in order to assess the immune status of patients before, during, and after immunotherapy. Ultimately, such tools would be routinely used to adjust the dose and immunization regimen to the needs of the patient. In the absence of such established biomarkers, the clinical efficacy of AIT can only be established in the context of large scale, placebo-controlled clinical studies.

References

Akdis CA, Blaser K. Mechanisms of allergen-specific immunotherapy. Allergy. 2000;55(6):522–30.

Akdis CA, Blesken T, Akdis M, Wuthrich B, Blaser K. Role of interleukin 10 in specific immunotherapy. J Clin Invest. 1998;102(1):98–106.

Akdis M, Verhagen J, Taylor A, Karamloo F, Karagiannidis C, Crameri R, Thunberg S, Deniz G, Valenta R, Fiebig H, et al. Immune responses in healthy and allergic individuals are characterized by a fine balance between allergen-specific T regulatory 1 and T helper 2 cells. J Exp Med. 2004;199(11):1567–75.

Dreborg S, Lee TH, Kay AB, Durham SR. Immunotherapy is allergen-specific: a double-blind trial of mite or timothy extract in mite and grass dual-allergic patients. Int Arch Allergy Immunol. 2012;158(1):63–70.

Durham SR, Walker SM, Varga EM, Jacobson MR, O'Brien F, Noble W, Till SJ, Hamid QA, Nouri-Aria KT. Long-term clinical efficacy of grass-pollen immunotherapy. N Engl J Med. 1999;341(7):468–75.

Durham SR, Emminger W, Kapp A, de Monchy JG, Rak S, Scadding GK, Wurtzen PA, Andersen JS, Tholstrup B, Riis B, et al. SQ-standardized sublingual grass immunotherapy: confirmation of disease modification 2 years after 3 years of treatment in a randomized trial. J Allergy Clin Immunol. 2012;129(3):717–25.

Francis JN, Till SJ, Durham SR. Induction of IL-10^{+}CD4^{+}CD25^{+} T cells by grass pollen immunotherapy. J Allergy Clin Immunol. 2003;111(6):1255–61.

Francis JN, James LK, Paraskevopoulos G, Wong C, Calderon MA, Durham SR, Till SJ. Grass pollen immunotherapy: IL-10 induction and suppression of late responses precedes IgG4 inhibitory antibody activity. J Allergy Clin Immunol. 2008;121(5):1120–5.

Group BDW. Biomarkers and surrogate endpoints: preferred definitions and conceptual framework. Clin Pharmacol Ther. 2001;69(3):89–95.

Horak F, Zieglmayer P, Zieglmayer R, Lemell P, Devillier P, Montagut A, Melac M, Galvain S, Jean-Alphonse S, Van Overtvelt L, et al. Early onset of action of a 5-grass-pollen 300-IR sublingual immunotherapy tablet evaluated in an allergen challenge chamber. J Allergy Clin Immunol. 2009;124(3):471–7.

James LK, Shamji MH, Walker SM, Wilson DR, Wachholz PA, Francis JN, Jacobson MR, Kimber I, Till SJ, Durham SR. Long-term tolerance after allergen immunotherapy is accompanied by selective persistence of blocking antibodies. J Allergy Clin Immunol. 2011;127(2):509–16.

Keilholz U, Weber J, Finke JH, Gabrilovich DI, Kast WM, Disis ML, Kirkwood JM, Scheibenbogen C, Schlom J, Maino VC, et al. Immunologic monitoring of cancer vaccine therapy: results of a workshop sponsored by the Society for Biological Therapy. J Immunother. 2002;25(2):97–138.

Lee M, Lee BW, Vichyanond P, Wang JY, Bever HV. Sublingual immunotherapy for house dust mite allergy in Southeast Asian children. Asian Pac J Allergy Immunol. 2013;31(3):190–7.

Lichtenstein LM, Ishizaka K, Norman PS, Sobotka AK, Hill BM. IgE antibody measurements in ragweed hay fever. Relationship to clinical severity and the results of immunotherapy. J Clin Invest. 1973;52(2):472–82.

Moingeon P, Batard T, Fadel R, Frati F, Sieber J, Van Overtvelt L. Immune mechanisms of allergen-specific sublingual immunotherapy. Allergy. 2006;61(2):151–65.

Niederberger V, Horak F, Vrtala S, Spitzauer S, Krauth MT, Valent P, Reisinger J, Pelzmann M, Hayek B, Kronqvist M, et al. Vaccination with genetically engineered allergens prevents progression of allergic disease. Proc Natl Acad Sci U S A. 2004;101(Suppl 2):14677–82.

Passalacqua G, Durham SR. Allergic rhinitis and its impact on asthma update: allergen immunotherapy. J Allergy Clin Immunol. 2007;119(4):881–91.

Pauli G, Larsen TH, Rak S, Horak F, Pastorello E, Valenta R, Purohit A, Arvidsson M, Kavina A, Schroeder JW, et al. Efficacy of recombinant birch pollen vaccine for the treatment of birch-allergic rhinoconjunctivitis. J Allergy Clin Immunol. 2008;122(5):951–60.

Prentice RL. Surrogate endpoints in clinical trials: definition and operational criteria. Stat Med. 1989;8(4):431–40.

Resch Y, Michel S, Kabesch M, Lupinek C, Valenta R, and Vrtala S. Different IgE recognition of mite allergen components in asthmatic and nonasthmatic children. J Allergy Clin Immunol. 2015.

Scadding G, Durham S. Mechanisms of sublingual immunotherapy. J Asthma. 2009;46(4):322–34.

Shamji MH, Durham SR. Mechanisms of immunotherapy to aeroallergens. Clin Exp Allergy. 2011;41(9):1235–46.

Shamji MH, Wilcock LK, Wachholz PA, Dearman RJ, Kimber I, Wurtzen PA, Larche M, Durham SR, Francis JN. The IgE-facilitated allergen binding (FAB) assay: validation of a novel flow-cytometric based method for the detection of inhibitory antibody responses. J Immunol Methods. 2006;317(1–2):71–9.

Shamji MH, Ljorring C, Francis JN, Calderon MA, Larche M, Kimber I, Frew AJ, Ipsen H, Lund K, Wurtzen PA, et al. Functional rather than immunoreactive levels of IgG4 correlate closely with clinical response to grass pollen immunotherapy. Allergy. 2012;67(2):217–26.

Shyur SD, Jan RL, Webster JR, Chang P, Lu YJ, Wang JY. Determination of multiple allergen-specific IgE by microfluidic immunoassay cartridge in clinical settings. Pediatr Allergy Immunol. 2010;21(4 Pt 1):623–33.

Skrindo I, Lupinek C, Valenta R, Hovland V, Pahr S, Baar A, Carlsen KH, Mowinckel P, Wickman M, Melen E, et al. The use of the MeDALL-chip to assess IgE sensitization: a new diagnostic tool for allergic disease? Pediatr Allergy Immunol. 2015;26(3):239–46.

van Neerven RJ, Wikborg T, Lund G, Jacobsen B, Brinch-Nielsen A, Arnved J, Ipsen H. Blocking antibodies induced by specific allergy vaccination prevent the activation of CD4+ T cells by inhibiting serum-IgE-facilitated allergen presentation. J Immunol. 1999;163(5):2944–52.

Van Overtvelt L, Wambre E, Maillere B, von Hofe E, Louise A, Balazuc AM, Bohle B, Ebo D, Leboulaire C, Garcia G, et al. Assessment of Bet v 1-specific CD4+ T cell responses in allergic and nonallergic individuals using MHC class II peptide tetramers. J Immunol. 2008;180(7):4514–22.

Author Biography

Prof. Dr. Jiu-Yao Wang PhD is currently a Distinguished Professorof Pediatrics and Director of the Allergy and Clinical Immunology Research (ACIR) Center at National Cheng Kung University Medical Center, Tainan, Taiwan. After clinical residency and sub-specialty training in pediatricallergy, clinical immunology and rheumatology from 1984-1987 at the National Taiwan University Hospital, Taipei, he was appointed as Lecturer (1989), AssociateProfessor (1992), and Professor (2000) at the College of Medicine, National Cheng Kung University, Tainan, Taiwan. He obtained his Doctor of Philosophy (D.Phil.)degree from the MRC Immunochemistry Unit, University of Oxford, United Kingdomin 1996. For more than 25 years, Dr. Wang's lab has been focused on the study of thegenetic, environmental, and immunological factors involved in the innate immunity of allergic and asthmatic children in Taiwan. These studies have produced more than 110SCI journal papers that have been cited more than 2,000 times. He has received several academic awards, including the 37th Ten Outstanding Young Persons in Medical Research Award (1999) sponsored by Taiwan Junior Commercial Chamber,Taiwan, and Gold Medal of K.T. Li Foundation for the Development of Science and Technology (2014). He has been the president of the Taiwan Academy of Pediatric Allergy, Asthma, and Clinical Immunology (TAPAACI), Board Member and Congress Chair of the 2013 Asian Pacific Academy of Allergy, Asthma and Clinical Immunology (APAAACI), and currently serves as a National Delegate and Board Member of the Educational Council of the World Allergy Organization (WAO).

Chapter 10
Immune Responses to Allergens in Atopic Disease: Considerations for Bioinformatics

Wayne R. Thomas

Abstract Allergies are mediated by several immunological mechanisms that make different contributions in different patients. Measurements of T_H2 and regulatory T cell responses are poorly standardized and are highly influenced by cell culture conditions and allergen preparations. There are, however, purified allergens and new techniques that can facilitate direct ex vivo measurements and purified allergens for use. IgE antibodies can be accurately and reproducibly measured using pure allergens that represent the important specificities and allergen arrays can be used to identify cross-reactivities, and hence, reduce false positives. IgE binding to most allergens is dominated by a few components that usually bind with the same pattern for most patients, although there are exceptions that are now recognized to provide important information. Measurements with allergen arrays can also reduce the underestimation of antibody titers due to underrepresentation of the important allergen components. Furthermore, IgE titers should be analysed preferably as continuous variables or if cut offs are used for thresholds the titers should reflect a high probability of substantive atopy and these titers differ for different allergens. The IgE value that has historically been used to indicate reaction sensitivity, 0.35 IU/mL, is the historic reaction sensitivity for IgE measurements and is a trivial amount of antibody for many allergens but a significant amount for others. Positive skin test responses can be readily elicited in subjects with trivially low IgE titers and no history of allergic disease and are thus a poor measure of atopy. The type of immune responses elicited by different allergens, as shown by IgE and IgG subclass titers and, from limited data, T cell responses to different allergens differ in memory cell phenotypes and cytokine profiles. There is much from these considerations that can be applied to improve the design and analysis of bioinformatic studies.

W.R. Thomas (✉)
Telethon Kids Institute, University of Western Australia, 100 Roberts Road, Subiaco, WA 6008, Australia
e-mail: Wayne.Thomas@telethonkids.org.au

A. Tao and E. Raz (eds.), *Allergy Bioinformatics*, Translational Bioinformatics 8,
DOI 10.1007/978-94-017-7444-4_10

Keywords Immune response • IgE titer • T cell responses • Cytokine profiles • Allergen provocation

10.1 Introduction

While the participation of T_H2 cells and IgE antibodies in atopy and atopic diseases needs no introduction, the relative roles of IgE and T_H2-cell cytokines in the immunopathology of allergy are not well understood, and are the influences of regulatory and ancillary non-T_H2 responses. Type 1 allergic responses occur quickly with, for inhalant allergy, smooth muscle constriction, edema, and mucous secretion produced within 30 min of encountering the specific antigen. For many people this is followed by late reactions and chronic tissue damage. In skin tests, the late phase reactions show as swelling and erythema occurring after 2–6 h and are associated with mixed inflammatory cell infiltration. This can result not only from the secretion of T_H2 cell cytokines but also from the release of TNF-α (Grimbaldeston et al. 2006), IL-33 (Hsu et al. 2010) and lipid mediators (Pettipher 2008) from mast cells. Separate from the acute responses, T_H2 cytokines have direct chronic effects on inflammation, such as IL-5-mediated tissue eosinophilia, IL-13-mediated fibrosis and mucin induction. Insights into the mechanism of human atopic disease are now being obtained with new types of therapy and with gene expression and genome association studies.

Therapeutic trials with the anti-IgE monoclonal antibody omalizumab and monoclonal antibody antagonists of T_H2 cytokines are now showing the importance of the different responses in different patients. Direct T cell-mediated disease has been demonstrated by the ability of anti-IL-13 antibody (Scheerens et al. 2014) to inhibit late phase responses and improve lung function with little effect on IgE or immediate responses. Subjects that respond best to anti-cytokine treatment have high levels of periostin (Corren et al. 2011), a T_H2-cytokine-induced tissue matrix protein. The beneficial effects of omalizumab for the treatment of immediate hypersensitivity have been repeatedly demonstrated although typically only 30 % of subjects obtain a worthwhile benefit (Holgate 2013). This monoclonal antibody can also reduce periostin levels (Parulekar et al. 2014), possibly by inhibiting the ability of IgE to enhance antigen presentation to T_H2 cells.

Studies of gene expression in T cell responses to allergen extracts have shown the activation of many T_H2 and other genes in inflammation-linked pathways (Bosco et al. 2006; Zhao et al. 2012), but there has been little consensus or follow-up and none have been examined responses to purified allergen components or directly compared allergens from different sources. Studies examining epithelial cells of asthmatics have been more informative, with the standout discovery being the identification of periostin expression (Woodruff et al. 2007), a valuable biomarker of asthmatic responses (Takayama et al. 2006).

Four of the five most consistent associations found in genome-wide association studies (GWAS), namely IL-33, thymic stromal lymphopoietin (TSLP), IL-1RL1, the IL-33 receptor (ST2) and HLA-DQ, strongly implicate T_H2 responses (Ober and Yao 2011) with the remaining 17q21 association implicating a gene for tissue remodeling. Other less universal associations with IL13, IL2RB, and SMAD3 similarly implicate T_H2 cells and another, RORA, group 2 innate lymphoid cells. Two others, SLC22A5 and PYHIN1, implicate epithelial cells and macrophages, respectively. Although these associations point to potentially important elements in atopy, they are limited because, from different estimates, they only account for 4–24 % of the known inheritance of asthma (Lee et al. 2011). Whole genome sequencing might reveal some missing polymorphisms but the missing inheritance is most likely due to gene-gene interactions and even interactions of only two genes can create problems for the current computational platforms. Another anomaly is that the genes identified as important for asthma by GWAS have little overlap with those important for atopy. Accordingly, a meta-analysis for GWAS for atopy revealed the importance of TLR6, STAT6, HLA-DQB1, IL-1RL1, LPP, IL-2, and HLA-B (Bonnelykke et al. 2013). These genes are important for immune function in general, and STAT6 for T_H2 responses, but only IL-1R1 and HLA-DQ were associated with asthma.

A feature of the analyses of therapy, gene expression, and genetic associations is that they have paid only limited attention to the size of the immune responses associated with atopic disease and none to the different types of immune responses induced by different allergens. Thus, there is a considerable scope for further studies with more informed designs and more informative measurements for quantitative analyses. Points that are critical for a better understanding of how atopy is measured and the varying nature of atopic responses to different allergens will be presented below.

10.2 Measurement of Atopy

10.2.1 IgE Antibody Binding

IgE binding has been most widely measured by the proprietary immunoabsorption test first released in 1974 as the Phadebas RAST and now the Phadia ImmunoCAP test. Absolute (gravimetric) measures of IgE binding could be derived from RAST tests but direct calibration was introduced in 1990 with the attachment of anti-IgE antibodies to the immunoabsorbent matrix to capture IgE standards (Yman 2001). The measurements are given in International Units (IU) of IgE based on a WHO reference preparation with 1 IU IgE now known to be 2.42 ng. All the early tests and current standard tests use unfractionated extracts made from sources of the allergens to bind IgE antibody. Since an excess of allergen on the solid phase matrix is essential for the quantitation of these assays, the underrepresentation of allergen components and a lack of knowledge of the components present in extracts are important issues. Phadia tests with selected purified or recombinant allergens are now available but for the most part use components that are already well represented in extracts. Other proprietary assays like the Hycore allergen fluorescence system from HYTEC and Immulite 2000 3gAllergyT system also give absolute measurements.

Microarrays or allergen chips that give a profile of IgE binding to some of the allergen components from different allergens are now available. They have considerable clinical utility in distinguishing cross-reactivities that confound diagnosis and in some cases the profiles show if the suspected allergen is likely to cause mild or severe disease (Canonica et al. 2013). The Phadia ImmunoCAP ISAC measures IgE antibodies to a panel of 112 components from 51 allergen sources. Technically, thousands of components could be arrayed on a chip but the number has been limited by their known clinical utility and performance in multiplex assay conditions. Although stated as arbitrary, the ISAC Standardized Units (ISU) have been calibrated against standards characterized by ImmunoCAP assays and a good concordance of titers determined with microarrays and ImmunoCAP has been found for grass allergen components (Jahn-Schmid et al. 2003). A chip developed from the European Union-funded project MeDALL has multiplexed 170 allergen components coated in minute quantities (Lupinek et al. 2014). It has good titration characteristics for IgE concentrations up to 300 ng/mL but significant inhibition, not found for ImmunoCAP (Squire et al. 1986), occurs with physiological levels of IgG antibody. Other old (Jahn-Schmid et al. 2003) and new formats (Suzuki et al. 2011), however, do not show any indication of IgG interference so high-density quantitative arrays appear feasible.

Other quantitative IgE binding assays have been validated. One method that can be applied to microtiter-plate ELISA and has been used extensively for international reference sera for infectious disease (Holder et al. 1995), is to perform parallel assays of a test antigen and a heterologous standard. Providing the antibodies have similar affinity, as shown by titration curves, titers can be interpolated from the standard. Humanized monoclonal anti-Der p 2 antibodies have been developed for standards with standard binding of the allergens to the solid phase being accomplished by capture with the equal concentrations of purified monoclonal antibodies for each allergen (Ichikawa et al. 1999; Trombone et al. 2002) or, for recombinant components, with the use of monoclonal antibodies to epitope-tags (Hales et al. 2004, 2006). When conducted with Fel d 1 (Hales et al. 2013a; Ichikawa et al. 1999), Der p 1&2 (Trombone et al. 2002) and with a panel of house dust mite (HDM) components (Hales et al. 2006), the titers show good concordance with ImmunoCAP and for Fel d 1 also with titers determined by radioimmunoassay (Ichikawa et al. 1999). Unpublished work-up of the assays used by Hales et al. (2013a, 2006) showed no effect from pre-absorption of allergic sera with protein G to remove IgG antibodies.

10.2.2 IgE Antibody Predictive Values

The titer of 0.35 IU/mL of IgE antibody (2.42 ng/mL), which is frequently taken as a measure of atopy, is the historic lower limit of IgE that could be reliability detected by RAST test (Yman 2001) and has no biological significance. Indeed, IgE antibody titers of over 0.35 IU/mL to single protein antigens of *Haemophilus influenzae* and *Streptococcus pneumoniae* bacteria are found in 40–50 % of non-atopic subjects (Hollams et al. 2010) and, as shown for *H. influenzae*, these bacteria induce T_H1-biased responses with little IL-5 but with detectable IL-13 (Epton et al. 2002a). The limitation of the 0.35 IU/mL cut-off is particularly evident in population samples as shown in an unselected Finnish cohort (Haahtela and Jaakonmaki 1981) where allergic symptoms were found in only 40 % of subjects with detectable IgE antibody compared to 30 % of subjects without detectable IgE. Titers, over 17.5 IU/mL were, however, highly associated with allergic symptoms and titers below 3.5 IU/mL with asymptomatic subjects.

More recent European compilations found that, typically, 10 % of subjects with 0.35 IU/mL of IgE antibody to an aeroallergen would have been diagnosed with allergy by a physician with the probability increasing linearly with the log of the titer until about 3.5 IU/mL where it attained 80–100 % (Soderstrom et al. 2003). There are similarly shaped probability curves for the diagnosis of milk, egg, fish, and peanut allergy in children except that they plateau at over 17.5 IU/mL where 75–90 % probabilities are evident. A diagnosis of atopy in infants is however made with lower IgE titers (Urisu et al. 2014).

The probability that 5-year old children will have asthma symptoms shows a similar association with IgE titers. For HDM the probability of having allergic disease for subjects with an IgE titer of 0.35 IU/mL is 15 % and this slowly increases to a plateau level of only about 55 %, even for children with very high titers for example 300 IU/mL (Simpson et al. 2005). The probability curves for children with cat and dog allergies are however sharper reaching 85 % and 90 % disease prevalences at the higher antibody concentrations, thus showing greater penetration of disease. Analysis of children with seasonal rhinoconjunctivitis showed that the relationship with grass pollen allergen titers and disease resembled that for the cat and dog, with the curves reaching an 80 % disease prevalence (Marinho et al. 2007).

Some studies provide an important perspective on what might be expected from investigations that use unselected cross-sectional population samples. Women in a pregnancy cohort in Michigan USA, a state with similar prevalences of allergies to most temperate environments, showed a 46 % incidence of having an IgE titer of over 0.35 IU/mL to an allergen in the absence of symptoms compared to the incidence of 66 % in symptomatic subjects (Abraham et al. 2007). When analysed for different allergens, IgE was found in 10–20 % of the asymptomatic subjects, depending on the allergen, compared to 20–50 % for the symptomatic subjects. For cat, dog, and horse allergy, where some self-awareness of symptoms would be expected, 14.5 % of asymptomatic subjects had titers of over

0.35 IU/mL. In Europe, titers of asymptomatic subjects have indicated cut-offs of approximately 10 IU/mL for pollens and HDM (Pastorello et al. 1995), but 25–40 % of symptomatic subjects had titers below this value and 25–30 % of asymptomatic subjects had titers above. The relationship of IgE titer with asymptomatic sensitization also should consider that it often precedes symptomatic allergy (Bodtger et al. 2003).

10.2.3 Skin Prick Tests

Skin prick test (SPT) reactions are hypersensitivity reactions and, thus, a direct measure of allergenicity. They have been demonstrated with many purified and recombinant allergens (Niederberger et al. 2014), including minor allergens, so it is important to consider their relationship with allergic disease and atopic sensitization. For aeroallergens, SPTs can be readily demonstrated in subjects that have less than 0.35 IU/mL of IgE antibody (Haahtela and Jaakonmaki 1981). For example, in a teenage cohort, only 84 % of subjects with positive SPTs to grass pollen had over 0.35 IU/mL of IgE antibody and this number fell to 62 % for cat extracts. Conversely, only 3 % of subjects with detectable IgE were SPT negative. A contemporary investigation showed that, depending on the allergen, between 2.5 and 19.4 % of the IgE titers of SPT positive subjects were between 0.1 and 0.35 IU/mL. The association of SPT with allergic disease is also only weakly positive. In a cohort of unselected Finnish teenagers (Haahtela et al. 1980) that conducted SPTs with 12 inhalant allergens, only 61 % of the SPT positive subjects compared to 23 % of the negative subjects had allergic symptoms, and conversely 72 % of the symptomatic and 33 % of the asymptomatic subjects had positive SPT reactions. Contemporary multicenter studies with wider panels of aeroallergens (Haahtela et al. 2014) showed similar results, although subjects with larger SPT wheals had better associations with allergic disease. Similar analyses for food allergens have established larger wheal sizes to improve the specificity of the prediction of egg, milk and peanut allergies (Niggemann et al. 2005).

10.2.4 Allergen Provocation

Advances in protein isolation and recombinant DNA technology have allowed provocation studies that can measure the response of the target tissues using molecularly defined allergen components. This has provided dose response characteristics that can be used to define the importance of different allergen components.

Nasal provocation with Bet v 1 showed that symptoms were produced by low doses of allergen, usually under 1 μg in moderately allergic subjects (Godnic-Cvar et al. 1997; Niederberger et al. 2001; Tresch et al. 2003). The response to the provocation correlated with the IgE titers but the subjects with low titers of 2–3 IU/mL still produced significant reactions (Tresch et al. 2003). Similar doses (10–100 μg) of Art v 1, the serodominant mugwort allergen, induced nasal symptoms with both natural and recombinant allergens and there were weak correlations with serum IgE and skin test reactivity (Schmid-Grendelmeier et al. 2003). A comparison of IgE titers and nasal provocation tests using different allergen components (Niederberger et al. 2001) found that a regimen with a top dose of 0.6 μg/mL resulted in consistent nasal responses to Bet v 1, Bet v 2, Phl p 1, Phl p 2, and Phl p 5, but that the responses to provocation with Phl p 2 and Bet v 2 far exceeded those expected from their IgE titers. Indeed, some subjects with highly positive skin and nasal provocation tests did not have detectable IgE antibody, even for allergens other than Phl p 2 and Bet v 2.

Conjunctival challenge with Bet v 1 (Arquint et al. 1999) showed that 0.15 μg of allergen could consistently induce symptoms. Similarly, natural date profilin, which cross-reacts extensively with profilins from many pollens and foods, produced large conjunctival reactions in grass-allergic subjects with as little as 5 μg/mL of allergen (Nunez et al. 2012).

Bronchial challenge with purified natural Der p 1&2 from HDM has been shown to cause immediate hypersensitivity symptoms, as measured by lung function tests, with doses of only 0.1 μg of allergen, although responses were more prevalent with doses increasing up to 1 μg. The responses were the same as those induced by HDM extracts containing the equivalent amount of allergen (Van Der Veen et al. 2001). The size of the immediate changes in the forced expiratory responses did not correlate with IgE titers and maximal responses were found in subjects with titers of only 3 IU/mL while typical IgE titers of HDM allergic subjects are 50–200 IU/ml (Van Der Veen et al. 1998). Differing from the immediate responses, larger late phase responses were found to HDM extract, indicating the importance of other allergen components, but this interpretation could be a function of the experimental design because in this study subjects that produced large immediate responses to low doses of pure allergen were not given higher doses that might have been required to elicit the late phase responses. Bronchial responses of grass pollen sensitized subjects have also been demonstrated with μg doses of profilin (Ruiz-Garcia et al. 2011), an allergen component that induces IgE titers of only a few IU/ml compared to the 50–100 IU/mL titers seen with the serodominant grass allergens. Its presence in amounts 1/10–1/100 less than these allergens, however, would be considered clinically significant.

Allergic responses to challenge with food allergen components have been accomplished using the apple PR-10 protein Mal d 1 (Bolhaar et al. 2005) in patients with birch pollen-related apple allergy. A double-blind placebo-controlled food challenge (DBPCFC) produced allergic symptoms with 10 μg, whereas a genetically engineered mutant with half the IgE binding capacity required tenfold more allergen for the same response. Oral provocation with profilin showed that even lower doses-induced oral and systemic, respiratory tract symptoms in food-allergic patients in a region with high grass exposure (Alvarado et al. 2014). In this study, 0.074 μg profilin induced reactions in 22 % of patients and 7.2 μg profilin induced reactions in 50 % of patients, with some requiring rescue medication.

Provocation with single allergen components does not take into consideration the importance of a single component within a mixture where components can cooperatively cross-link IgE receptors. Cooperation has been demonstrated in an in vitro model using synthetic allergens and monoclonal antibodies (Handlogten et al. 2014) but, importantly, this was observed only when IgE titers were low. When high amounts of allergen were used, however, co-cross-linking was markedly inhibitory. Given, the simplicity of the experiments and the availability of recombinant allergens, it is surprising that no similar real-life experiments could be found for human allergy.

10.3 IgG Responses

IgG antibody has often been thought to have the capacity to protect from disease. Its production in allergy-tolerant beekeepers has been cited as an example, but a recent analysis has shown that their high IgG4 titers are accompanied by a paucity of IgE (Varga et al. 2013). A protective association in the presence of high titers of IgE antibodies was shown for HDM allergic children admitted to a hospital emergency department for asthma exacerbation (Hales et al. 2006). These children had very little IgG1 or IgG4 antibody in conspicuous contrast to children with stable asthma and similar IgE antibody titers. Further investigation (Hales et al. 2009) found that the lowest IgG1 or IgG4 anti-HDM titers were in children with recurrent and frequent asthma. A similar conclusion for cat allergy was drawn from a study of 5-year olds from two birth cohorts where a positive association between anti-cat IgE antibody titers and wheezing was found, but it was attenuated by IgG1 antibodies and not by IgG4 antibodies (Custovic et al. 2011). Persistent milk allergy is also associated with IgE without IgG1 and IgG4 antibodies (Ahrens et al. 2012). IgG antibody might be a marker for a less allergenic response or could act directly by blocking or sequestering allergen and by mediating negative Fc gamma receptor interactions.

10.3.1 IgE and IgG Antibody Patterns to Different Allergens

Different allergens induce distinctive patterns of IgE and IgG antibody (Table 10.1). HDM induce high IgE, IgG1, and IgG4 antibody titers in HDM atopic subjects but rarely induce IgG in non-atopic subjects (Hales et al. 2006). The IgG titers are higher for serodominant allergens with typical titers of 10 μg/mL for IgG1 and 1 μg/mL for IgG4, although it is complicated because IgG titers are low for subjects with recurrent asthma (Hales et al. 2006, 2009). Grass (Jutel et al. 2005; Rossi and Monasterolo 2004), tree (Harfast et al. 1998; Mobs et al. 2010) and

Table 10.1 IgE and IgG binding of common allergens[a]

Allergen	IgE IU/ml	IgG1 μg/ml	IgG4 μg/ml	IgG in non-sensitized
House dust mite	80 (5–150)	10 (0–70)	1 (0–7)	No[b]
Grass Pollen	6 (3–300)	0.15 (0.01–0.3)	0.1 (0–0.6)	No
Birch Pollen	15 (3–60)	0.2 (0–0.6)	0.2 (0–0.3)	No
Ragweed Pollen	25 (6–83)	0.2 (0.1–0.3)	0.3 (0.25–0.45)	No
Cat[c]	10 (0–100)	2.0 (0–10)	0.2 (0–3)	Yes[d]
Alternaria alternaria	23 (1–50)	0.3 (0–0.6)	0.15 (0–0.3)	No
Bee venom	5 (2–11)	9 (5–17)	2 (1–11)	Yes in beekeepers
Peanut	40 (1–200)	20 (10–40)	8 (0.5–55)	Low/absent
Milk	16 (0–200)	4 (2–8)	3 (1–6)	Low/absent
Egg	6.0 (1–90)	8 (1–30)	0.5 (0.03–4.0)	Yes[e]

aThe values are median titres and approximations of interquartile ranges (See text for references)
[b]No means that it is unusual to find IgG in subjects without IgE antibodies
[c]Cat median IgE titres vary from 18.2 (2–100) in asthmatic children to 2.9 (<0.35–20) for adults with rhinoconjunctivitis
[d]Yes but more prevalent in sensitized
[e]Yes but higher in sensitized children

weed (Creticos et al. 2014; Peebles et al. 1998) pollens induce IgE titers similar to HDM but induce very little IgG1 antibody and only low, under 1 μg/mL, titers of IgG4, which is seen only in atopic subjects. IgE and IgG responses to fungi, as exemplified by those to Alternaria, are similar to the responses to pollens (Lizaso et al. 2008; Vailes et al. 2001). The IgE titers to cat allergens vary greatly from patient to patient and differences have been found for different clinical presentations. Subjects with rhinoconjunctivitis have repeatedly been shown to have low IgE titers, often lower than 1 IU/mL (Gronlund et al. 2008; Hales et al. 2013a; Linden et al. 2011; Smith et al. 2004; Worm et al. 2011) while asthmatic children show higher titers, similar to those seen with pollens (Gronlund et al. 2008). A clear difference with other common allergens is that many non-atopic subjects have high IgG1 and IgG4 antibody responses to cat allergens (Hales et al. 2013a; Platts-Mills et al. 2001). Cat allergen Fel d 1 induces IgG1 and IgG4 in allergic and non-allergic subjects while the other cat allergens only induce IgG4 and in much lesser amounts in the non-allergic subjects (Hales et al. 2013a). The IgG titers to cat in atopic patients, typically 2 μg/mL for IgG1 and 0.2 μg/mL for IgG4, are less than those to HDM but higher than that seen to pollens and fungal spores. IgE titers to bee venom allergens in atopic subjects, usually 2–10 IU/mL, are lower than those to HDM and pollen but the IgG1 and IgG4 titers are higher, similar to HDM (Varga et al. 2013) and are also found in nonallergic beekeepers, although not in people with limited exposure. For food allergies, peanut allergy is associated with high IgE titers to Ara h 1&2 (Eller and Bindslev-Jensen 2013; Hong et al. 2012), similar to HDM, and high anti-peanut IgG1 and IgG4 titers that are associated with IgE production (Hong et al. 2012; Tay et al. 2007). The IgE titers found in milk-allergic (Shek et al. 2005) and egg-allergic (Ahrens et al. 2010) subjects are similar to those seen to pollen allergens but the anti-milk and anti-egg responses differ in that, although they both induce high IgG1 and IgG4 antibody titers, the IgG is associated with IgE titers for milk (Shek et al. 2005) but not to the same degree for egg (Ahrens et al. 2010; Tay et al. 2007).

10.4 T Cells

10.4.1 Associations of T Cell Responses with Allergic Symptoms

The in vitro proliferation of peripheral blood mononuclear cell (PBMC) T cells stimulated with grass and birch pollen extracts was shown to be similar for symptomatic sensitized subjects and asymptomatic sensitized subjects but higher than that of non-sensitized subjects (Assing et al. 2006). However, the T cells from the symptomatic subjects released higher levels of T_H2 cytokines. A similar distinction between symptomatic and asymptomatic subjects could be made for HDM allergy when PBMC were stimulated with HDM extracts (Till et al. 1997), where IL-5 was induced only in symptomatic subjects. It should however be noted that for both the pollen and HDM, the IgE titers were much lower in asymptomatic people. Similar associations were found when comparing anti-HDM responses in asthmatic and non-asthmatic subjects in a birth cohort (Heaton et al. 2005). T cell stimulation also shows a potential for diagnosing milk allergy since casein has been shown to not only stimulate T cells from PBMC of allergic subjects to secrete IL-4 and IL-13 but also to do so in a manner inversely proportional to the dose of milk that the donors could tolerate (Michaud et al. 2014).

10.4.2 Measuring Anti-Allergen T Cell Responses

Allergen stimulation assays measure recall responses of T cells that have been sensitized to allergen presented by dendritic cells in the lymph nodes. Here, the T cells become either central, effector, or resident memory cells (Turner et al. 2014). Central memory cells migrate through the blood, reentering lymph nodes through high endothelial venules, while effector memory cells circulate via the blood and tissues, entering lymph nodes via the afferent lymph, having being programmed via dendritic cells to home to the tissue exposed to the sensitizing antigen. Resident memory cells do not recirculate but persist in the tissues originally exposed to antigen. They have been studied in human skin and surgically removed lungs (Purwar et al. 2011) as well as from biopsies from lung and small bowel transplantation patients (Turner et al. 2014). Studies of PBMC from allergic and nonallergic individuals show they have effector and central cell responses to the pollen allergen Bet v 1, but mainly only central memory responses to HDM Der p 2 (Wambre et al. 2011). There is no information on resident memory cells.

While allergen-specific T_H2 responses can be detected in PBMC from allergic subjects, it is difficult to compare experiments from different laboratories and measurements of immunoregulatory responses also differ. For example, IL-10 release induced by purified Der p 1 (Macaubas et al. 1999) and Der p 2 (Hales et al. 2002) has been shown to be higher for PBMC from HDM allergic subjects than nonallergic subjects, while other studies of subjects sensitized to a variety of allergens using different culture methods have indicated the reverse (Akdis et al. 2004). More recent investigations have shown that IL-10 does not regulate T cell responses in Der p 1-stimulated in vitro cultures (Maggi et al. 2007) and that Der p 1 stimulated PBMC from non-atopic and atopic asthmatic subjects produce the same IL-10 responses (Wang et al. 2009). Some of discrepancies could be due to patient selection, differences in cell culture and the allergen preparations, which have not always been described (Akdis et al. 2004). In vitro culture systems lack the tissue-specific dendritic cells and in addition, the physiological milieu of cytokines can become non-physiologically depleted or concentrated in the course of the assays. Early detection of responses with techniques such as PCR might circumvent some of these problems, but not all regulation occurs via transcriptional control and tissue-specific cytokines would still be missing. Cell cytokine-secretion assays that detect cytokine-producing cells before they proliferate and differentiate would only detect effector memory cells. Methods that enrich T_H2 cells with markers such as CCR4 and CRT_H2 (Shamji et al. 2015) can help early detection, but the physiological relevance of responses in the absence of other cell types needs to be considered. Recent strategies such as the isolation of cells expressing CD154, the CD40 ligand which is transiently activated directly after allergen stimulation, are helping to focus studies on the immediate ex vivo responses (Campbell et al. 2010), and advances in multiparameter flow cytometry (O'Donnell et al. 2013), mass cytometry (Atkuri et al. 2015) and digital PCR (Hindson et al. 2011) are among the new techniques that can further improve this. The use of allergen-MHC tetramers (Van Hemelen et al. 2015) provides a method for directly isolating allergen responsive T cells, although the representative nature of the MHC restriction elements and the epitopes used need to be taken into account.

A major impediment to evaluating many of the studies of T cell responses is the type and quality of the allergen used. Extracts of the allergen source continue to be used despite being known to contain substances that modulate cytokine secretion, including not only microbial products but also nonallergenic proteins produced by the allergen source (Rockwood et al. 2013). Comparison of the gene expression of epithelial cells exposed to grass extract and purified Phl p 1 showed that the extract up-regulated 262 genes compared to 71 for Phl p 1. Only some genes were regulated by both stimuli with some being in the opposite direction (Roschmann et al. 2012). Purified natural and recombinant allergens from HDM (Hales et al. 2002), cat (Hales et al. 2013a), grass pollen (Wurtzen et al. 1998) and birch pollen

(Gafvelin et al. 2005; Mobs et al. 2010) have been shown to elicit ex vivo T cell cytokine responses, although the responses are not as large as those historically achieved with extracts.

10.4.3 T Cell Responses to Different Allergens

Direct comparisons using Der p 2 and Bet v 1 components found that they had not only sensitized different memory compartments but also induced cells to release different patterns of cytokines with a strong T_H2 bias for Bet v 1 and a mixed pattern of T_H2 cytokines with IL-10 and IFN-γ release for Der p 2 (Wambre et al. 2011). These results concur with prior studies of the allergens (Gafvelin et al. 2005; Hales et al. 2002; Mobs et al. 2010) although it should be emphasized that, although low, IFN-γ can be readily detected after Bet v 1 stimulation. The difference between HDM and birch not only provides an interesting avenue of research but also shows that measurements used to determine the degree of sensitization with one allergen might be different for another. There has been little analysis of differences between allergen components from one source. It was found that the less important HDM allergen component Der p 7 induced more IFN-γ than the serodominant Der p 1, while it induced the same proliferative and IL-5 cytokine release as Der p 1, even for subjects without anti-Der p 7 antibodies (Hales et al. 2000). A similar difference has now been revealed for cat allergy with the lipocalin allergens Fel d 4 and Fel d 7 but not Fel d 1 stimulating PBMC to release IL-10, with all of the allergens inducing similar levels of T_H2 cytokines (Hales et al. 2013a). Individual cytokine signatures for different allergen components, as well as a disconnect between IgE binding and T_H2 cytokine responses, have been reported for PBMC responses to the peptide epitopes of allergens. For example, in cockroach allergy, the epitopes of Bla g 5 epitope induced both IL-5 and IFN-γ, while Bla g 2 epitopes induced only IFN-γ (Oseroff et al. 2012), although, to be cautious, the study was confounded by biased contributions from individual subjects and no consideration was made of the physicochemical properties of the peptides.

10.5 Serodominant Allergen Components

Allergen components allow IgE binding assays to be conducted with known and optimal concentrations of reagents to provide accurate reproducible measurements. Indeed, the sum of the binding to different components often exceeds the binding measured with extracts, providing a more accurate assessment (see below). The pattern of binding to the components additionally provides information on the source of the sensitization, cross-reactivities, and, in a few cases, special types of sensitization. For common sources of allergen, the IgE titers to a few immunodominant components far exceed those to other IgE binding components. Since it is well established in mice that large allergic responses to one component can stimulate allergic responses to bystander antigens (Blumchen et al. 2006; Cadot et al. 2010; Cunningham et al. 2012; Eisenbarth et al. 2004; Kulis and Burks 2015), low responses to bystander antigens would be expected to be prevalent and be of questionable significance.

For grass pollen, the summated IgE titers to separate components showed higher binding than that measured with extracts, and this was mostly due to binding to the group 1, 2, 5, and 6 allergens (Mari 2003; Rossi et al. 2001), with groups 1 & 2 and 5 & 6 being, respectively, biochemically related. These results were more significant because recombinant proteins were used that excluded IgE binding to cross-reactive carbohydrates. The importance of this pattern has recently been confirmed with the demonstration that a cocktail of these components bound 80 % of the IgE in 70 % of grass-allergic subjects and that subjects with proportionally larger contributions from other allergens had low total titers (Darsow et al. 2014).

For birch pollen, nearly all of the IgE binding to extracts in most subjects can be accounted for by binding to Bet v 1 (Moverare et al. 2002), although IgE from subjects allergic to other pollens bind cross-reactive allergens in birch extracts, and there is cross-reactivity with Bet v 1-like allergens found in tree pollen and foods.

IgE binding to a panel of HDM components showed that IgE binding to Der p 1&2 accounted for over 50 % of the binding to extracts and over 50 % of the summated binding to all the components (Hales et al. 2006). Most of rest of the IgE binding was accounted for by binding to the mid-tier components Der p 4, 5, and 7, which, when present, had titers proportional to Der p 1&2. Other components only sporadically had titers similar to the mid-tier components, and, from known comparative studies, it can be calculated that many of the other denominated components make little contribution (Thomas et al. 2010). More recently, the IgE binding to the group 21 allergens has been shown to be similar to that of the biochemically related group five components, and the pleotropin Der p 23 has been shown to bind IgE with titers similar to Der p 1&2 and with, despite some exceptions, a strong correlation with the titers to Der p 1&2 (Vrtala et al. 2014). The allergenicity of HDM proteins that continue to be described from genomic (Chan et al. 2015) and proteomic (An et al. 2013) studies requires absolute determinations of IgE binding and comparison with other allergens, noting the discrepancies already found between IgE binding and SPT tests (Chan et al. 2015) and between studies from different laboratories (An et al. 2013; Chan et al. 2015). Component-resolved diagnosis for HDM allergy has already had application demonstrating that Australian aboriginals, once the subject for genetic and epidemiological studies, showed IgE binding to the HDM component amylase and not to the serodominant allergens (Hales et al. 2007). Exposure to scabies, which is often endemic in many countries, might be the culprit. IgE binding to Der p 10 and Der p 11 shows some usefulness in that, although Der p 10 (tropomyosin) binding is sporadic, it defines a subpopulation that has a very broad specificity of binding to

HDM components (Resch et al. 2011), and Der p 11(paramyosin) is a very minor allergen for subjects with inhalant allergy but is a major allergen for atopic dermatitis patients (Banerjee et al. 2015).

The mite *Blomia tropicalis*, important in highly populated tropical and subtropical regions, has biochemically different serodominant components from the *Dermatophagoides* spp. The evolutionary related Blo t 5 & 21 are the primary sensitizers and provide a method to distinguish *B. tropicalis* sensitization from sensitization to cross-reactive *Dermatophagoides pteronyssinus* allergens (Thomas et al. 2010).

The summated IgE binding to a panel of Bla g 1, 2, 4, 5, and 7 cockroach components (Satinover et al. 2005) often gave higher IgE binding titers than those measured with extracts, although 36 % of subjects were positive to extracts and negative to the components. This could be due to minor allergens or cross-reactivities since these sera typically had low IgE titers. The Bla g 2 and 5 were the most important allergens, followed by Bla g 5, but they do not show the universal dominance found for Der p 1&2. More recently described cockroach arginine kinase and protease allergens (Pomes and Arruda 2014) could be important from their prevalence but the IgE binding titers have not been reported.

As shown with a panel of allergen components, Fel d 1 contributes over 50 % of the IgE binding for 60 % of cat-allergic subjects (Hales et al. 2013a). The remaining cat-allergic subjects usually have one of either Fel d 2, 4, 7, or 8 as their highest IgE binding component often accounting for the majority of the IgE binding. It is therefore possible that there are subgroups of cat allergies defined by responses to different serodominant allergens, and this appears important in severe cross-reactive allergy to mammalian allergens (Konradsen et al. 2014). Atopic dermatitis patients with cat allergy also show disproportionally high IgE binding to Fel d 2 & 4 (Wisniewski et al. 2013), which, like Der p 11 for HDM allergy, indicates that there are different types of sensitization for different diseases.

Considering food allergen components, IgE antibodies are mainly directed to casein in the case of milk allergy (Shek et al. 2005), and to conalbumin and ovalbumin for egg allergy (Caubet et al. 2012; Everberg et al. 2011). For peanut allergy, Ara h 1&2, and to a lesser extent Ara h 3, contribute most of the binding (Eller and Bindslev-Jensen 2013; Hong et al. 2012).

A recent comprehensive study of the allergen components in bee venom allergy showed that although the phospholipaseA2 Api m 1 was the highest IgE binding specificity, demonstrating binding IgE binding in 72.2 % of subjects and usually with high titer, the components Api m 2, 3, 4, 5, and 10 also made large contributions (Kohler et al. 2014) and that Api m 3 or Api m 10 were the only IgE binding components for in 5 % of the patients.

10.6 Influence of the Characteristics of Allergen Components on Immune Response and Innate Interactions

Immune responses are initiated following either tissue damage or interactions with innate immune receptors, called pattern recognition receptors that have evolved to recognize molecules produced by pathogens. The two families of receptors that are important for non-infectious antigens are the Toll-like receptors (TLR) and the C-type lectin receptors (CTLR). It has been proposed, without definitive evidence that these and other actions of innate immunity are important for allergenicity, and, since these differ for different allergens (Table 10.2), there will be different pathways to sensitization.

Members of the TLR family found on antigen-presenting cells have distinctive specificities for lipids, nucleic acids and some proteins, and polysaccharides. For allergens, the recognition of lipoproteins and proteolipids by TLR-2 and TLR-4 are important. Historically, TLR were found to direct T_H1 type responses for protective immunity, but they also direct T_H2 responses in the lung (Tan et al. 2010). The high occurrence of lipid binding amongst important allergens might be due to the TLR pathway. There are, however, vast differences in the type of lipid binding by different allergens (Thomas 2014). Fel d 1 has an extremely small lipid binding cavity that only fits small hormone-like lipids, and the same is true for the salivary lipocalins like Fel d 4 and Can f 6. Others, such as the von Ebner gland lipocalins Fel d 7 and Can f 1 and the ML-domain protein Der p 2, have structures that bind much larger lipids.

The CTLR bind glycosylated allergens or allergens bound to glycans and glycolipids. Some just capture antigen on antigen-presenting cells while others modify TLR-initiated signals to direct the nature of immune responses. DC-SIGN engagement, for example, can inhibit T_H1 responses by blocking IL-12 production and Dectin-2 can stimulate the release of the T_H2-stimulating cysteinyl leukotriene. The effects of CTLR signaling are, however, very dependent on the precise circumstances (Parsons et al. 2014; Wang et al. 2009), so only possible influences on allergenicity can be proposed. A very probable CTLR function would be to direct allergens to alternately activated antigen-presenting cells that have their inflammatory functions modified by T_H2 cytokines and have up-regulated CTLR. Table 10.2 shows the many glycosylated allergens and the carbohydrate ligands Phl p 1&2. Included in the list are Der p 2 and Phl p 5, previously thought to be unglycosylated but now known to have unusual forms of glycosylation (Halim et al. 2015).

Chitin can interact with both CTLR and TLR ligands and enhance T_H1, T_H2, and T_H17 responses (Da Silva et al. 2010). Chitin adjuvant activity could accordingly be responsible for the high allergenicity of the chitin-binding peritropin allergen Der p 23, and, although the chitinase-like allergens Der p 15 and Der p 18 are not potent sensitizers, they induce IgE titers that, unlike most other mite components, do not correlate with the titers to Der p 1&2 (Hales

Table 10.2 Interactions of allergens and innate immunity

Biochemical activity	Allergens	Mechanism	Evidence
Cysteine protease	Der p 1, Der f 1	Proposed cleavage of immunological receptors including CD23 and proteolysis mediated increase of mucosal permeability	Cysteine proteases have protease-dependent adjuvant activity for T_H2 responses but despite being ubiquitous they are not typically allergenic or even immunodominant allergens for other mite families. Activity within the reducing milieu of the endosome is feasible as found for parasites
Serine protease	Pen c 13, Pen c 18, Asp f 13, Asp f 18	Increase inflammation via protease activated receptors (PARs)	Reported as prevalent fungal allergens but the IgE binding titres are not documented. The abundant serine protease of HDM are weak allergens
Kallikrein	Can f 5	Kallikeins have proinflammatory properties from the release of bradykinin from kininogen	Can f 5 induces very high IgE titres in some people and is similar to the major human semen allergen
Glycosylation (biochemically demonstrated)	Der p 1, Der p 2, Fel d 1, Art v 1, Bla g 2, Phl p 1, Phl p 5, Ara h1, caseins, Der p 15	Focus allergens onto C-type lectin pattern recognition receptors to modulate Toll-like receptor signaling and to induce cysteinyl leukotrienes that activate type 2 innate immune cells	Plausible scenarios for inducing T_H2-biased signaling especially through DC-SIGN, Dectin-1 and Dectin-2. Dectin-2 dependent induction of cysteinyl leukotriene in mice has been demonstrated with HDM extracts. High expression of CTLRs is found on alternatively activated macrophages in T_H2 milieus
Proteolipid Lipidation	Not known	Bind innate receptors especially TLR-2 and TLR4 to activate dendritic cells	See lipoproteins
Carbohydrate ligand	Phl p 1, Phl p2	As for glycosylation	As for glycosylation
Lipoproteins Lipid Ligand (proposed)	Der p 2, Der f 2, Bet v 1, Der p 5, Der p 7, Fel d 1, Fel d 7, Fel d 8, Can f 1	As for proteolipids	Many important allergens are lipoproteins. The ability of Der p 2 to sensitise mice by mimicking the lipid binding MD2 protein of the TLR-4 complex has been demonstrated
Chitin binding	Der p 23, Der p 15, Der p 18, Blo t 12	Allergens attach to chitin thus having adjuvant properties	Der p 23 is a recently recognized potent HDM allergen although the other allergens Der p 15 and 18 that would be expected to bind chitin are weak sensitisers. Blo t 12, molecularly similar to Der p 15, can bind chitin to increase its allergenicity in mice

et al. 2013b). Zakzuk et al. (2014) have shown that chitin complexed to the Blo t 12 chitin-binding protein of *B. tropicalis* can increase its ability to sensitize mice, providing experimental evidence for its binding and adjuvanticity.

The cysteine protease activity of Der p 1 has engendered speculation about its importance in allergic sensitization, but its role remains unproven (Thomas 2013). Cysteine proteases are highly susceptible to oxidative inactivation, and no activity has been found in HDM preparations, so is not clear how the activity would manifest in the oxidative extracellular environment. Moreover, despite being ubiquitous enzymes, cysteine proteases are rarely immunodominant allergen components. Enzymatically activated cysteine protease does, however, have a propensity to induce T_H2 responses in mice and, as found for parasite cysteine proteases; they could act within the reducing milieu of the endosome. The serine protease allergens of HDM are poor allergens (Hales et al. 2006; Thomas et al. 2010) and serine proteases from other sources are rarely immunodominant. From the prevalence of IgE binding, it appears that some fungal serine proteases are immunodominant, as indicated for the *Penicillium* spp. group 13 and 18 allergens (Shen et al. 2007) and the *Cladosporium herbarum* Cla h 9 (Poll et al. 2009), but this has yet to be demonstrated with IgE binding titers. The kallikrein Can f 5 allergen of dogs is a serine protease that does induce high IgE titers in some people (Mattsson et al. 2010) and could act via its hydrolytic activity on the kallikrien substrate kininogen that releases bradykinin that triggers the release of nitric oxide and leukotrienes.

Many immunodominant allergens are lipoproteins, suggesting that the possession of TLR2 and TLR 4 signaling properties enhances allergenicity, as well other lipid mediated functions (Bublin et al. 2014). Der p 2, a homologue of the MD-2 protein, could mimic the MD-2 function of binding lipopolysaccharide and delivering it to TLR-4 to activate antigen-presenting cells. It has been shown to induce a T_H2-type TLR-4-dependent response in MD-2 knockout mice (Trompette et al. 2009) and Der f 2 has been shown to bind lipopolysaccharide at high affinity using the amino acids equivalent to those used by MD-2 (Ichikawa et al. 2009). There is, however, still no evidence for its importance in humans where MD-2 is found in abundance on myeloid cells, although interestingly unless induced by IFN-γ it is absent on epithelial cells (Roy et al. 2011).

While the importance of the potential innate interactions above have not been demonstrated, the poor allergenicity of HDM ferritin (Chan et al. 2015; Epton et al. 2002b) and arginine kinase (Hales et al. 2006, 2007), which are produced in large amounts (Batard et al. 2006), suggests that more than abundance or the presence of a component in an allergenic source is required to induce high allergic responses.

10.7 Conclusions

Bioinformatics provides powerful tools to dissect out the immune mechanisms that lead to atopy and allergy. An important advantage is that human atopy is highly prevalent so it should be possible to effectively stratify and refine gene expression and genetic associations, as proposed here particularly by the sensitizing allergen. HDM and grass pollen allergy are highly prevalent in many countries, thus, separate analyses can be made to remove the complications resulting from the inclusion of responses that are qualitatively and quantitatively different. In further analyses, the responses to different allergens can be compared rather than pooled so that the known differences might be exploited. Since the atopic responses to the common allergens are mainly directed to a few allergen components, and these can be used for reproducible measurements, quantitative trait GWAS can be applied to both IgE and IgG antibodies. The latter is associated with both sensitization and protection so the results could provide important information on the development of disease. A critical area for development is the establishments of reproducible T cell assays, especially those that can determine the responses of T cells without introducing in vitro artifacts, and, as for basic immunological investigations in mice, use defined pure allergens in known and reproducible quantities. The limited experiments to date comparing allergen components suggest that there is a wealth of information that can be gleaned by comparing the different types of responses to the different allergens. Much is being obscured in responses to extracts that are mixtures of unknown composition, often, like HDM extracts, having no relationship to natural exposure. There are recent technological advances in mass cytometry and digital PCR with that have the potential to provide better ex vivo measures of T cells that will allow assays with allergen components. It should also be noted that anti-cytokine and anti-IgE therapy indicate that not all atopic subjects follow the same path to disease, so subgroups should be expected. Clearly, biomarkers discovered from bioinformatic analyses can be used to analyse therapeutic trials, as periostin has been already. The defining test for allergy still remains challenge with antigen by the natural route of exposure, and it has been shown that, even for minor food allergens, this can be accomplished with small doses of highly purified natural and recombinant proteins. A direct relationship between the type of responses induced by allergen components after provocation and their immunological and bioinformatic profiles can thus be established. The use of molecularly defined provocation tests has severely declined in recent years (Niederberger et al. 2014), possibly due to regulatory processes, so the development of guidelines for its application that are acceptable to governing bodies and allergy societies should be recommended.

References

Abraham CM, Ownby DR, Peterson EL, Wegienka G, Zoratti EM, Williams LK, Joseph CL, Johnson CC. The relationship between seroatopy and symptoms of either allergic rhinitis or asthma. J Allergy Clin Immunol. 2007;119(5):1099–104.

Ahrens B, Lopes de Oliveira LC, Grabenhenrich L, Schulz G, Niggemann B, Wahn U, Beyer K. Individual cow's milk allergens as prognostic markers for tolerance development? Clin Exp Allergy. 2012;42(11):1630–7.

Ahrens B, Lopes de Oliveira LC, Schulz G, Borres MP, Niggemann B, Wahn U, Beyer K. The role of hen's egg-specific IgE, IgG and IgG4 in the diagnostic procedure of hen's egg allergy. Allergy. 2010;65(12):1554–7.

Akdis M, Verhagen J, Taylor A, Karamloo F, Karagiannidis C, Crameri R, Thunberg S, Deniz G, Valenta R, Fiebig H, et al. Immune responses in healthy and allergic individuals are characterized by a fine balance between allergen-specific T regulatory 1 and T helper 2 cells. J Exp Med. 2004;199(11):1567–75.

Alvarado MI, Jimeno L, De La Torre F, Boissy P, Rivas B, Lazaro MJ, Barber D. Profilin as a severe food allergen in allergic patients overexposed to grass pollen. Allergy. 2014;69(12):1610–6.

An S, Chen L, Long C, Liu X, Xu X, Lu X, Rong M, Liu Z, Lai R. Dermatophagoides farinae allergens diversity identification by proteomics. Mol Cell Proteomics. 2013;12(7):1818–28.

Arquint O, Helbling A, Crameri R, Ferreira F, Breitenbach M, Pichler WJ. Reduced in vivo allergenicity of Bet v 1d isoform, a natural component of birch pollen. J Allergy Clin Immunol. 1999;104(6):1239–43.

Assing K, Nielsen CH, Poulsen LK. Immunological characteristics of subjects with asymptomatic skin sensitization to birch and grass pollen. Clin Exp Allergy. 2006;36(3):283–92.

Atkuri KR, Stevens JC, Neubert H. Mass cytometry: a highly multiplexed single-cell technology for advancing drug development. Drug Metab Dispos. 2015;43(2):227–33.

Banerjee S, Resch Y, Chen KW, Swoboda I, Focke-Tejkl M, Blatt K, Novak N, Wickman M, van Hage M, Ferrara R, et al. Der p 11 is a major allergen for house dust mite-allergic patients suffering from atopic dermatitis. J Invest Dermatol. 2015;135(1):102–9.

Batard T, Hrabina A, Bi XZ, Chabre H, Lemoine P, Couret MN, Faccenda D, Villet B, Harzic P, Andre F, et al. Production and proteomic characterization of pharmaceutical-grade Dermatophagoides pteronyssinus and Dermatophagoides farinae extracts for allergy vaccines. Int Arch Allergy Immunol. 2006;140(4):295–305.

Blumchen K, Gerhold K, Schwede M, Niggemann B, Avagyan A, Dittrich AM, Wagner B, Breiteneder H, Hamelmann E. Effects of established allergen sensitization on immune and airway responses after secondary allergen sensitization. J Allergy Clin Immunol. 2006;118(3):615–21.

Bodtger U, Poulsen LK, Malling HJ. Asymptomatic skin sensitization to birch predicts later development of birch pollen allergy in adults: a 3-year follow-up study. J Allergy Clin Immunol. 2003;111(1):149–54.

Bolhaar ST, Zuidmeer L, Ma Y, Ferreira F, Bruijnzeel-Koomen CA, Hoffmann-Sommergruber K, van Ree R, Knulst AC. A mutant of the major apple allergen, Mal d 1, demonstrating hypo-allergenicity in the target organ by double-blind placebo-controlled food challenge. Clin Exp Allergy. 2005;35(12):1638–44.

Bonnelykke K, Matheson MC, Pers TH, Granell R, Strachan DP, Alves AC, Linneberg A, Curtin JA, Warrington NM, Standl M, et al. Meta-analysis of genome-wide association studies identifies ten loci influencing allergic sensitization. Nat Genet. 2013;45(8):902–6.

Bosco A, McKenna KL, Devitt CJ, Firth MJ, Sly PD, Holt PG. Identification of novel Th2-associated genes in T memory responses to allergens. J Immunol. 2006;176(8):4766–77.

Bublin M, Eiwegger T, Breiteneder H. Do lipids influence the allergic sensitization process? J Allergy Clin Immunol. 2014;134(3):521–9.

Cadot P, Meyts I, Vanoirbeek JA, Vanaudenaerde B, Bullens DM, Ceuppens JL. Sensitization to inhaled ryegrass pollen by collateral priming in a murine model of allergic respiratory disease. Int Arch Allergy Immunol. 2010;152(3):233–42.

Campbell JD, Buchmann P, Kesting S, Cunningham CR, Coffman RL, Hessel EM. Allergen-specific T cell responses to immunotherapy monitored by CD154 and intracellular cytokine expression. Clin Exp Allergy. 2010;40(7):1025–35.

Canonica GW, Ansotegui IJ, Pawankar R, Schmid-Grendelmeier P, van Hage M, Baena-Cagnani CE, Melioli G, Nunes C, Passalacqua G, Rosenwasser L, et al. A WAO-ARIA-GA^{2}LEN consensus document on molecular-based allergy diagnostics. World Allergy Organ J. 2013;6(1):17.

Caubet JC, Bencharitiwong R, Moshier E, Godbold JH, Sampson HA, Nowak-Wegrzyn A. Significance of ovomucoid-and ovalbumin-specific IgE/IgG(4) ratios in egg allergy. J Allergy Clin Immunol. 2012;129(3):739–47.

Chan TF, Ji KM, Yim AK, Liu XY, Zhou JW, Li RQ, Yang KY, Li J, Li M, Law PT, et al. The draft genome, transcriptome, and microbiome of Dermatophagoides farinae reveal a broad spectrum of dust mite allergens. J Allergy Clin Immunol. 2015;135(2):539–48.

Corren J, Lemanske RF, Hanania NA, Korenblat PE, Parsey MV, Arron JR, Harris JM, Scheerens H, Wu LC, Su Z, et al. Lebrikizumab treatment in adults with asthma. N Engl J Med. 2011;365(12):1088–98.

Creticos PS, Esch RE, Couroux P, Gentile D, D'Angelo P, Whitlow B, Alexander M, Coyne TC. Randomized, double-blind, placebo-controlled trial of standardized ragweed sublingual-liquid immunotherapy for allergic rhinoconjunctivitis. J Allergy Clin Immunol. 2014;133(3):751–8.

Cunningham PT, Elliot CE, Lenzo JC, Jarnicki AG, Larcombe AN, Zosky GR, Holt PG, Thomas WR. Sensitizing and Th2 adjuvant activity of cysteine protease allergens. Int Arch Allergy Immunol. 2012;158(4):347–58.

Custovic A, Soderstrom L, Ahlstedt S, Sly PD, Simpson A, Holt PG. Allergen-specific IgG antibody levels modify the relationship between allergen-specific IgE and wheezing in childhood. J Allergy Clin Immunol. 2011;127(6):1480–5.

Da Silva CA, Pochard P, Lee CG, Elias JA. Chitin particles are multifaceted immune adjuvants. Am J Respir Crit Care Med. 2010;182(12):1482–91.

Darsow U, Brockow K, Pfab F, Jakob T, Petersson CJ, Borres MP, Ring J, Behrendt H, Huss-Marp J. Allergens. Heterogeneity of molecular sensitization profiles in grass pollen allergy–implications for immunotherapy? Clin Exp Allergy. 2014;44(5):778–86.

Eisenbarth SC, Zhadkevich A, Ranney P, Herrick CA, Bottomly K. IL-4-dependent Th2 collateral priming to inhaled antigens independent of Toll-like receptor 4 and myeloid differentiation factor 88. J Immunol. 2004;172(7):4527–34.

Eller E, Bindslev-Jensen C. Clinical value of component-resolved diagnostics in peanut-allergic patients. Allergy. 2013;68(2):190–4.

Epton MJ, Hales BJ, Thompson PJ, Thomas WR. T cell cytokine responses to outer membrane proteins of Haemophilus influenzae and the house dust mite allergens Der p 1 in allergic and non-allergic subjects. Clin Exp Allergy. 2002a;32(11):1589–95.

Epton MJ, Smith W, Hales BJ, Hazell L, Thompson PJ, Thomas WR. Non-allergenic antigen in allergic sensitization: responses to the mite ferritin heavy chain antigen by allergic and non-allergic subjects. Clin Exp Allergy. 2002b;32(9):1341–7.

Everberg H, Brostedt P, Oman H, Bohman S, Moverare R. Affinity purification of egg-white allergens for improved component-resolved diagnostics. Int Arch Allergy Immunol. 2011;154(1):33–41.

Gafvelin G, Thunberg S, Kronqvist M, Gronlund H, Gronneberg R, Troye-Blomberg M, Akdis M, Fiebig H, Purohit A, Horak F, et al. Cytokine and antibody responses in birch-pollen-allergic patients treated with genetically modified derivatives of the major birch pollen allergen Bet v 1. Int Arch Allergy Immunol. 2005;138(1):59–66.

Godnic-Cvar J, Susani M, Breiteneder H, Berger A, Havelec L, Waldhor T, Hirschwehr R, Valenta R, Scheiner O, Rudiger H, et al. Recombinant Bet v 1, the major birch pollen allergen, induces hypersensitivity reactions equal to those induced by natural Bet v 1 in the airways of patients allergic to tree pollen. J Allergy Clin Immunol. 1997;99(3):354–9.

Grimbaldeston MA, Metz M, Yu M, Tsai M, Galli SJ. Effector and potential immunoregulatory roles of mast cells in IgE-associated acquired immune responses. Curr Opin Immunol. 2006;18(6):751–60.

Gronlund H, Adedoyin J, Reininger R, Varga EM, Zach M, Fredriksson M, Kronqvist M, Szepfalusi Z, Spitzauer S, Gronneberg R, et al. Higher immunoglobulin E antibody levels to recombinant Fel d 1 in cat-allergic children with asthma compared with rhinoconjunctivitis. Clin Exp Allergy. 2008;38(8):1275–81.

Haahtela T, Burbach GJ, Bachert C, Bindslev-Jensen C, Bonini S, Bousquet J, Bousquet-Rouanet L, Bousquet PJ, Bresciani M, Bruno A, et al. Clinical relevance is associated with allergen-specific wheal size in skin prick testing. Clin Exp Allergy. 2014;44(3):407–16.

Haahtela T, Heiskala M, Suoniemi I. Allergic disorders and immediate skin test reactivity in Finnish adolescents. Allergy. 1980;35(5):433–41.

Haahtela T, Jaakonmaki I. Relationship of allergen-specific IgE antibodies, skin prick tests and allergic disorders in unselected adolescents. Allergy. 1981;36(4):251–6.

Hales BJ, Bosco A, Mills KL, Hazell LA, Loh R, Holt PG, Thomas WR. Isoforms of the major peanut allergen Ara h 2: IgE binding in children with peanut allergy. Int Arch Allergy Immunol. 2004;135(2):101–7.

Hales BJ, Chai LY, Hazell L, Elliot CE, Stone S, O'Neil SE, Smith WA, Thomas WR. IgE and IgG binding patterns and T-cell recognition of Fel d 1 and non-Fel d 1 cat allergens. J Allergy Clin Immunol Pract. 2013a;1(6):656–65.

Hales BJ, Elliot CE, Chai LY, Pearce LJ, Tipayanon T, Hazell L, Stone S, Piboonpocanun S, Thomas WR, Smith WA. Quantitation of IgE binding to the chitinase and chitinase-like house dust mite allergens Der p 15 and Der p 18 compared to the major and mid-range allergens. Int Arch Allergy Immunol. 2013b;160(3):233–40.

Hales BJ, Hazell LA, Smith W, Thomas WR. Genetic variation of Der p 2 allergens: effects on T cell responses and immunoglobulin E binding. Clin Exp Allergy. 2002;32(10):1461–7.

Hales BJ, Laing IA, Pearce LJ, Hazell LA, Mills KL, Chua KY, Thornton RB, Richmond P, Musk AW, James AL, et al. Distinctive immunoglobulin E anti-house dust allergen-binding specificities in a tropical Australian Aboriginal community. Clin Exp Allergy. 2007;37(9):1357–63.

Hales BJ, Martin AC, Pearce LJ, Laing IA, Hayden CM, Goldblatt J, Le Souef PN, Thomas WR. IgE and IgG anti-house dust mite specificities in allergic disease. J Allergy Clin Immunol. 2006;118(2):361–7.

Hales BJ, Martin AC, Pearce LJ, Rueter K, Zhang G, Khoo SK, Hayden CM, Bizzintino J, McMinn P, Geelhoed GC, et al. Anti-bacterial IgE in the antibody responses of house dust mite allergic children convalescent from asthma exacerbation. Clin Exp Allergy. 2009;39(8):1170–8.

Hales BJ, Shen H, Thomas WR. Cytokine responses to Der p 1 and Der p 7: house dust mite allergens with different IgE-binding activities. Clin Exp Allergy. 2000;30(7):934–43.

Halim A, Carlsson MC, Madsen CB, Brand S, Moller SR, Olsen CE, Vakhrushev SY, Brimnes J, Wurtzen PA, Ipsen H, et al. Glycoproteomic analysis of seven major allergenic proteins reveals novel post-translational modifications. Mol Cell Proteomics. 2015;14(1):191–204.

Handlogten MW, Deak PE, Bilgicer B. Two-allergen model reveals complex relationship between IgE crosslinking and degranulation. Chem Biol. 2014;21(11):1445–51.

Harfast B, van Hage-Hamsten M, Lilja G. IgG subclass antibody responses to birch pollen in sibling pairs discordant for atopy. Pediatr Allergy Immunol. 1998;9(4):208–14.

Heaton T, Rowe J, Turner S, Aalberse RC, de Klerk N, Suriyaarachchi D, Serralha M, Holt BJ, Hollams E, Yerkovich S, et al. An immunoepidemiological approach to asthma: identification of in-vitro T-cell response patterns associated with different wheezing phenotypes in children. Lancet. 2005;365(9454):142–9.

Hindson BJ, Ness KD, Masquelier DA, Belgrader P, Heredia NJ, Makarewicz AJ, Bright IJ, Lucero MY, Hiddessen AL, Legler TC, et al. High-throughput droplet digital PCR system for absolute quantitation of DNA copy number. Anal Chem. 2011;83(22):8604–10.

Holder PK, Maslanka SE, Pais LB, Dykes J, Plikaytis BD, Carlone GM. Assignment of Neisseria meningitidis serogroup A and C class-specific anticapsular antibody concentrations to the new standard reference serum CDC1992. Clin Diagn Lab Immunol. 1995;2(2):132–7.

Holgate ST. Stratified approaches to the treatment of asthma. Br J Clin Pharmacol. 2013;76(2):277–91.

Hollams EM, Hales BJ, Bachert C, Huvenne W, Parsons F, de Klerk NH, Serralha M, Holt BJ, Ahlstedt S, Thomas WR, et al. Th2-associated immunity to bacteria in teenagers and susceptibility to asthma. Eur Respir J. 2010;36(3):509–16.

Hong X, Caruso D, Kumar R, Liu R, Liu X, Wang G, Pongracic JA, Wang X. IgE, but not IgG4, antibodies to Ara h 2 distinguish peanut allergy from asymptomatic peanut sensitization. Allergy. 2012;67(12):1538–46.

Hsu CL, Neilsen CV, Bryce PJ. IL-33 is produced by mast cells and regulates IgE-dependent inflammation. PloS one. 2010;5(8):e11944.

Ichikawa K, Iwasaki E, Baba M, Chapman MD. High prevalence of sensitization to cat allergen among Japanese children with asthma, living without cats. Clin Exp Allergy. 1999;29(6):754–61.

Ichikawa S, Takai T, Yashiki T, Takahashi S, Okumura K, Ogawa H, Kohda D, Hatanaka H. Lipopolysaccharide binding of the mite allergen Der f 2. Genes Cells. 2009;14(9):1055–65.

Jahn-Schmid B, Harwanegg C, Hiller R, Bohle B, Ebner C, Scheiner O, Mueller MW. Allergen microarray: comparison of microarray using recombinant allergens with conventional diagnostic methods to detect allergen-specific serum immunoglobulin E. Clin Exp Allergy. 2003;33(10):1443–9.

Jutel M, Jaeger L, Suck R, Meyer H, Fiebig H, Cromwell O. Allergen-specific immunotherapy with recombinant grass pollen allergens. J Allergy Clin Immunol. 2005;116(3):608–13.

Kohler J, Blank S, Muller S, Bantleon F, Frick M, Huss-Marp J, Lidholm J, Spillner E, Jakob T. Component resolution reveals additional major allergens in patients with honeybee venom allergy. J Allergy Clin Immunol. 2014;133(5):1383–9.

Konradsen JR, Nordlund B, Onell A, Borres MP, Gronlund H, Hedlin G. Severe childhood asthma and allergy to furry animals: refined assessment using molecular-based allergy diagnostics. Pediatr Allergy Immunol. 2014;25(2):187–92.

Kulis M, Burks AW. Effects of a pre-existing food allergy on the oral introduction of food proteins: findings from a murine model. Allergy. 2015;70(1):120–3.

Lee SH, Park JS, Park CS. The search for genetic variants and epigenetics related to asthma. Allergy Asthma Immunol Res. 2011;3(4):236–44.

Linden CC, Misiak RT, Wegienka G, Havstad S, Ownby DR, Johnson CC, Zoratti EM. Analysis of allergen specific IgE cut points to cat and dog in the Childhood Allergy Study. Ann Allergy Asthma Immunol. 2011;106(2):153–8.

Lizaso MT, Tabar AI, Garcia BE, Gomez B, Algorta J, Asturias JA, Martinez A. Double-blind, placebo-controlled Alternaria alternata immunotherapy: in vivo and in vitro parameters. Pediatr Allergy Immunol. 2008;19(1):76–81.

Lupinek C, Wollmann E, Baar A, Banerjee S, Breiteneder H, Broecker BM, Bublin M, Curin M, Flicker S, Garmatiuk T, et al. Advances in allergen-microarray technology for diagnosis and monitoring of allergy: the MeDALL allergen-chip. Methods. 2014;66(1):106–19.

Macaubas C, Sly PD, Burton P, Tiller K, Yabuhara A, Holt BJ, Smallacombe TB, Kendall G, Jenmalm MC, Holt PG. Regulation of T-helper cell responses to inhalant allergen during early childhood. Clin Exp Allergy. 1999;29(9):1223–31.

Maggi L, Santarlasci V, Liotta F, Frosali F, Angeli R, Cosmi L, Maggi E, Romagnani S, Annunziato F. Demonstration of circulating allergen-specific CD4+ CD25 high Foxp3+ T-regulatory cells in both nonatopic and atopic individuals. J Allergy Clin Immunol. 2007;120(2):429–36.

Mari A. Skin test with a timothy grass (Phleum pratense) pollen extract vs. IgE to a timothy extract vs. IgE to rPhl p 1, rPhl p 2, nPhl p 4, rPhl p 5, rPhl p 6, rPhl p 7, rPhl p 11, and rPhl p 12: epidemiological and diagnostic data. Clin Exp Allergy. 2003;33(1):43–51.

Marinho S, Simpson A, Soderstrom L, Woodcock A, Ahlstedt S, Custovic A. Quantification of atopy and the probability of rhinitis in preschool children: a population-based birth cohort study. Allergy. 2007;62(12):1379–86.

Mattsson L, Lundgren T, Olsson P, Sundberg M, Lidholm J. Molecular and immunological characterization of Can f 4: a dog dander allergen cross-reactive with a 23 kDa odorant-binding protein in cow dander. Clin Exp Allergy. 2010;40(8):1276–87.

Michaud B, Aroulandom J, Baiz N, Amat F, Gouvis-Echraghi R, Candon S, Foray AP, Couderc R, Bach JF, Chatenoud L, et al. Casein-specific IL-4-and IL-13-secreting T cells: a tool to implement diagnosis of cow's milk allergy. Allergy. 2014;69(11):1473–80.

Mobs C, Slotosch C, Loffler H, Jakob T, Hertl M, Pfutzner W. Birch pollen immunotherapy leads to differential induction of regulatory T cells and delayed helper T cell immune deviation. J Immunol. 2010;184(4):2194–203.

Moverare R, Westritschnig K, Svensson M, Hayek B, Bende M, Pauli G, Sorva R, Haahtela T, Valenta R, Elfman L. Different IgE reactivity profiles in birch pollen-sensitive patients from six European populations revealed by recombinant allergens: an imprint of local sensitization. Int Arch Allergy Immunol. 2002;128(4):325–35.

Niederberger V, Eckl-Dorna J, Pauli G. Recombinant allergen-based provocation testing. Methods. 2014;66(1):96–105.

Niederberger V, Stubner P, Spitzauer S, Kraft D, Valenta R, Ehrenberger K, Horak F. Skin test results but not serology reflect immediate type respiratory sensitivity: a study performed with recombinant allergen molecules. J Invest Dermatol. 2001;117(4):848–51.

Niggemann B, Rolinck-Werninghaus C, Mehl A, Binder C, Ziegert M, Beyer K. Controlled oral food challenges in children—when indicated, when superfluous? Allergy. 2005;60(7):865–70.

Nunez R, Carballada F, Lombardero M, Jimeno L, Boquete M. Profilin as an aeroallergen by means of conjunctival allergen challenge with purified date palm profilin. Int Arch Allergy Immunol. 2012;158(2):115–9.

O'Donnell EA, Ernst DN, Hingorani R. Multiparameter flow cytometry: advances in high resolution analysis. Immune Netw. 2013;13(2):43–54.

Ober C, Yao TC. The genetics of asthma and allergic disease: a 21st century perspective. Immunol Rev. 2011;242(1):10–30.

Oseroff C, Sidney J, Tripple V, Grey H, Wood R, Broide DH, Greenbaum J, Kolla R, Peters B, Pomes A, et al. Analysis of T cell responses to the major allergens from German cockroach: epitope specificity and relationship to IgE production. J Immunol. 2012;189(2):679–88.

Parsons MW, Li L, Wallace AM, Lee MJ, Katz HR, Fernandez JM, Saijo S, Iwakura Y, Austen KF, Kanaoka Y, et al. Dectin-2 regulates the effector phase of house dust mite-elicited pulmonary inflammation independently from its role in sensitization. J Immunol. 2014;192(4):1361–71.

Parulekar AD, Atik MA, Hanania NA. Periostin, a novel biomarker of TH2-driven asthma. Curr Opin Pulm Med. 2014;20(1):60–5.

Pastorello EA, Incorvaia C, Ortolani C, Bonini S, Canonica GW, Romagnani S, Tursi A, Zanussi C. Studies on the relationship between the level of specific IgE antibodies and the clinical expression of allergy: I. Definition of levels distinguishing patients with symptomatic from patients with asymptomatic allergy to common aeroallergens. J Allergy Clin Immunol. 1995;96(5 Pt 1):580–7.

Peebles RS Jr, Liu MC, Adkinson NF Jr, Lichtenstein LM, Hamilton RG. Ragweed-specific antibodies in bronchoalveolar lavage fluids and serum before and after segmental lung challenge: IgE and IgA associated with eosinophil degranulation. J Allergy Clin Immunol. 1998;101(2 Pt 1):265–73.

Pettipher R. The roles of the prostaglandin D(2) receptors DP(1) and CRTH2 in promoting allergic responses. Br J Pharmacol. 2008;153(1):S191–9.

Platts-Mills T, Vaughan J, Squillace S, Woodfolk J, Sporik R. Sensitisation, asthma, and a modified Th2 response in children exposed to cat allergen: a population-based cross-sectional study. Lancet. 2001;357(9258):752–6.

Poll V, Denk U, Shen HD, Panzani RC, Dissertori O, Lackner P, Hemmer W, Mari A, Crameri R, Lottspeich F, et al. The vacuolar serine protease, a cross-reactive allergen from Cladosporium herbarum. Mol Immunol. 2009;46(7):1360–73.

Pomes A, Arruda LK. Investigating cockroach allergens: aiming to improve diagnosis and treatment of cockroach allergic patients. Methods. 2014;66(1):75–85.

Purwar R, Campbell J, Murphy G, Richards WG, Clark RA, Kupper TS. Resident memory T cells (T(RM)) are abundant in human lung: diversity, function, and antigen specificity. PloS one. 2011;6(1):e16245.

Resch Y, Weghofer M, Seiberler S, Horak F, Scheiblhofer S, Linhart B, Swoboda I, Thomas WR, Thalhamer J, Valenta R, et al. Molecular characterization of Der p 10: a diagnostic marker for broad sensitization in house dust mite allergy. Clin Exp Allergy. 2011;41(10):1468–77.

Rockwood J, Morgan MS, Arlian LG. Proteins and endotoxin in house dust mite extracts modulate cytokine secretion and gene expression by dermal fibroblasts. Exp Appl Acarol. 2013;61(3):311–25.

Roschmann KI, van Kuijen AM, Luiten S, Jonker MJ, Breit TM, Fokkens WJ, Petersen A, van Drunen CM. Comparison of Timothy grass pollen extract-and single major allergen-induced gene expression and mediator release in airway epithelial cells: a meta-analysis. Clin Exp Allergy. 2012;42(10):1479–90.

Rossi RE, Monasterolo G. Evaluation of recombinant and native timothy pollen (rPhl p 1, 2, 5, 6, 7, 11, 12 and nPhl p 4)-specific IgG4 antibodies induced by subcutaneous immunotherapy with timothy pollen extract in allergic patients. Int Arch Allergy Immunol. 2004;135(1):44–53.

Rossi RE, Monasterolo G, Monasterolo S. Measurement of IgE antibodies against purified grass-pollen allergens (Phl p 1, 2, 3, 4, 5, 6, 7, 11, and 12) in sera of patients allergic to grass pollen. Allergy. 2001;56(12):1180–5.

Roy S, Sun Y, Pearlman E. Interferon-gamma-induced MD-2 protein expression and lipopolysaccharide (LPS) responsiveness in corneal epithelial cells is mediated by Janus tyrosine kinase-2 activation and direct binding of STAT1 protein to the MD-2 promoter. J Biol Chem. 2011;286(27):23753–62.

Ruiz-Garcia M, Garcia Del Potro M, Fernandez-Nieto M, Barber D, Jimeno-Nogales L, Sastre J. Profilin: a relevant aeroallergen? J Allergy Clin Immunol. 2011;128(2):416–8.

Satinover SM, Reefer AJ, Pomes A, Chapman MD, Platts-Mills TA, Woodfolk JA. Specific IgE and IgG antibody-binding patterns to recombinant cockroach allergens. J Allergy Clin Immunol. 2005;115(4):803–9.

Scheerens H, Arron JR, Zheng Y, Putnam WS, Erickson RW, Choy DF, Harris JM, Lee J, Jarjour NN, Matthews JG. The effects of lebrikizumab in patients with mild asthma following whole lung allergen challenge. Clin Exp Allergy. 2014;44(1):38–46.

Schmid-Grendelmeier P, Holzmann D, Himly M, Weichel M, Tresch S, Ruckert B, Menz G, Ferreira F, Blaser K, Wuthrich B, et al. Native Art v 1 and recombinant Art v 1 are able to induce humoral and T cell-mediated in vitro and in vivo responses in mugwort allergy. J Allergy Clin Immunol. 2003;111(6):1328–36.

Shamji MH, Bellido V, Scadding GW, Layhadi JA, Cheung DK, Calderon MA, Asare A, Gao Z, Turka LA, Tchao N, et al. Effector cell signature in peripheral blood following nasal allergen challenge in grass pollen allergic individuals. Allergy. 2015;70(2):171–9.

Shek LP, Bardina L, Castro R, Sampson HA, Beyer K. Humoral and cellular responses to cow milk proteins in patients with milk-induced IgE-mediated and non-IgE-mediated disorders. Allergy. 2005;60(7):912–9.

Shen HD, Tam MF, Tang RB, Chou H. Aspergillus and Penicillium allergens: focus on proteases. Curr Allergy Asthma Rep. 2007;7(5):351–6.

Simpson A, Soderstrom L, Ahlstedt S, Murray CS, Woodcock A, Custovic A. IgE antibody quantification and the probability of wheeze in preschool children. J Allergy Clin Immunol. 2005;116(4):744–9.

Smith W, Butler AJ, Hazell LA, Chapman MD, Pomes A, Nickels DG, Thomas WR. Fel d 4, a cat lipocalin allergen. Clin Exp Allergy. 2004;34(11):1732–8.

Soderstrom L, Kober A, Ahlstedt S, de Groot H, Lange CE, Paganelli R, Roovers MH, Sastre J. A further evaluation of the clinical use of specific IgE antibody testing in allergic diseases. Allergy. 2003;58(9):921–8.

Squire EN Jr, McBride DC, Ledoux R, Nelson HS. Assessment of IgE-mediated sensitivity during immunotherapy: comparison of the RAST with and without adsorption of IgG to titrated prick skin tests. Ann Allergy. 1986;56(1):22–7.

Suzuki K, Hiyoshi M, Tada H, Bando M, Ichioka T, Kamemura N, Kido H. Allergen diagnosis microarray with high-density immobilization capacity using diamond-like carbon-coated chips for profiling allergen-specific IgE and other immunoglobulins. Anal Chim Acta. 2011;706(2):321–7.

Takayama G, Arima K, Kanaji T, Toda S, Tanaka H, Shoji S, McKenzie AN, Nagai H, Hotokebuchi T, Izuhara K. Periostin: a novel component of subepithelial fibrosis of bronchial asthma downstream of IL-4 and IL-13 signals. J Allergy Clin Immunol. 2006;118(1):98–104.

Tan AM, Chen HC, Pochard P, Eisenbarth SC, Herrick CA, Bottomly HK. TLR4 signaling in stromal cells is critical for the initiation of allergic Th2 responses to inhaled antigen. J Immunol. 2010;184(7):3535–44.

Tay SS, Clark AT, Deighton J, King Y, Ewan PW. Patterns of immunoglobulin G responses to egg and peanut allergens are distinct: ovalbumin-specific immunoglobulin responses are ubiquitous, but peanut-specific immunoglobulin responses are up-regulated in peanut allergy. Clin Exp Allergy. 2007;37(10):1512–8.

Thomas WR. Innate affairs of allergens. Clin Exp Allergy. 2013;43(2):152–63.

Thomas WR. Allergen ligands in the initiation of allergic sensitization. Curr Allergy Asthma Rep. 2014;14(5):432.

Thomas WR, Hales BJ, Smith WA. House dust mite allergens in asthma and allergy. Trends Mol Med. 2010;16(7):321–8.

Till S, Dickason R, Huston D, Humbert M, Robinson D, Larche M, Durham S, Kay AB, Corrigan C. IL-5 secretion by allergen-stimulated CD4 + T cells in primary culture: relationship to expression of allergic disease. J Allergy Clin Immunol. 1997;99(4):563–9.

Tresch S, Holzmann D, Baumann S, Blaser K, Wuthrich B, Crameri R, Schmid-Grendelmeier P. In vitro and in vivo allergenicity of recombinant Bet v 1 compared to the reactivity of natural birch pollen extract. Clin Exp Allergy. 2003;33(8):1153–8.

Trombone AP, Tobias KR, Ferriani VP, Schuurman J, Aalberse RC, Smith AM, Chapman MD, Arruda LK. Use of a chimeric ELISA to investigate immunoglobulin E antibody responses to Der p 1 and Der p 2 in mite-allergic patients with asthma, wheezing and/or rhinitis. Clin Exp Allergy. 2002;32(9):1323–8.

Trompette A, Divanovic S, Visintin A, Blanchard C, Hegde RS, Madan R, Thorne PS, Wills-Karp M, Gioannini TL, Weiss JP, et al. Allergenicity resulting from functional mimicry of a Toll-like receptor complex protein. Nature. 2009;457(7229):585–8.

Turner DL, Gordon CL, Farber DL. Tissue-resident T cells, in situ immunity and transplantation. Immunol Rev. 2014;258(1):150–66.

Urisu A, Ebisawa M, Ito K, Aihara Y, Ito S, Mayumi M, Kohno Y, Kondo N. Japanese Guideline for Food Allergy 2014. Allergol Int. 2014;63(3):399–419.

Vailes LD, Perzanowski MS, Wheatley LM, Platts-Mills TA, Chapman MD. IgE and IgG antibody responses to recombinant Alt a 1 as a marker of sensitization to Alternaria in asthma and atopic dermatitis. Clin Exp Allergy. 2001;31(12):1891–5.

Van Der Veen MJ, Jansen HM, Aalberse RC, van der Zee JS. Der p 1 and Der p 2 induce less severe late asthmatic responses than native Dermatophagoides pteronyssinus extract after a similar early asthmatic response. Clin Exp Allergy. 2001;31(5):705–14.

Van Der Veen MJ, Lopuhaa CE, Aalberse RC, Jansen HM, van der Zee JS. Bronchial allergen challenge with isolated major allergens of Dermatophagoides pteronyssinus: the role of patient characteristics in the early asthmatic response. J Allergy Clin Immunol. 1998;102(1):24–31.

Van Hemelen D, Mahler V, Fischer G, Fae I, Reichl-Leb V, Pickl W, Jutel M, Smolinska S, Ebner C, Bohle B, et al. HLA class II peptide tetramers vs allergen-induced proliferation for identification of allergen-specific CD4 T cells. Allergy. 2015;70(1):49–58.

Varga EM, Kausar F, Aberer W, Zach M, Eber E, Durham SR, Shamji MH. Tolerant beekeepers display venom-specific functional IgG4 antibodies in the absence of specific IgE. J Allergy Clin Immunol. 2013;131(5):1419–21.

Vrtala S, Huber H, Thomas WR. Recombinant house dust mite allergens. Methods. 2014;66(1):67–74.

Wambre E, Bonvalet M, Bodo VB, Maillere B, Leclert G, Moussu H, Von Hofe E, Louise A, Balazuc AM, Ebo D, et al. Distinct characteristics of seasonal (Bet v 1) vs. perennial (Der p 1/Der p 2) allergen-specific CD4(+) T cell responses. Clin Exp Allergy. 2011;41(2):192–203.

Wang LH, Lin YH, Yang J, Guo W. Insufficient increment of CD4+ CD25+ regulatory T cells after stimulation in vitro with allergen in allergic asthma. Int Arch Allergy Immunol. 2009;148(3):199–210.

Wisniewski JA, Agrawal R, Minnicozzi S, Xin W, Patrie J, Heymann PW, Workman L, Platts-Mills TA, Song TW, Moloney M, et al. Sensitization to food and inhalant allergens in relation to age and wheeze among children with atopic dermatitis. Clin Exp Allergy. 2013;43(10):1160–70.

Woodruff PG, Boushey HA, Dolganov GM, Barker CS, Yang YH, Donnelly S, Ellwanger A, Sidhu SS, Dao-Pick TP, Pantoja C, et al. Genome-wide profiling identifies epithelial cell genes associated with asthma and with treatment response to corticosteroids. Proc Natl Acad Sci U S A. 2007;104(40):15858–63.

Worm M, Lee HH, Kleine-Tebbe J, Hafner RP, Laidler P, Healey D, Buhot C, Verhoef A, Maillere B, Kay AB, et al. Development and preliminary clinical evaluation of a peptide immunotherapy vaccine for cat allergy. J Allergy Clin Immunol. 2011;127(1):89–97.

Wurtzen PA, van Neerven RJ, Arnved J, Ipsen H, Sparholt SH. Dissection of the grass allergen-specific immune response in patients with allergies and control subjects: T-cell proliferation in patients does not correlate with specific serum IgE and skin reactivity. J Allergy Clin Immunol. 1998;101(2 Pt 1):241–9.

Yman L. Standardization of in vitro methods. Allergy. 2001;56(Suppl 67):70–4.

Zakzuk J, Benedetti I, Fernandez-Caldas E, Caraballo L. The influence of chitin on the immune response to the house dust mite allergen Blo T 12. Int Arch Allergy Immunol. 2014;163(2):119–29.

Zhao Y, Wang H, Gustafsson M, Muraro A, Bruhn S, Benson M. Combined multivariate and pathway analyses show that allergen-induced gene expression changes in CD4+ T cells are reversed by glucocorticoids. PloS one. 2012;7(6):e39016.

Author Biography

Professor Wayne Thomas has been a Senior Research Fellow of the Australian National Health and Medical Research Council since 1987, a Senior Principal Research Fellow at the (recently renamed) Telethon Kids Institute, and Professor at the University of Western Australia since 1998. He investigated fundamental T cell biology at the Clinical Research Centre in London (1974–1983) and at the Walter and Eliza Hall Institute in Melbourne Australia during which time he was the first to demonstrate antigen-induced production of interferon gamma from CD4 T cells. His best-known accomplishments, however, have been in pioneering the development of recombinant allergens and to characterize house dust mite allergens. These have been used to explore formulations for immunotherapy that have led to the use of peptides representing allergens for immunotherapy, which have now been successfully used in clinical trials. More recent studies have extended knowledge of the spectrum of important cat allergens and how the responses that they induce are affected by single allergen immunotherapy with Fel d 1. He has also studied modified allergens and the interaction of allergens with the innate immune system. Since 2007, in conjunction with Prof Belinda Hales, he has used defined recombinant bacterial and viral antigens to describe associations of antimicrobial immune responses in early life with the development of allergy and asthma leading to the exploration of possible mechanisms of the association between infection and allergy.

Chapter 11
Antigenicity, Immunogenicity, Allergenicity

Jianguo Zhang and Ailin Tao

Abstract The term "*immune*" pertains to the body keeping itself free from diseases, not to trigger any diseases. In this regard, it makes sense for us to divide *antigenicity* into *immunogenicity* and *allergenicity*. This distinction allows for the characterization of all types of modern antigens, i.e., to evaluate and modify a priori the allergenicity of an antigen before it is applied to humans. In this chapter, we also formulated the hypothesis that "Balanced Stimulation by Whole Antigens" is essential for immune development. This hypothesis revives the practicality of the "Hygiene Hypothesis" and can provide a fundamental solution to curb the increasing prevalence of allergic disease, namely, early exposure, at 0–1 year old or earlier, in utero, of representative allergens/protein antigens with immunogenicity retained or improved and allergenicity attenuated or eliminated.

Keywords Antigenicity · Immunogenicity · Allergenicity · Immune response · Balanced stimulation

11.1 Introduction

Allergic diseases are caused by an inappropriate initiation of Type 2 (T_H2) immune responses to innocuous environmental antigens that affect the upper airway mucosa (rhinitis), lung (asthma), the gut (food allergy), and the skin (dermatitis) (Julia et al. 2015). Over the last two to three decades, the prevalence of allergic

J. Zhang · A. Tao (✉)
Guangdong Provincial Key Laboratory of Allergy and Clinical Immunology,
The State Key Clinical Specialty in Allergy, The State Key Laboratory of Respiratory Disease,
The Second Affiliated Hospital of Guangzhou Medical University, 250# Changgang Road East,
Guangzhou, China
e-mail: Aerobiologiatao@163.com

A. Tao and E. Raz (eds.), *Allergy Bioinformatics*, Translational Bioinformatics 8,
DOI 10.1007/978-94-017-7444-4_11

Table 11.1 Composition of the immune system

Immune organs		Immunocytes	Immune molecules	
Central immune organ	Peripheral immune organ		Membrane surface molecules	Secretory molecule
Thymus	Spleen	Stem cell line	TCR	Immunoglobulin
		Lymphocyte	BCR	
Marrow	Lymph node	Mononuclear phagocyte	CD molecule	Complement
Bursa of Fabricius (birds)	Mucosa-associated lymphoid tissue	Other APC (dendritic cell, endothelial cell, etc.)	Adhesion molecule	Cytokines
	Skin-associated lymphoid tissue	Other immune cells (granulocyte, mast cell, platelet, erythrocyte, etc.)	MHC, etc.	

diseases has significantly increased and this has often been explained by a decline in infections during early life. It is thought that those who have had bacterial and viral infections during childhood are able to direct their maturing immune system (Table 11.1) toward a T_H1 type and counterbalance any pro-allergic responses of T_H2 cells (Yazdanbakhsh et al. 2002). The induction of a robust anti-inflammatory regulatory network by early life exposure to allergens offers a solution to the inverse association of allergen exposure with allergic disorders (Du Toit et al. 2008; Wu et al. 2014).

11.2 Differentiation of Antigenicity, Immunogenicity and Allergenicity

In textbooks, an antigen, also called an immunogen in some references, is a substance that binds to a specific antibody or is any molecule or molecular fragment that can be bound by a major histocompatibility complex (MHC) and presented to a T cell receptor (TCR). Two features, antigenicity and immunogenicity, are generally used to describe each antigen. Immunogenicity is the ability to induce a humoral and/or cell-mediated immune response. Antigenicity is the ability to specifically combine with the final products of the immune response (i.e., secreted antibodies and/or surface receptors on T cells) (Owen et al. 2013). Although all molecules that are immunogenic are also antigenic, the reverse is not true.

If we carefully contemplate these two characteristics that are used to define an antigen, we discover that immunogenicity and antigenicity are tightly related and are always duplicated. Antibodies are produced as the result of immune induction, not from thin air. And antigens cannot trigger an immune response unless they bind with their corresponding antibodies or receptors. The above two concepts simply and repeatedly describe a single generality of all antigens, yet this alone does not allow us to completely characterize various antigens and cannot help us to understand antigens in various guises. However, different antigens produce different immune responses as they encounter their antibodies or receptors. Using this feature, antigens can be more accurately defined by the difference in the type of immune responses they induce.

According to the classical definition of immunology, the major function of the immune system, as in the integrated anatomic system and other systems, is to avoid disease in the human body. The immune system has its own mechanisms for maintaining a general physiological balance in life by co-operating with other systems of the body. Here, we attempt to redefine and differentiate antigenicity into immunogenicity and allergenicity. We refer to antigenicity as the ability of an antigen to induce an immunological response when it is encountered by the human body. Antigenicity involves two types of immune characteristics, immunogenicity, and allergenicity. Immunogenicity refers to the ability of an antigen to trigger normal and protective immune responses after being encountered by the human body. We describe the immunogenicity of an antigen using the following three aspects: (1) the ability to defend the immune system (**immunological defense**), which is the ability to repel an exogenous antigen and to fight against infection; (2) the ability to keep the immune system stable (**immunological homeostasis**), which is the ability of the body to recognize and eliminate damaged tissue, inflammation and/or senescent cells, and (3) the ability to kill and to remove abnormally mutated cells so as to monitor and inhibit the growth of malignancies in the body (**immunological surveillance**). Thus, immunogenicity reflects the strength of these three functions.

Allergenicity refers to the ability of an antigen to induce an abnormal immune response, which is an overreaction and different from a normal immune response in that it does not result in a protective/prophylaxis effect but instead causes physiological function disorder or tissue damage.

To further simplify, each antigen carries immunogenic and allergenic properties:

$$\text{Antigenicity} = \text{Immunogenicity} + \text{Allergenicity}$$

11.3 How to Measure *Allergenicity*?

Allergenicity, like immunogenicity, also exhibits antigen specificity. Due to different antigen/antibody specificities, each antigen has different levels of allergenicity and immunogenicity and each individual has a different immune system, thus, antigens can be allergens in individuals of different ages and different immune statuses. In general, the most potent allergens are proteins, with polysaccharides ranking second.

Allergy usually is characterized by T_H2 (T helper 2) responses, which are described by increases in the levels of interleukin (IL)-4 and other T_H2-type cytokines (IL-5, IL-9, IL-13, and IL-21, etc.), activation and expansion of CD4+ T_H2 cells, induction of plasma cells secreting IgE, and activation of eosinophils, mast cells, and basophils, all of which can produce several types of T_H2-type cytokines (Anthony et al. 2007). Hence, the levels and duration of the T_H2 response define allergenicity.

Researchers can also ascertain a protein's identity by scrutinizing its history of medical use or by searching the literature to see whether any adverse reactions have been reported or by doing experiments to investigate the allergenicity and immunogenicity of the candidate protein(s) to be encountered by the human body. Literature reviewing also provides bioinformatics data that can be used to evaluate immunogenicity and allergenicity.

11.4 Important Factors for Early Exposure and Hygiene Hypothesis

Allergies are a major cause of chronic disease in all countries of the world with the incidence of reported cases significantly increasing over the past two to three decades. This increase has often been explained by some experts as the Hygiene Hypothesis (Cramer et al. 2012; Liu and Murphy 2003; Maizels et al. 2014; Sherriff and Golding 2002), that is, a decline in infections during early life could predispose children to be susceptible to allergy and that exposure to microbial products such as endotoxin can reduce the risk for allergic sensitization during early childhood. However, on the contrary, allergic sensitization among adults and in the elderly increased with increasing endotoxin levels (Min and Min 2015). The same observation has been made for the development of food allergy. Early consumption of peanuts in infancy is associated with a low prevalence of peanut allergy (Du Toit et al. 2008, 2015). Higher maternal intake of peanut, milk, and wheat during early pregnancy was associated with a decreased incidence of mid-childhood allergy and asthma (Bunyavanich et al. 2014). The role of diet, therefore, has been highlighted as a key factor that influences immune homeostasis and the development of allergic diseases (Julia et al. 2015). There is no benefit to delaying the introduction of any potentially allergenic food, such as milk, eggs, peanuts, or fish food beyond 6 months of age to prevent food allergy (Chin et al. 2014). Regarding the development of inhalant allergy, a similar conclusion has been drawn but with an exception for mites. It was demonstrated that early exposure to high levels ($\geq$10 μg/g dust) of dust mite allergen was associated with an increased risk of asthma and late-onset wheeze at age 7 years compared with exposure to low levels (< 0.05 μg/g dust) of dust mite allergen (Celedon et al. 2007). Also, pet exposure during the first year of life and an increasing number of siblings were both associated with a lower prevalence of allergic rhinitis and asthma in school children (Hesselmar et al. 2008).

Some experts have argued that appropriately targeted allergic hypersensitivity evolved to elicit anticipatory responses and to promote avoidance of suboptimal environmental substances (Palm et al. 2012). However, if allergen avoidance really benefited the immune system, there would not be a need for the immune system to establish immune memory. On the other hand, different allergens belong to different groups (see Chap. 5 in this book). Single-allergen avoidance means avoiding an entire group of allergens, making complete avoidance impossible. Moreover, complete avoidance can cause malnutrition, mental retardation and lost enjoyment of life (Wang 2010). Even worse, avoiding the consumption of certain substances may trigger defects in the immune system. In fact, our meta-analysis concluded that allergen avoidance may not always be successful in preventing allergic symptoms (Wu et al. 2014), especially for newborns.

A study of the prevalence of allergy in adults demonstrated that infection with pulmonary TB contributes significantly to atopy, particularly allergic rhinitis symptoms (Lin et al. 2013). A pilot experiment showed the cross-reactivity between antigens from roundworm *Ascaris lumbricoides* (AL) and house dust mite (HDM) allergens (Acevedo et al. 2009). Another experiment with larger samples further demonstrated that AL-antigens can inhibit up to 92 % of HDM-specific IgE-reactivity among allergic subjects, while only up to 54 % of AL-specific IgE-reactivity among ascariasis subjects was inhibited by HDM allergens (Valmonte et al. 2012), suggesting that AL antigens have broader and higher allergenicity than HDM allergens and noting that the latter would sensitize up to 70 % of the allergic population (He et al. 2014). A further study from Hagel I et al. indicates that it only took a mild infection with *A. lumbricoides* (0–5000 eggs/g feces) to significantly elevate the levels of IL-13, IL-6, IL-10 as well as the levels of IFN-γ and no mention of IgE or IgG in this group, while in the moderately infected group (> 5001–50,000 eggs/g feces), IL-13 and IL-10 are very significantly increased but no increase of IgE, IgG, IFN-γ, or IL-6 was observed in comparison with those in the urban nonparasitized control group. This result indicated that the protective response against allergy development by *A. lumbricoides* relies on IL-10 but is independent of the production of IFN-γ, IgE, or IgG. These observations are complicated by concurrent infections with *Giardia duodenalis* and *A. lumbricoides*. A study evaluated the effect of *A. lumbricoides* on *G. duodenalis* infection and T_H1/T_H2 type immune mechanisms toward this parasite in 251 rural parasitized and 70 urban nonparasitized school children (Hagel et al. 2011). In the group of children mildly infected with *A. lumbricoides*, the levels of IgG, IgE, IL-13, IL-6, IL-10, IFN-γ, and IL-6 are all very significantly increased, while in the group of children moderately infected with *A. lumbricoides*, only IL-13 and IL-10 are highly elevated and IFN-γ is significantly increased but with no significant effects on the levels of IgG, IgE, or IL-6. These results suggest that *A. lumbricoides* can modulate the immune responses by affecting both T_H1 and T_H2 type immunity (Hagel et al. 2011). Therefore, the conditions regarding the severity of the *A. lumbricoides* infection and whether or not co-infection with other species of parasites has occurred are very important in reaching the real conclusion. This information would have been helpful in a study (Palmer et al. 2002) that demonstrated in a cross-sectional sample of 2164 children that *A. lumbricoides* infection is associated with increased risk of childhood asthma and atopy in rural China but it did not contain detailed data on the above-mentioned two parameters. This could explain why previous studies of birth cohorts with participants in (sub)-urban environments that examined similar associations have yielded inconsistent results. In conclusion, the protective immune response induced by parasites in humans is dependent on the particular parasite (Ek et al. 2012) and, therefore, discussion of the protective effects without the antibody and cytokine measurements (IgG, IgE, IL-13, IL-6, IL-10, IFN-γ and IL-6) or investigation into the presence of co-infection is limited and insufficient. It is tempting to conclude that *A. lumbricoides* has high immunogenicity and low allergenicity and that this type of circumstance could significantly contribute to the maturation of our immune systems.

Pet exposure during the first year of life and an increased number of siblings were both associated with a lower prevalence of allergic rhinitis and asthma in school children (Hesselmar et al. 2008). Moreover, there is no benefit to delaying the introduction of any potentially allergenic food, such as milk, eggs, peanuts, or fish food beyond 6 months of age to prevent food allergy (Chin et al. 2014). In any case, it should be emphasized that early exposures to prevent the development of allergy should be with allergens, probiotics and non-infectious microbes (Douwes et al. 2006) but without exposure to bio-contaminants (such as biomass smoke), as these can obviously reduce lung function in young adults compared to exposure to smoke from liquefied petroleum gas (Kurmi et al. 2013).

Furthermore, the phenomenon of allergic sensitization is overrepresented among first-born or only children and less frequent in children from large families and those attending day care, suggesting that the frequent exchange of infections may protect children from allergic sensitization (Yazdanbakhsh et al. 2002). A study of gut commensals demonstrates that different rates of microbial colonization and infections with different bacterial types (*Clostridia* vs. *Lactobacilli*) would predispose children to allergy or no allergy (Sepp et al. 1997). This is similar to the situation seen with parasitic infections in children. **The protective effect of infections strictly depends on the specific species and the microbial/parasitic burden**. Thus, it is tempting to think that for immune system development and homeostasis, there are microbial and parasitic friends and foes that are very distinct and explicit. Therefore, **to better characterize the antigen, it is crucial to know how harmful (allergenic) and how beneficial (immunogenic) the antigen is to the immune system**.

Regarding mechanisms, early exposure to soil, house dust, and decaying plants increases gut microbial diversity and decreases serum immunoglobulin E levels, thus enhancing innate immunity (Zhou et al. 2015). Exposure to a non-hygienic environment did not induce significant airway neutrophilia, yet it altered the number of immunologically active cells in the lung and reduced subsequent allergic inflammation (George et al. 2006). Further studies suggested that early exposure to unhygienic conditions and infections is associated with different expression of Toll-like receptors (Majak et al. 2009) and early exposure to a farm environment seems to influence methylation patterns in distinct genes (Michel et al. 2013), therefore, epigenetic mechanisms may contribute to the development of asthma and other allergies.

11.5 Balanced Stimulation by Whole Antigens for Immune System Development

Based on what has been stated for early exposure factors in the previous sections, allergen number reduction results (see Chap. 5 in this book), and the Hygiene Hypothesis, we hypothesized that **Balanced Stimulation by Whole Antigens is necessary for healthy immune system development**. This hypothesis contains three essential parts: (1) Administration of all types of allergens in very early life contributes to the healthy maturation of the immune system and protects children from allergy development; those infants who miss exposure to one or some types of allergens during the key period of immune system expansion may develop atopy to these substances when they grow up. After a diagnosis of allergy, the affected cases would be treated by immunotherapy with these allergens. (2) Regarding mechanisms, the maternal immune status is a key factor in whether the fetus is primed for a T_H1 or T_H2 response. A balanced level of T_H1 cytokines (IFN-γ, IL-10) provided by maternal T cells drives the direction of the homeostatic development of the initial T_H0 cells of the fetus that are then further educated for tolerance during infancy and improved by a balanced stimulation with all types of allergens, even if the developing immune status is T_H2-biased. Conversely, an unbalanced stimulation that lacks of one or more types of allergens in the first year of life would negatively influence the evolving balance and/or enhance any existing T_H2-biased immune status, thus allowing the development of allergic disease. Furthermore, maternal milk, a tight link between mothers and their children, contains free dietary and environmental allergens, IgM/IgG/IgA, tolerogenic factors (such as interleukin 10, transforming growth factor-β (TGF-β), lactoferrin, antioxidants, etc.), gut growth factors (such as cortisol, thyroxine, epidermal growth factor, TGF-β, etc.) and microbiota-influencing factors (such as prebiotics, oligosaccharides, casein, etc.). These factors can be transferred to the infant during breastfeeding. During childhood and adolescence (Fig. 11.1), tolerance develops to dietary and inhalant allergens and reinforces the immune system memory to these antigens (Julia et al. 2015). (3) There is a necessity for lasting memory T cells to be restimulated in order to sustain their immortality and immunotolerance capability. It is the exposure of antigens in a certain space-time continuum that stimulate and reinforce the development of the immune system. During the naïve stage, early exposure to superantigens with attenuated allergenicity could potentially strengthen and confer immune system tolerance to their allergenicity-untouched natural counterparts—this is similar to the process of allergen-specific immunotherapy. A good example comes from the progression of smallpox vaccination (Fig. 11.2). Smallpox vaccines were originally made with whole smallpox virus that then evolved over generations to being made with the cowpox virus, which was actually a type of allergenicity attenuation that resulted in a vaccine that could therefore be safely administered to humans for protection against the smallpox virus (Fig. 11.2). It is tempting to speculate that other infectious diseases (SARS, AIDS, Ebola, etc.) could be eradicated by vaccination with their allergenicity-attenuated counterparts.

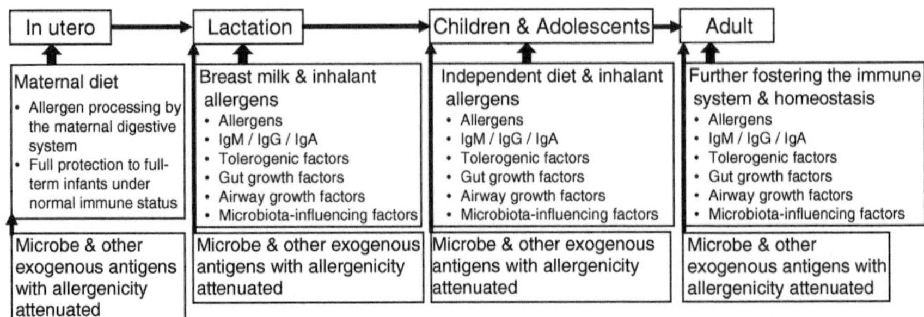

Fig. 11.1 Possible mechanisms of mother-to-offspring transfer of protection against allergy. Adapted from the reference (Julia et al. 2015). Maternal milk, a tight link between mothers and their children, contains free dietary and environmental allergens, IgM/IgG/IgA, tolerogenic factors (such as interleukin 10, transforming growth factor-β (TGF-β), lactoferrin, antioxidants, etc.), gut growth factors (such as cortisol, thyroxine, epidermal growth factor, TGF-β, etc.) and microbiota-influencing factors (such as prebiotics, oligosaccharides, casein, etc.). These factors can be transferred to the infant during breastfeeding. During childhood and adolescence, tolerance develops to dietary and inhalant allergens and reinforces the immune system memory to these antigens, otherwise the body could become allergic to these allergens/antigens along with the gradual induction of immune tolerance. Allergic disease would march onward if the adult immune system is not able to be tolerant to the allergens. In any case, the adult immune system also needs to be fostered by sustained antigen stimulation to avoid any damage to immune homeostasis. Nevertheless, the allergic status can be modified/reduced and the immune system enhanced/reinforced by immunotherapy with microbial and other exogenous antigens that can attenuate allergenicity

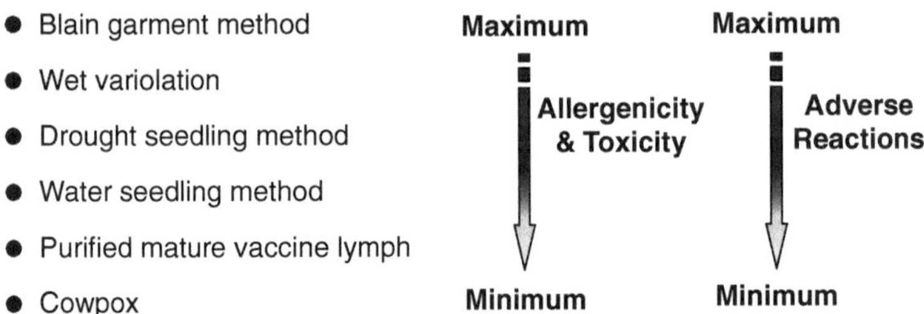

Fig. 11.2 Smallpox vaccines evolved from an original primitive type to fostered and purified vaccine lymph. The real essence lies in attenuation of the allergenicity

11.6 What Is the Future? Can We Ever Win?

Increasing allergic reactions have been described as the "Allergy March," which is the progression of atopic manifestations persisting over years and is characterized by a typical sequence of clinical symptoms from colic during infancy (tummy pains, including bad stomach aches, vomiting and diarrhea, itchiness on the baby's face, lips and buttocks, etc.), to eczema when the child is under two-three years old (itchy skin as well as reactions to certain foods and allergens in the air), to rhinitis and then asthma. Nevertheless, we can safely heal this kind of disease by etiological immunotherapy with allergenicity-attenuated vaccines.

We have seen through vaccination that we can strengthen our immune system (Table 11.1) enough to protect us against all types of infectious diseases. It is conceivable that we can also protect from allergy development by early exposure to allergenicity-attenuated vaccines that have been genetically engineered from the environmental antigens. Thus, it is possible that **Immune Giants** could be created that would train our immune systems to be ready to handle all types of environmental antigens, no matter whether they are allergens or infectious microbes (Fig. 11.1). Regardless, the control of exposure to environmental antigens and undesirable commensal microorganisms will always be an important and challenging part of human health.

References

Acevedo N, Sanchez J, Erler A, Mercado D, Briza P, Kennedy M, Fernandez A, Gutierrez M, Chua K, Cheong N, et al. IgE cross-reactivity between Ascaris and domestic mite allergens: the role of tropomyosin and the nematode polyprotein ABA-1. Allergy. 2009;64(11):1635–43.

Anthony RM, Rutitzky LI, Urban JF Jr, Stadecker MJ, Gause WC. Protective immune mechanisms in helminth infection. Nat Rev Immunol. 2007;7(12):975–87.

Bunyavanich S, Rifas-Shiman SL, Platts-Mills TA, Workman L, Sordillo JE, Camargo CA Jr, Gillman MW, Gold DR, Litonjua AA. Peanut, milk, and wheat intake during pregnancy is associated with reduced allergy and asthma in children. J Allergy Clin Immunol. 2014;133(5):1373–82.

Celedon JC, Milton DK, Ramsey CD, Litonjua AA, Ryan L, Platts-Mills TA, Gold DR. Exposure to dust mite allergen and endotoxin in early life and asthma and atopy in childhood. J Allergy Clin Immunol. 2007;120(1):144–9.

Chin B, Chan ES, Goldman RD. Early exposure to food and food allergy in children. Can Fam Physician. 2014;60(4):338–9.

Cramer C, Link E, Koletzko S, Lehmann I, Heinrich J, Wichmann HE, Bauer CP, Berg AV, Berdel D, Herbarth O, et al. The hygiene hypothesis does not apply to atopic eczema in childhood. Chem Immunol Allergy. 2012;96:15–23.

Douwes J, van Strien R, Doekes G, Smit J, Kerkhof M, Gerritsen J, Postma D, de Jongste J, Travier N, Brunekreef B. Does early indoor microbial exposure reduce the risk of asthma? The prevention and incidence of asthma and mite allergy birth cohort study. J Allergy Clin Immunol. 2006;117(5):1067–73.

Du Toit G, Katz Y, Sasieni P, Mesher D, Maleki S, Fisher H, Fox A, Turcanu V, Amir T, Zadik-Mnuhin G, et al. Early consumption of peanuts in infancy is associated with a low prevalence of peanut allergy. J Allergy Clin Immunol. 2008;122(5):984–91.

Du Toit G, Roberts G, Sayre PH, Bahnson HT, Radulovic S, Santos AF, Brough HA, Phippard D, Basting M, Feeney M, et al. Randomized trial of peanut consumption in infants at risk for peanut allergy. N Engl J Med. 2015;372(9):803–13.

Ek C, Whary M, Ihrig M, Bravo L, Correa P, Fox J. Serologic evidence that ascaris and toxoplasma infections impact inflammatory responses to Helicobacter pylori in Colombians. Helicobacter. 2012;17(2):107–15.

George CL, White ML, Kulhankova K, Mahajan A, Thorne PS, Snyder JM, Kline JN. Early exposure to a nonhygienic environment alters pulmonary immunity and allergic responses. Am J Physiol Lung Cell Mol Physiol. 2006;291(3):L512–22.

Hagel I, Cabrera M, Puccio F, Santaella C, Buvat E, Infante B, Zabala M, Cordero R, Di Prisco M. Co-infection with *Ascaris lumbricoides* modulates protective immune responses against *Giardia duodenalis* in school Venezuelan rural children. Acta Trop. 2011;117(3):189–95.

He Y, Liu X, Huang Y, Zou Z, Chen H, Lai H, Zhang L, Wu Q, Zhang J, Wang S, et al. Reduction of the number of major representative allergens: from clinical testing to 3-dimensional structures. Mediat Inflamm. 2014;2014:291618.

Hesselmar B, Aberg N, Aberg B, Eriksson B, Björkstén B. Does early exposure to cat or dog protect against later allergy development? Clin Exp Allergy. 2008;29(5):611–7.

Julia V, Macia L, Dombrowicz D. The impact of diet on asthma and allergic diseases. Nat Rev Immunol. 2015;15(5):308–22.

Kurmi OP, Devereux GS, Semple WC, Steiner S, Simkhada MF, Lam P, Hubert KB, Ayres JG. Reduced lung function due to biomass smoke exposure in young adults in rural Nepal. Eur Respir J. 2013;41(1):25–30.

Lin CT, Gopala K, Manuel AM. The impact of pulmonary tuberculosis treatment on the prevalence of allergic rhinitis. Ear Nose Throat J. 2013;92(8):358–99.

Liu AH, Murphy JR. Hygiene hypothesis: fact or fiction? J Allergy Clin Immunol. 2003;111(3):471–8.

Maizels RM, McSorley HJ, Smyth DJ. Helminths in the hygiene hypothesis: sooner or later? Clin Exp Immunol. 2014;177(1):38–46.

Majak P, Brzozowska A, Bobrowska-Korzeniowska M, Stelmach I. Early exposure to unhygienic conditions and infections is associated with expression of different Toll-like receptors. J Investig Allergol Clin Immunol. 2009;19(4):260–5.

Michel S, Busato F, Genuneit J, Pekkanen J, Dalphin JC, Riedler J, Mazaleyrat N, Weber J, Karvonen AM, Hirvonen MR, et al. Farm exposure and time trends in early childhood may influence DNA methylation in genes related to asthma and allergy. Allergy. 2013;68(3):355–64.

Min KB, Min JY. Exposure to household endotoxin and total and allergen-specific IgE in the US population. Environ Pollut. 2015;199:148–54.

Owen JA, Punt J, Stranford SA, Jones PP (2013). Kuby immunology, 7th edn (Susan Winslow).

Palm N, Rosenstein R, Medzhitov R. Allergic host defences. Nature. 2012;484(7395):465–72.

Palmer LJ, Celedon JC, Weiss ST, Wang B, Fang Z, Xu X. *Ascaris lumbricoides* infection is associated with increased risk of childhood asthma and atopy in rural China. Am J Respir Crit Care Med. 2002;165(11):1489–93.

Sepp E, Julge K, Vasar M, Naaber P, Bjorksten B, Mikelsaar M. Intestinal microflora of Estonian and Swedish infants. Acta Paediatr. 1997;86(9):956–61.

Sherriff A, Golding J. Hygiene levels in a contemporary population cohort are associated with wheezing and atopic eczema in preschool infants. Arch Dis Child. 2002;87(1):26–9.

Valmonte G, Cauyan G, Ramos J. IgE cross-reactivity between house dust mite allergens and *Ascaris lumbricoides* antigens. Asia Pac Allergy. 2012;2(1):35–44.

Wang J. Management of the patient with multiple food allergies. Current Allergy Asthma Rep. 2010;10(4):271–7.

Wu H, Guo Y, Wang J, Zhang J, Wang S, Zhang X, Tao A. The importance of allergen avoidance in high risk infants and sensitized patients: a meta-analysis study. Allergy Asthma Immunol Res. 2014;6(6):525–34.

Yazdanbakhsh M, Kremsner PG, van Ree R. Allergy, parasites, and the hygiene hypothesis. Science. 2002;296(5567):490–4.

Zhou D, Zhang H, Bai Z, Zhang A, Bai F, Luo X, Hou Y, Ding X, Sun B, Sun X, et al. Exposure to soil, house dust and decaying plants increases gut microbial diversity and decreases serum immunoglobulin E levels in BALB/c mice. Environ Microbiol. 2015;. doi:10.1111/1462-2920.12895.

Author's Biography

Jianguo Zhang is a Chief Physician and Director of the Otolaryngology Department of the Second Affiliated Hospital of Guangzhou Medical University. He is engaged in otolaryngology diagnosis and research and has directed several clinical trials on the diagnosis and immunotherapy of allergic rhinitis. Prof. Zhang has received funding for many research projects and published dozens of research articles. His research focuses on novel immunotherapeutic methods and allergenicity evaluation and modification for toxins and transgenic candidate genes. Prof. Zhang significantly contributed to the hypothesis "Balanced Stimulation by Whole Antigens." He is also interested in the integrated application of multiple surgical treatments and developed the Sleep Apnea Treatment Center. Prof. Zhang was given the Anti-SARS Advanced Individual of Guangdong and Guangzhou Award for his excellent contribution to SARS control in 2003 and received the Guangdong "Outstanding Teachers" Award for his exceptional teaching.

Ailin Tao is a Professor at Guangzhou Medical University, Director of Guangdong Provincial Key Laboratory of Allergy and Clinical Immunology, Principal Investigator of the State Key Laboratory of Respiratory Disease, Deputy Director of the State Key Clinical Specialty in Allergy of the Second Affiliated Hospital of Guangzhou Medical University, Member of the State Committee for Transgenic Safety Assessment, Standing Committee Member of Allergy Branch of Guangdong Medical Association, Member of Guangdong Provincial Committee for Transgenic Safety Assessment. Email: Aerobiologiatao@163.com. Professor Ailin TAO earned his doctorate degree from the State Key Laboratory of Crop Genetic Improvement of Huazhong Agricultural University in 2002, followed by a postdoctoral training at Postdoctoral Station of Basic Medicine in Shantou University Medical College, majoring in allergen proteins. His most recent research has been on allergy bioinformatics, allergy, and clinical immunology and disease models, such as allergic asthma, allergic rhinitis, infection and inflammation induced by allergy, inflammatory, and protracted diseases caused by antigens or superantigens. He has gained experience in the field of allergology including the mechanisms of immune tolerance, allergy triggering factors, and chronic inflammation pathways and allergenicity evaluation and modification for food and drugs. He proposed some new concepts including "Representative Major Allergens," "Allergenicity Attenuation" of immunotoxin and allergens, "broad-spectrum immunomodulator" as well as the theoretical hypothesis of "Balanced Stimulation by Whole Antigens." Prof. TAO's laboratory focuses on the diagnosis of allergic disease and the medical evaluation of food and drug allergenicity and its modification. Prof. TAO has now constructed a system for the prediction, quantitative assessment and simultaneous modification of epitope allergenicity, which has been applied to more than 20 allergens, and he also developed a bioinformatics software program for allergen epitope prediction, SORTALLER (http://sortaller.gzhmu.edu.cn), which performed significantly better than the other existing software, reaching a perfect balance of high specificity (98.4 %) and sensitivity (98.6 %) for discriminating allergenic proteins from several independent datasets of protein sequences of diverse sources. Furthermore, this program has a Matthews correlation coefficient as high as 0.970, a fast running speed and can rapidly predict a set of amino acid sequences with a single click. The software has been frequently used by researchers from many institutions in China and over 30 countries worldwide, thus becoming the number one allergen epitope prediction software program. Prof TAO has set up an allergen database ALLERGENIA (http://ALLERGENIA.gzhmu.edu.cn) that has several advantages over other databases, such as a wide selection of nonredundant allergens, excellent astringency and accuracy, and user-friendly analytical functions.

Chapter 12
Bioinformatic Classifiers for Allergen Sequence Discrimination

Yuyi Huang and Ailin Tao

Abstract With the rapid development of biotechnology and sequencing techniques, the volume of various protein and sequence data has grown exponentially. However, compared to the rapid breakthrough in large-scale sequencing, management and mining of these data has made little progress. Proteins are a collection of macromolecules that play essential roles in organisms. Prediction of the functional classification of a protein is a major task of bioinformatics. Recently, with the rapid developments in bioinformatics, computational classification of proteins has revolutionized protein research by guiding experimental design.

Keywords Allergen · Bioinformatics · Classifiers · Protein · Prediction

12.1 Introduction

12.1.1 The Primary Structure of Proteins

Proteins are a collection of macromolecules that play essential roles in organisms. Predicting the functional classification of a protein is a major task of bioinformatics since the primary structure of a protein determines its characteristics

Y. Huang · A. Tao (✉)
Guangdong Provincial Key Laboratory of Allergy and Clinical Immunology,
The State Key Clinical Specialty in Allergy, The State Key Laboratory of Respiratory Disease,
The Second Affiliated Hospital of Guangzhou Medical University,
250# Changgang Road East, Guangzhou 510260, Guangdong Province,
People's Republic of China
e-mail: taoailin@gzhmu.edu.cn

Y. Huang
e-mail: chvipdata@126.com

A. Tao and E. Raz (eds.), *Allergy Bioinformatics*, Translational Bioinformatics 8,
DOI 10.1007/978-94-017-7444-4_12

and function (Wu et al. 2013). Protein structure is dependent on its amino acid sequence and it is widely recognized that this sequence can be used to predict the function of a protein (Popov et al. 2014).

All proteins are made of a set of 20 amino acids, each of which is assigned a unique letter, and can be represented as AA = [A, C, D, E, F, G, H, I, K, L, M, N, P, Q, R, S, T, V, W, and Y]. Different amino acid alignments make different proteins. When two amino acids link through esterification between the carboxyl group of the first amino acid and the amino group of the next, the connection is called a peptide bond. The linear structure formed by linked amino acids is called a peptide, and when a peptide becomes longer and consists of several amino acids, it is called polypeptide. One end of a polypeptide has a free amino group and is called the amino terminus and the other end has a free carboxyl group and is called the carboxyl terminus. In a sense, proteins are basically a strand or several strands of polypeptides. For example, ferredoxin and myoglobin are composed of a single polypeptide chain, insulin is composed of two polypeptide chains that are connected by disulfide bonds, and hemoglobin is composed of four polypeptides that are connected non-covalently. An amino acid sequence is translated from messenger RNAs, which were transcribed from DNA, and thus ultimately, it is the DNA sequence that determines the characteristics and function of proteins.

12.2 Sequence Data Mining and Classification

With the rapid development of biotechnology and sequencing techniques, the volume of various protein and sequence data has expanded greatly but management and mining of these data has made little progress. Data itself is not knowledge. Some point mutations or even large-scale deletions may exert little effect on the biological function of a certain protein. These amino acid fragments are called noise sequences. Other amino acids or sequences are largely conserved, however, once replaced or deleted, severe consequences may occur that lead to partial or complete loss of protein function and these fragments are called characteristic sequences (Zhang et al. 2012). This difference is the basis of sequence data mining and classification.

The traditional way to understand the function of a protein is through experimentation that can be expensive and time consuming; therefore, it is not possible to test every protein experimentally. Meanwhile, various sequence databases have been filled by an ocean of sequence data waiting to be mined for useful information. This data could have great clinical significance in understanding the potential correlation between the sequence and function of proteins as well as predict protein function from its sequence through machine learning and data mining. Nobel Prize winner W. Gilbert had pointed out that traditional biology seeks answers by experiments, while emerging approaches would be based on full knowledge of the genome electronically stored in databases. The next generation of biology research should be initiated by theories, with the scientist starting from theoretical assumptions and then testing the assumptions by experimentation (Whelan et al. 2013). In recent years, with the rapid development in bioinformatics, computational classification of proteins has revolutionized protein research by guiding experimental design (Yu et al. 2015; Jones et al. 2014; Nevado and Perez-Enciso 2015; Srinivasan et al. 2013).

12.3 Prediction of Allergen Classification

12.3.1 Introduction

Sequence data is a discrete set of ordered data containing genetic information as an ordered series of a certain length. In fact, conservation and divergence of the sequence data contain very valuable information. They reflect the adaptive changes of the genetic information during the long period of evolution, thus representing the "memory" of a species. If we can quantitatively describe such characteristic "memory," we would be able to predict the unknown; therefore, analysis of the signature of the sequence data becomes the key to predicting sequence classification. Prediction of allergen protein classification is basically performed in two ways, the feature extraction-based method and the similarity-based method. The feature extraction-based method extracts the characteristic parameters of the amino acid sequence based on conservation and divergence data. The amino acid sequence of a protein can be abstractly described by the characteristic parameters, which can then be used for machine learning and classification. The similarity-based method collects a sufficiently large database of allergen sequences, which is a prerequisite for the sequence similarity model to work. The sequences with high similarities can be retrieved and, through certain approaches, correlations among the sequences can be identified and thereby an unknown function of a sequence can be predicted.

12.3.2 Allergen Classification

Allergen classification can be fulfilled by several methods as follows.

12.3.3 Similarity-Based Method

The criteria proposed by FAO/WHO (FAO/WHO 2001) to distinguish allergens and nonallergens are based on similarity. A protein is identified as an allergen if (i) it contains a 6-aa sequence of an already known allergen, or (ii) it shows at least 35 % identity with an already known allergen in an 80-aa segment. Such definition requires a pre-established database for sequence alignment. That is, a database of known allergens would be required before similarity testing using BLAST (Altschul et al. 1997), FASTA (Pearson 1988), or other alignment tools. The criteria above belong to the classical model of allergen classification, which in practice may be adjusted.

Construction of a similarity model includes the following three steps: (i) Establish an allergen sequence database, excluding repeated and redundant data to ensure accuracy and applicability; (ii) Submit the target sequence to the database and search for similar sequences using BLAST or FASTA; and (iii) Filter the results according to existing or customized standards for classification.

Accuracy of the similarity model depends on the accuracy of the allergen sequence database, thus, only if the sequences in the database are confirmed to be allergens, proteins having similarities with the database sequences would be allergens. So it is of utmost importance to establish an accurate sequence database, examples of which are described below.

Allermatch

Allermatch (Fiers et al. 2004) is an online allergen prediction tool using FAO/WHO criteria. The prediction steps are as follows: (1) Amino acid sequences of known allergens are obtained in FASTA format; (2) The submitted sequence is read into overlapping 80-aa slices, also in FASTA format; (3) Each of the slices is aligned with sequences of known allergens in the database; and (4) Using the FAO/WHO standard, determine whether the target sequence is a potential allergen or not. As mentioned above, the accuracy of the similarity model depends on accuracy of the allergen sequence database that is used. In this regard, the Allermatch database has collected 730 known allergens from WHO-IUIS or SwissProt.

WebAllergen

WebAllergen (Riaz et al. 2005) is another online allergen prediction service. It provides a database of 664 known allergens. Users can submit their sequences to align with the sequences in the database or other allergen sequence databases of their own. The server converts submitted allergens into allergen motifs, which are aligned with the target sequence to be classified.

12.3.4 Feature Extraction-Based Method

Prediction of allergen classification using the feature extraction-based method consists of the following steps: (1) Construct a sequence database of allergens and non-allergens using data that is accurate, objective, and typical; (2) Extract characteristic parameters according to conservation and divergence of the sequence among related protein family so that the amino acid sequence of the protein can be classified by characteristic parameters; (3) Design a classifier by matching learning; and (4) Predict the classification of an unknown protein using the classifier. Due to the diversity and complexity of the allergen sequences in sequence length and order, conservation and divergence of each site of the amino acid sequences differs, making it difficult to find a characteristic parameter representative of known allergens and to construct a proper classifier. However, many advances have been made in this field as described below:

EVALLER
EVALLER (Martinez Barrio et al. 2007) is an interactive web program based on the filtered length-adjusted allergen peptides (DFLAP) algorithm. DFLAP is characterized by two features: (1) highly specific allergen peptides can be selected by comparison to nonallergens and (2) an allergen discrimination model is established using the support vector machine (SVM) model. This algorithm shows very high sensitivity and specificity in identification of potential allergens. The EVALLER web server accepts FASTA protein sequences as input for allergenicity assessment and the results are presented as both text and graphics so that the user can conveniently evaluate accuracy of the results.

AllerHunter
AllerHunter (Muh et al. 2009) is a web-based program for the assessment of potential allergenicity and cross-reactivity. This program integrates FAO/WHO and the SVM learning model and has been tested using 1356 known allergens and nonallergens. This system works by, first, evaluating the potential allergenicity and cross-reactivity of a protein sequence using the SVM learning model, with a likelihood score as output, and then, the output is further evaluated by FAO/WHO criteria, resulting in a comprehensive assessment of the allergenicity.

AlgPred
AlgPred (Saha and Raghava 2006) integrates several different methods to identify allergens. Its test set and training set include 578 strands of allergen sequences and 700 strands of nonallergen sequences. These sequences were divided into five clusters by BLAST similarity search (E value 8E-4, identity 26 %), where no similarity exists between different clusters and each cluster contains a nearly equal number of sequences. A classifier is then designed using a corresponding machine learning approach. By grouping the sequences into clusters, the similarity within the training set can be reduced without reducing the total number of sequences, so that accuracy of classification prediction can be improved.

Allerdictor
Allerdictor (Dang and Lawrence 2014) proposes a sequence-based method for allergenicity prediction. The basic idea of this program is to represent a protein by its k-mer feature. Due to the feature construction nature and adoption of a linear SVM model, linear prediction can be achieved. Allerdictor uses both naïve Bayes (NB) and SVM classifiers. The SVM model performs better than NB when dealing with highly similar sequences between allergens and nonallergens. The length of k-mer peptide is the most important parameter of the Allerdictor model. The performance of Allerdictor varies when using different k values. A k value of 5 or 6 shows the highest performance, with a value of $k = 5$ indicating a near-perfect false positive rate (FP), while a value of $k = 6$ shows higher sensitivity while maintaining a low FP rate.

SORTALLER

SORTALLER (Zhang et al. 2012) is an online allergen classifier based on an allergen family featured peptide (AFFP) dataset and normalized BLAST E values, which established the featured vectors for the support vector machine (SVM) model. AFFPs are allergen-specific peptides panned from irredundant allergens and harbor perfect information with noise fragments eliminated because of their similarity to nonallergens. SORTALLER performed significantly better than other existing software and reached a perfect balance with high specificity (98.4 %) and sensitivity (98.6 %) for discriminating allergenic proteins from several independent datasets of protein sequences of diverse sources, also highlighted by a Matthews correlation coefficient (MCC) as high as 0.970, a fast running speed, and rapid predicting of a batch of amino acid sequences with a single click.

12.4 Conclusions

With the continuous development of genetically modified crops and products, the allergenicity problem is becoming increasingly severe. Bioinformatics studies have made much progress in predicting the allergenicity of highly conserved sequences with breakthroughs in sequence acquisition, processing, storage, distribution, analysis, interpretation, etc. However, precise distinction between potential allergens and nonallergens is still far from perfect. Thus, assessment of allergenicity and cross-reactivity of proteins by the comprehensive application of mathematics, computer science, and biology tools will be the direction of this area of research in the coming years.

References

Altschul SF, Madden TL, Schaffer AA, Zhang J, Zhang Z, Miller W, Lipman DJ. Gapped BLAST and PSI-BLAST: a new generation of protein database search programs. Nucleic Acids Res. 1997;25(17):3389–402.

Dang HX, Lawrence CB. Allerdictor: fast allergen prediction using text classification techniques. Bioinformatics. 2014.

FAO/WHO. Evaluation of allergenicity of genetically modified foods. Report of a Joint FAO/WHO Expert Consultation on Allergenicity of Foods Derived from Biotechnology. 2001;22–5.

Fiers MW, Kleter GA, Nijland H, Peijnenburg AA, Nap JP, van Ham RC. Allermatch, a webtool for the prediction of potential allergenicity according to current FAO/WHO Codex alimentarius guidelines. BMC Bioinf. 2004;5:133.

Jones P, Binns D, Chang HY, Fraser M, Li W, McAnulla C, McWilliam H, Maslen J, Mitchell A, Nuka G, et al. InterProScan 5: genome-scale protein function classification. Bioinformatics. 2014;30(9):1236–40.

Martinez Barrio A, Soeria-Atmadja D, Nister A, Gustafsson MG, Hammerling U, Bongcam-Rudloff E. EVALLER: a web server for in silico assessment of potential protein allergenicity. Nucleic Acids Res. 2007;35(Web Server issue):694–700.

Muh HC, Tong JC, Tammi MT. AllerHunter: a SVM-pairwise system for assessment of allergenicity and allergic cross-reactivity in proteins. PLoS ONE. 2009;4(6):e5861.
Nevado B, Perez-Enciso M. Pipeliner: software to evaluate the performance of bioinformatics pipelines for next-generation resequencing. Molecular ecology resources. 2015;15(1):99–106.
Pearson WR. FASTA: Improved tools for biological sequence comparison. Proc Natl Acad Sci. 1988;85(8):2444–8.
Popov I, Nenov A, Petrov P, Vassilev D. Bioinformatics in Proteomics: A Review on Methods and Algorithms. Biotechnol Biotechnol Equip. 2014;23(1):1115–20.
Riaz T, Hor HL, Krishnan A, Tang F, Li KB. Web Allergen: a web server for predicting allergenic proteins. Bioinformatics. 2005;21(10):2570–1.
Saha S, Raghava GP. AlgPred: prediction of allergenic proteins and mapping of IgE epitopes. Nucleic Acids Res. 2006;34(Web Server issue):W202–9.
Srinivasan SM, Vural S, King BR, Guda C. Mining for class-specific motifs in protein sequence classification. BMC Bioinformatics. 2013;14:96.
Whelan FJ, Yap NV, Surette MG, Golding GB, Bowdish DM. A guide to bioinformatics for immunologists. Front Immunol. 2013;4:416.
Wu FX, Ressom H, Dunn MJ. Focus on bioinformatics in proteomics. Proteomics. 2013;13(2):219–20.
Yu Guoxian, Rangwala H, Domeniconi C, Zhang Guoji, Zhang Zili. Predicting Protein Function Using Multiple Kernels. IEEE/ACM Trans Comput Biol Bioinform. 2015;12(1):219–33.
Zhang L, Huang Y, Zou Z, He Y, Chen X, Tao A. SORTALLER: predicting allergens using substantially optimized algorithm on allergen family featured peptides. Bioinformatics. 2012;28(16):2178–9.

Author's Biography

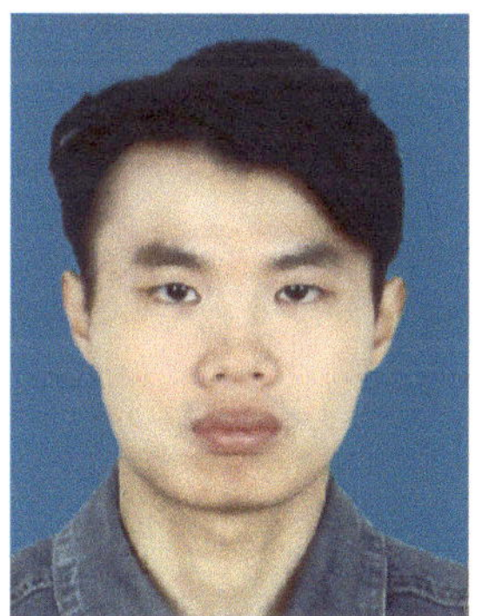

Yuyi Huang is a research associate working at Guangdong Provincial Key Laboratory of Allergy and Clinical Immunology, the State Key Laboratory of Respiratory Disease, the State Key Clinical Specialty in Allergy of the Second Affiliated Hospital of Guangzhou Medical University. Email: chvipdata@126.com.

He got his master degree of Immunology and Molecular Biology from Guangzhou Medical University in 2010. He is currently working at Guangzhou Medical University. He has experience of alleviation of the allergenic potential of allergens. His main research is focused on allergen bioinformatics. He has developed an allergen database ALLERGENIA and an allergen predicting software SORTALLER.

Ailin Tao is a Professor of Guangzhou Medical University, Director of Guangdong Provincial Key Laboratory of Allergy and Clinical Immunology, Principal Investigator of the State Key Laboratory of Respiratory Disease, Deputy Director of the State Key Clinical Specialty in Allergy of the Second Affiliated Hospital of Guangzhou Medical University, Member of the State Committee for Transgenic Safety Assessment, Standing Committee Member of Allergy Branch of Guangdong Medical Association, Member of Guangdong Provincial Committee for Transgenic Safety Assessment, and Master Tutor.

Prof. Ailin TAO earned his doctorate degree from the State Key Laboratory of Crop Genetic Improvement of Huazhong Agricultural University in 2002, followed by a postdoctoral training at Postdoctoral Station of Basic Medicine in Shantou University Medical College, majoring in allergen proteins. Recent studies of his lab mainly focus on allergy bioinformatics, allergy and clinical immunology, diseases models such as allergic asthma, allergic rhinitis, infection and inflammation induced by allergy, and inflammatory and protracted diseases caused by antigen or superantigen. He has accumulated some experience in the allergology field including the mechanism on immune tolerance, allergy triggering factors and chronic inflammation pathways, and allergenicity evaluation and modification for food and drugs. He proposed some new concepts such as "Representative Major Allergens," "Allergenicity Attenuation" of immunotoxin and allergens, and "broad-spectrum immunomodulator." He put forward the theoretical hypothesis of "Balanced Stimulation by Whole Antigens." He focuses on the diagnosis of allergic disease, and the allergenicity medical evaluation and modification for food and drugs. He constructed the method system for prediction, quantitative assessment, and modification in parallel on allergenicity epitopes, which has been applied for more than 20 cases of allergens. He has developed a bioinformatics software for allergen discriminating, SORTALLER (http://sortaller.gzhmu.edu.cn), which performed significantly better than other existing software and reached a perfect balance with high specificity (98.4 %) and sensitivity (98.6 %) for discriminating allergenic proteins from several independent datasets of protein sequences of diverse sources, also highlighting with the Matthews correlation coefficients as high as 0.970, fast running speed and rapidly predicting a batch of amino acid sequences with a single click. The software has been frequently used by researchers from lots of institutions in more than 20 Chinese cities and over 30 countries worldwide, thus becoming the TOP 1 in this field. He has set up an allergen database ALLERGENIA (http://allergenia.gzhmu.edu.cn), which has obvious international advantages in five aspects: wide allergen coverage, nonredundancy, astringency, accuracy, and friendly and usable analysis functions.

Chapter 13
Strategies for the Modification and Evaluation of Allergenicity

Zhaoyu Liu, Huifang Chen and Ailin Tao

Abstract Allergic disease is a global public health problem with increasing attention being paid to the evaluation and modification of allergenicity. The major aim of this chapter is to provide an overview of the field of allergenicity evaluation and modification focusing on bioinformatics evaluation, in vitro evaluation, and in vivo evaluation, which are the most commonly used methods for allergenicity evaluation, followed by information on allergenicity modification, which is required to reduce the allergenicity of the allergen proteins. Site-directed mutagenesis and error-prone PCR, which are used to modify allergenicity, are elucidated and summarized in this chapter. The field of allergenicity evaluation and modification is expected to lay the foundation for the future safe application of allergen reagents.

Keywords Allergenicity · Evaluation · Site-directed mutagenesis · Animal model · Histamine

13.1 Introduction

With advances in society and changes in dietary habits, the incidence of allergic diseases has been continually increasing. Epidemiological surveys show that 20–30 % of the global population is affected by a variety of allergic diseases to

Z. Liu · H. Chen · A. Tao (✉)
Guangdong Provincial Key Laboratory of Allergy and Clinical Immunology,
The State Key Clinical Specialty in Allergy, the State Key Laboratory of Respiratory Disease,
The Second Affiliated Hospital of Guangzhou Medical University,
250# Changgang Road East, Guangzhou, China
e-mail: Aerobiologiatao@163.com

Z. Liu
e-mail: sysu1924@gmail.com

A. Tao and E. Raz (eds.), *Allergy Bioinformatics*, Translational Bioinformatics 8,
DOI 10.1007/978-94-017-7444-4_13

different degrees (Furuta et al. 2005). Allergic diseases are chronic diseases caused by a wide range of factors including foods, dust mites, pollens, drugs, etc. (Geier et al. 2011; Scurlock and Burks 2004). Most allergic diseases are induced by closely related proteins or protein components. Therefore, bioinformatics and molecular biology methods used to evaluate the allergenicity of proteins and to reduce or even eliminate their allergenicity by site-directed mutagenesis have important practical significance and application value. The following are areas of particular importance for human health as it relates to allergic diseases:

1. **Reducing the incidence of allergic disease**

Evaluation of the allergenicity of proteins can help the allergic patient to reduce or completely avoid exposure to the sensitizing allergens and, thus, reduce the triggering of allergic diseases (Marini et al. 1996). Our meta-analysis demonstrated that allergen avoidance for newborns did not reduce the subsequent incidence of allergic diseases but did significantly reduce the incidence of asthma and wheezing in high-risk infants. However, previously sensitized patients who had reduced their exposure to known allergens did not subsequently show improvement in their lung functions. Hence, we argued that allergen avoidance may not always be successful in preventing allergic reactions (Wu et al. 2014).

2. **Improving the safety and efficacy of allergen-specific immunotherapy**

Allergen-specific immunotherapy is not only a very effective treatment in reducing allergic symptoms to a particular allergen (Bublin et al. 2010), but also provides a unique opportunity to prevent the development of allergic reactions to other potential allergens in the long-term course of the disease (Mobs et al. 2010). Traditional immunotherapeutic reagents are often crude extracts with an unknown composition. Because the use of this type of reagent could induce severe side effects and even life-threatening anaphylaxis due to their complex and unknown components, hypoallergenic allergen derivatives with reduced allergenicity have been recombinantly engineered to reduce side effects during allergen-specific immunotherapy (SIT).

3. **Increasing the safety and efficacy of drugs in clinical applications**

By evaluating and modifying the allergenicity of protein drugs (such as immunotoxins, etc.) early in their development, increases in the safety and security of the drug can be made well before any use in clinical applications, thus reducing the number of adverse effects and risks associated with drug development (Dean et al. 1994; Galbiati et al. 2010).

4. **Promoting the safety of Genetically Modified (GM) food**

In the field of agriculture, improvements to food crops focus on certain excellent agronomic traits, e.g., disease resistance, insect resistance, stress resistance, and improvement of the food quality. Some as yet unidentified transforming genes could be similar to allergen genes and the products of these transgenes could eventually enter into the human world or be consumed as food, so the safety of transgenes is a growing public concern (Goodman and Tetteh 2011; Kuiper et al. 2001). In 1996, with the awareness that the *2S* gene can increase protein content, the agriculture company Pioneer Hi Bred transformed the *2S* gene into soybeans. However, immunological identification results showed that the 2S protein is an important allergen and that this transgenic soybean would likely cause allergic reactions (Nordlee et al. 1996; Saalbach et al. 1994) and, thus, this transgenic soybean or its products cannot be publicly sold for consumption. Therefore, evaluation and modification of the allergenicity of all transgenic genes in the GM process not only can avoid the potential allergenic problem, but also can improve the safety of GM foods.

The strategy for assessing protein allergenicity and the subsequent modification of protein allergenicity as discussed in this chapter is as follows: first, evaluate the allergenicity of the proteins using bioinformatics and/or experimental methods; second, predict the allergic epitopes and the modification sites through bioinformatics methods and obtain the mutant sequences that maintain the dimensional structure and biological activity but reduce the allergenicity; third, modify the allergen proteins through site-directed mutagenesis and re-evaluate their allergenicity.

13.2 Evaluation of Protein Allergenicity

Allergic disease is a global public health problem, with allergen detection and evaluation increasingly receiving much attention (Pawankar 2014). The allergenicity evaluation of the candidate proteins can be achieved by sequence alignment with allergen (Chrysostomou and Seker 2014; Kimber et al. 2003; Thomas et al. 2005; van Esch et al. 2011). Sequence alignment, using computer technology and allergen databases, identifies and compares the sequence identities and positivities between the target protein and the allergen, and then predicts the potential allergenicity of the protein (Gibson 2006). In vitro evaluation methods include enzyme-linked immunosorbent assay (ELISA) and histamine provocation tests (Thomas et al. 2004). In vivo evaluation methods include challenge tests in animal models (Aldemir et al. 2009; Ladics et al. 2010). Although there are many methods to evaluate protein allergenicity, there is not yet just one specific method that can effectively evaluate protein allergenicity. Therefore, development of a combination of evaluation methods is necessary to establish a fast, efficient, and accurate allergenicity evaluation procedure.

13.2.1 Bioinformatics Methods for Evaluation of Allergenicity

With the rapid increase in the identification of allergen protein sequences and related data, the application of bioinformatics to the field of allergy is becoming more and more important (Brusic and Petrovsky 2003). Using bioinformatics methods, evaluation of the allergenicity of the target proteins can be quickly performed through identity predicting and cross-reactivity predicting (Aalberse 2007; Schein et al. 2007). This section will give a synopsis of the use of bioinformatics in allergenicity evaluation.

Allergen proteins always share some common characteristics, such as digestion resistance, resistance to acid/alkali, thermal stability, and certain specific amino acid sequences (Brusic and Petrovsky 2003). Allergen proteins have conserved structures and functions and, therefore, analysis of sequence similarities between target proteins and known allergens could be an efficient method for preliminary evaluation of the allergenicity of proteins.

The most commonly used protein allergen databases are the International Union of Immunological Societies (IUIS) (King et al. 1994), Allergome (Mari and Scala 2006), Allergenia (http://allergenia.gzhmu.edu.cn), the Structural Database of Allergenic Proteins (SDAP) (Hileman et al. 2002; Ivanciuc et al. 2003), and Allergenonline (Richard et al. 2007). These databases use a local alignment strategy and are the commonly used sequence comparing methods (Do and Katoh 2008; Mount 2007). BLAST, with a fast search speed, is more effectively used on searches of large databases. FASTA searches are more sensitive to local matches, and therefore, FASTA is the preferred method to use when evaluating protein allergenicity.

In recent years, in combination with artificial neural network algorithms, a variety of software for allergen identification have been developed based on FASTA and BLAST sequence alignment. FAO/WHO, EVALLER, AlgPred (amino acid), AlgPred (dipeptide), AlgPred (ARPs BLAST), AllerHunter, and SORTALLER are the mostly common used allergen discrimination software (Martinez Barrio et al. 2007; Muh et al. 2009; Saha and Raghava 2006; Zhang et al. 2012). The sensitivity and specificity of these allergen discrimination softwares are shown in Table 13.1. The FAO/WHO is a method recommended by the UN Food and

Table 13.1 Comparison of different prediction methods

Methods	SE (%)	SP (%)	ACC (%)	MCC
FAO/WHO[a]	99.2	9.6	54.4	0.198
EVALLER[b]	86.6	99.0	92.8	0.863
AlgPred (amino acid)[c]	92.4	80.2	86.3	0.731
AlgPred (dipeptide)[c]	88.8	88.2	88.5	0.770
AlgPred (ARPs BLAST)[c]	81.8	98.0	89.9	0.809
AllerHunter[d]	82.2	99.2	90.7	0.826
SORTALLER[e]	98.6	98.4	98.5	0.970

Note: *SE* sensitivity; *SP* specificity; *ACC* accuracy; *MCC* Matthews correlation coefficients
[a]Making use of the script file in accordance with the FAO/WHO guidelines (2001)
[b]http://www.slv.se/en-gb/Group1/Food-Safety/e-Testing-of-protein-allergenicity/e-Test-allergenicity/
[c]www.imtech.res.in/raghava/algpred/
[d]http://tiger.dbs.nus.edu.sg/AllerHunter/
[e]http://sortaller.gzhmu.edu.cn/

Agriculture Organization and the World Health Organization. This method has two criteria for identifying potential protein allergens: (1) the presence of at least six contiguous amino acids identical to a known allergen or (2) more than 35 % identity in the amino acid sequence of the expressed protein, using a window of 80 amino acids and a suitable gap penalty (using Clustal alignment programs or equivalent alignment programs) (Hviid 2007; Ladics 2008). Based on prior allergen identifying software, our lab has developed a new algorithm and bioinformatics evaluation software called SORTALLER (Zhang et al. 2012). SORTALLER uses a neural network algorithm with the advantages of having a high running speed and high accuracy as well as being more convenient, since multiple sequences can be evaluated at the same time. The sensitivity and specificity of SORTALLER are both more than 98 %, and it has now been linked to the Internet (http://sortaller.gzhmu.edu.cn) for public use. Regarding the specific principles, evaluating criterion, and application of these bioinformatics evaluating methods, please refer to Chap. 14 in this book.

13.2.2 Laboratory Methods for the Evaluation of Allergenicity

Although protein allergenicity can be quickly evaluated by bioinformatics methods, the accuracy of bioinformatics assessments depends on the allergen protein database and the corresponding algorithm and is affected by relatively high rates of false positives and false negatives (Thomas et al. 2005). In addition, current sequence similarity analyses mainly focus on comparing the primary amino acid sequence and predictions using conformational epitopes are relatively unreliable. Therefore, further evaluation of allergenicity using laboratory methods should also be done to obtain more accurate results.

Laboratory allergenicity evaluations are divided into serological and cytological methods (Thomas et al. 2009). Serological methods are mainly based on the binding capacity of the protein with specific serum immunoglobulin (IgE, IgG, etc.), while cytological methods depend on the production of inflammatory mediators after stimulation of the immunocyte with the protein (Thomas et al. 2009). In addition, these laboratory assessments are divided into in vitro evaluation and in vivo tests (Thomas et al. 2009). In vitro assessments include ELISA, mast cell and whole blood stimulation tests, etc. In vivo methods use the allergenic proteins in animal models or to directly stimulate the human body and then evaluate the allergenicity of the protein based on the presence or absence of allergic symptoms and/or the levels of specific IgE. In vivo allergenicity assessments include skin tests, provocation tests, and animal models. Provocation tests and skin tests use a small amount of allergen protein applied directly to humans followed by observation of symptoms to assess the potential allergenicity of the target protein. Animal models have been developed to establish a corresponding model for assessing the potential allergenicity of the target protein and to evaluate the allergenicity of the proteins based on symptoms and production of antibody and cytokines. Rats, mice, and guinea pigs are the commonly used animals because they are similar to humans in their immune responding mechanisms. Furthermore, they have advantages such as that they are small and inexpensive and have short breeding cycles.

1. **IgE binding assays for the evaluation of allergenicity**

IgE is one of five immunoglobulin types in human serum and its concentration is normally very low, about 10–100 U/mL; however, patients with allergic or parasitic diseases have significantly higher levels. In 1966, Johansson (Sweden) and Ishizakas (Japan) first isolated IgE from ragweed allergic patients' serum and demonstrated the relationship between IgE and allergic diseases. IgE is mainly synthesized by B cells located in the mucosa-associated lymphatic tissue of respiratory and gastrointestinal tracts and is regulated by the levels of T_H1/T_H2 cytokines in vivo. It is widely accepted that an imbalance between T_H1/T_H2 cells leads to the development of allergic diseases. T_H2 cells synthesize high levels of interleukin 4 (IL-4), IL-5, and IL-13, which promote B cells to release allergen-specific IgE (Jeon et al. 2015). Detection of allergen-specific IgE in biological fluid is one of the methods of evaluation of proteins before and after modification. Common methods for detection of allergen-specific IgE are described below:

(a) **Enzyme-Linked Immunosorbent Assay (ELISA)**

Purified allergen is coated onto high binding plates to capture allergen-specific IgE in biological fluids. Among the most commonly used methods are the sandwich ELISA, capture ELISA, and competition ELISA. A sandwich ELISA measures allergens between two layers of antibodies (captured and detection antibodies). The allergen to be tested must contain at least two different antigenic sites capable of binding antibodies. The advantage of the sandwich ELISA is that the sample does not have to be purified before analysis, and the assay can be very sensitive (up to 2–5 times more sensitive than direct or indirect ELISAs).

(b) **Western Blot (WB)**

Western blot (WB) is a method that combines the advantages of electrophoresis with the immunological method of antigen–antibody binding. The samples tested can be crude extracts from cells or tissue or recombinant proteins that were separated though SDS-PAGE gel electrophoresis and then transferred to a NC or PVDF membrane. Specific antibodies are then used to detect the presence of the corresponding antigen. Non-specific antigen-binding sites can be blocked using agents such as bovine serum albumin (BSA) before adding the probe antibody. This method is now widely used in the study of gene expression at the protein level, the detection of antibody activity, the evaluation of allergenicity of recombinant allergenic proteins, and for screening of specific antibodies in patient serum. The disadvantage of WB is that it is a time-consuming operation and is, therefore, often combined with the ELISA method to improve efficiency.

(c) **The ImmunoCAP system**

The ImmunoCAP system was developed and is manufactured by the Swedish company Phadia and is an automatic detection system for the diagnosis of allergic diseases. This system measures total and specific IgE in human serum. It has been widely recognized worldwide and was named as "the gold standard in allergy detection" by numerous allergy experts because of its high specificity and sensitivity.

(d) **Radioallergosorbent Test (RAST)**

After Johansson's group isolated IgE from allergic patients in 1967, they introduced the radioallergosorbent test (RAST) as a method to measure IgE in serum (Gleich and Jones 1975). The main steps include adding test serums and control serums to a microtiter plate that has been coated with purified allergen, followed by the addition of an isotope-labeled anti-IgE antibody. Bound radioactivity is detected and used to calculate the amount of specific IgE by a standard curve or judged as positive when the radioactivity of the specimen is higher than the healthy control by 3 times the standard deviation (3SD). There is a high correlation between RAST and skin testing and physicians are able to use both of these methods in the diagnosis and treatment of allergic disorders (Tandy et al. 1996). However, there are several disadvantages in the RAST method including costly reagents, time-consuming, risk of radioactive contamination/environmental pollution, and short expiration date of the radioisotopes, as well as the complications resulting from different kits using different reference serums that are inaccurate when compared with each other and that are affected by high concentrations of specific IgG in the serum.

2. **Basophil/Mast cell activation tests for the evaluation of allergenicity**

(a) **Whole Blood Histamine Release Assay (WBHR)**

Histamine is an important allergic inflammatory mediator, found pre-stored in the metachromatic granules within mast cells and basophils, and can be quickly released after stimulation. Mast cells reside in various tissues of human organs, while basophils are found in the blood (Jensen et al. 2014). The whole blood histamine release assay (WBHR) is an important screening tool for the detection of allergens or for evaluation of the allergenicity of allergens, and has a high correlation with specific IgE assays and skin prick testing (SPT). WBHR is not affected by high levels of total IgE or by the presence of anti-allergic drugs like β-agonists, steroids, or anti-histamines. When the patient's skin is not suitable for SPT, or in the case of infants that cannot be tested by SPT, the WBHR is a simple method and a valuable diagnostic tool for screening allergens (Nolte et al. 1990). The required sample volume is small, only 3 mL of blood is needed for complete screening against the common allergens. The advantage of WBHR is not only in clinical testing, but also in scientific research in the evaluation of allergenic proteins. All recombinant proteins, mutant proteins, and vaccines could be evaluated for their allergenicity using this method.

(b) **Mast Cell Activation Test for the evaluation of allergenicity**

Mast cells are secretory cells that resident in multiple tissues and organs and are best known for their involvement in many physiological and pathological processes, including allergic reactions, renovation of damaged tissue, chronic inflammation, tumor immunity, and others (Kinoshita et al. 1999). Mast cells play an important role in allergic and inflammatory reactions such as rhinitis, asthma, urticaria, and anaphylactic shock. Mast cells can be activated by re-exposure to allergens and release a series of immunoregulatory mediators including histamine, β-hexosaminidase, tryptase, chymase, vascular endothelial growth factor, and several cytokines and chemokines. These mediators initiate early- and late-phase inflammatory and allergic responses including increased vascular permeability, tissue swelling, bronchial contraction, leukocyte recruitment, glandular secretion, etc. Activation of mast cells (primary mast cells or mast cell lines) in vitro is an important method for the detection of allergies and for the assessment of the allergenicity of allergens with/without modification. Primary mast cells can be isolated from surgical tissue, obtained from tonsillectomy, circumcision, and resections of lung, colon, and other cancers. The characteristics of primary mast cells are their close proximity to the environment of the physiological condition but tissue-resident mast cells usually have large heterogeneity. Enzymatic digestion is the method frequently used for isolating tissue mast cells, often resulting in cells with low viability and less than desired purity. Mast cells can be derived from $CD34^+$ stem cells isolated from bone marrow or umbilical cord blood by the addition of certain cytokines (Schmetzer et al. 2014) and these mature mast cells can be used in allergen stimulation assays. This method can overcome the problems resulting from the isolation of primary tissue-resident mast cells but it requires a lengthy production time and is costly (Holm et al. 2008). So in some regard, mast cell lines possess some advantages for use in in vitro activation tests for allergenicity assessment. Currently, the most commonly used mast cell lines include Laboratory of Allergic Diseases 2 (LAD2) (Kirshenbaum et al. 2014), Human Mast Cell-1 (HMC-1) (Xia et al. 2011), and Rat Basophil Leukemia (RBL-2H3) (Wan et al. 2014).

Mast cells are packed with 50–500 secretory granules, which account for over 40 % of its cell volume, and are "ready to go" immediately after cells are activated by an appropriate stimulus. The granules are pre-formed and newly synthesized in the Golgi, and, when activated, the secreted products are released extracellularly to cause inflammation. Many substances can activate mast cells and cause the degranulation reaction such as allergens, IgE, ionophore, compound 48/80, substance P, C5a, C3a, and so on. The mast cell stimulation assay is an important step in assessing the potential allergenicity of GM allergens by analyzing the inflammatory mediators in the mast cell supernatants and comparing the supernatants stimulated with wild-type (WT) allergen.

(c) **Detection methods for various basophil/mast cell mediators**

Histamine Detection

Histamine is derived from histidine decarboxylation, a reaction catalyzed by L-histidine decarboxylase. It is a type of biogenic amine with a small molecular weight and a short half-life. Histamine is found pre-formed in granules in mast cells and basophils. It can be released immediately after stimulation with allergens and causes bronchial smooth muscle contraction, increased mucus secretion, telangiectasia, increased vascular permeability, and provokes allergic diseases. Both IgE-mediated and non-IgE-mediated stimuli can cause histamine release. Mast cells can also be activated to release histamine via toll-like receptors (TLRs) but this depends on the cell phenotype (Meng et al. 2013). Quantification of histamine can be performed in several ways and many commercial ELISA kits are ready to use and accompanied by clear manufacturer's guidelines. Another method to detect histamine was previously described (see WBHR) and is dependent on glass fiber-coated plates and fluorometric detection using a Hisreader-501 device. Unfortunately, the short half-life and instability of histamine make it difficult to detect. Also, the complexity of the ELISA test for the quantitative analysis of histamine and the narrow detection range make these assays not ideal, and thus, other more sufficient tools are needed for the development of valid mast cell activation tests.

β-hexosaminidase Detection

β-hexosaminidase is one of the main mediators released by mast cells and basophils after stimulation. The release of β-hexosaminidase is considered as a common indicator of allergic disease. Reagents required for the β-hexosaminidase assay can be made in the laboratory except for the substrate 4-nitrophenyl N-acetyl-β-D-glucosaminidase (Huang et al. 2015a). This method is also commonly used to validate an IgE-mediated allergic response.

Tryptase Detection

Tryptase is the most abundant serine protease and has been considered as a marker for mast cell/basophil activation. The serum level of tryptase is normally less than 12 μg/L and increases in anaphylactic reactions; however, negative results do not exclude the possibility of allergic disease. Up to now, the tryptase detection assay is not significantly used for screening food allergy. Mast cell tryptase is a tetrameric neutral serine protease with a molecular weight of 134 KDa and is composed of four monomers with a molecular weight of 32 KDa. Tryptase can be divided into two subtypes: α-tryptase and β-tryptase that share approximately 90 % sequence identity between each other. α-tryptase is mainly secreted by inactivated mast cells and presents as proenzyme in the blood, while β-tryptase is located in mast cell secretory granules and is released when mast cell activation occurs. β-tryptase is released from activated mast cells in parallel with histamine when systemic anaphylaxis occurs, its level peaking at 15–120 min with a half-life of 90–150 min, while histamine levels peak at 5 min with a half-life of 15–30 min. There is a longer window of detection time for β-tryptase than for histamine, which is significant for clinical investigations. High concentration of β-tryptase can be found in biological fluids collected 1–6 h after a suspected allergic reaction (Payne and Kam 2004).

3. Animal Models for the evaluation of allergenicity

Considering the increasing number of people suffering from allergic diseases, the development of preclinical approaches to evaluate the potential allergenicity of proteins has become more and more urgent. In the past decades, we have seen the evolution of animal models that can accurately simulate the human body response. Animal models are a safe and effective in vivo experimental tool that can be used to evaluate the allergenicity of proteins by testing specific antibodies and cytokines in serum, bronchoalveolar lavage fluid (BALF), nasal irrigation fluid, and other biological fluids, as well as for determining the severity of the allergic reactions. By far the most commonly used animal models include those for asthma, AR, food allergy, and atopic dermatitis (AD), and the most commonly used species are guinea pig, mouse, and rat.

(a) Animal species used for the evaluation of allergenicity in vivo

Early in the 1990s, European and American scholars started to use animal models in food allergy studies. Animals such as guinea pigs, mice, and rats have been used for assessment of the allergenicity of foods. In a study of cow's milk allergy, guinea pig was the appropriate model using oral sensitization (Kitagawa et al. 1995). Guinea pigs are also a suitable species that can be used in experimental models for shrimp and ovalbumin allergy. Nearly 100 % of sensitized guinea pigs can be induced to develop intestinal symptoms of food allergy after challenge. Due to the high level of homology in gene expression between murine and human, mice have been commonly used in allergy research. Rats also have many advantages compared with other species, especially in toxicity testing (Penninks and Knippels 2001). Due to their organ functions highly similar to humans, as well as readily to operate, Chinese miniature pig models have also been used for food/protein allergenicity assessment (Huang et al. 2010, 2015b).

Guinea pigs are one of the most widely used experimental animals for the study of allergic asthma/AR. The advantages of using guinea pigs are that they can be easily sensitized, with a stronger reaction than other types of animals after allergen challenge, and they develop type I hypersensitivity (Sutovska et al. 2015). Stable high titers of specific IgG and IgE are easily induced and symptoms of both of immediate and delayed reaction are seen. Their pathology and asthma symptoms might be similar to humans, but their reaction to OVA shows large individual differences. After sensitization, a portion of guinea pigs develop anaphylactic shock and die after OVA challenge, whereas other guinea pigs may not appear to have any or only minor asthmatic reactions. In order to prevent this occurrence, antihistamine drugs, such as diphenhydramine and pyrilamine, are usually given before the stimulation in order to prevent the death of guinea pigs from overreactions. Guinea pigs are also suitable for developing food allergy models using oral sensitization, followed by re-exposure to the same food allergens either by injection or taking orally. Outcomes include death or stress shock, indicating that guinea pigs become sensitized in a way similar to humans, thus, they can be a good species to use in animal models of milk allergy.

Mice are also often used to build classical models of asthma/AR and the commonly used strains are BALB/C, CBA/J, and C57BL/6. Owing to their small size and short breeding cycle, mice can save greatly on the supply of allergen and the cost of these experiments is relatively low for mouse models. In addition, with the emergence of a large number of related immunology and molecular biological reagents in recent years, and due to the more clearly understood immunologic and genetic background of mice, their use in animal models is widespread. Both BALB/c and C57BL/6 are the most commonly used species and are useful for allergic/infectious/transgenic animal models of asthma/AR, as well as for food allergy models. BALB/c mice are easily induced to produce airway hyperresponsiveness (AHR), easily sensitized with OVA, and produce high titers of IgE. It is not easy to induce AHR in C57BL/6 mice, but they are able to be sensitized to house dust mite (HDM) and, thus, can be used to create an allergic asthma model induced by HDM. Mice can be sensitized with allergens such as peanut lectin, ovalbumin (OVA), and BSA by intraperitoneal injection (i.p.), or orally sensitized with Cholera toxin (CT) as an adjuvant together with food allergens such as peanuts and milk, in which case the mice produce high levels of specific IgG1 and IgE. $Al(OH)_3$ as an adjuvant to OVA or other food allergen proteins can be used in food allergy models. Mice are also often used for the construction of transgenic animal models in studying skin allergy diseases.

Rats have some advantages when developing allergic-related animal models. First, pure strains are widely available and vigorously reproduce. Second, their cost and related reagents are relatively inexpensive. Three, sufficient amount of specimen can be collected. Their late-phase reaction and IgE-mediated allergic reactions are similar to humans. After stimulation, the symptoms of the immediate-phase and late-phase reaction of airway hypersensitivity are easily observed. The time of occurrence of the late-phase reaction more closely resembles that of clinical patients. Rats are also a suitable animal model for investigating food allergy: their size is suitable for kinetic analysis of serum-specific antibodies on a single individual; they can be orally sensitized without adjuvant; and sensitized rats can show allergy symptoms after challenge similar to humans. BN rats are a high immunoglobulin (especially IgE) response strain and are widely used in models of food allergy, which can provide much useful information when evaluating any potential allergenicity of novel foods or proteins. BN rats have been used in food allergy models both with and without the use of adjuvant (Abril-Gil et al. 2015).

Animal models, such as those described using guinea pigs, mice, and rats, have been commonly used for the assessment of the allergenicity of proteins, but an ideal animal model has not yet been developed (Abril-Gil et al. 2015; Piacentini et al. 2003; Vinje et al. 2009). Thus, investigators should choose a suitable animal model according to their research purposes.

(b) **Animal models for the evaluation of allergenicity**

(i) **Models of allergic rhinitis/asthma**

Asthma is a serious respiratory disease threatening human health that can develop from rhinitis, and together they have been described as "one airway, one disease." The morbidity of AR/asthma is rising, with characteristic recurrent and protracted symptoms that seriously affect the quality of life of patients as well as increasing financial burdens. Rhinitis/asthma is generally considered as chronic allergic airway inflammation and involves a variety of inflammatory cells, inflammatory mediators, and cytokines. Due to the complex processes leading to the occurrence and development of rhinitis/asthma, there are various restrictions in studying the pathogenesis of human disease in clinical practice and clinical trials (Rice et al. 2008). Prediction of the outcome of clinical treatment requires animal models that can simulate human disease. Rhinitis/asthma-related animal models are commonly used to explore the pathophysiology, etiology, pathogenesis, and treatment methods of allergic respiratory diseases, as well as to evaluate the allergenicity of aeroallergens such as pollen, mite, bacteria, animal fur, and so on. Different species of animals differ in their sensitivity to different allergens, and rodents, including guinea pigs, rats, and mice, are the most commonly used animals in building rhinitis/asthma-related animal models.

Although there are various limitations in using each type of animal as has been described, nevertheless, they play an important role in the study of AR/asthma and the evaluation of protein allergenicity. To validate the success of our animal models, we record and calculate the scores of sneezing, scratching, and other nasal symptoms after allergen challenge. After the models were successfully built, the animals are challenged with allergen. After 24–48 h, animals are then euthanized to collect specimens for various purposes: serum collection for detection of specific IgE and associated inflammatory factors, collection of nasal lavage fluid/BALF for counting the total number of nucleated cells and eosinophils, as well as testing of the supernatant for the detection of cytokines, collection of nasal mucosa tissue/lung tissue for observation of the infiltration of eosinophils and mast cells, and observation of thickening of bronchial smooth muscle by HE staining.

(ii) **Models of food allergy**

Food allergies are adverse to human health and the mechanism is still unclear. An animal model of food allergy is mainly used to assess the potential allergenicity of novel foods/proteins, especially GM foods (Boyce et al. 2011).

Clinical manifestations of food allergy differ widely between individuals; from abdominal discomfort, diarrhea, itchy skin, lips, and mucosal edema to life-threatening laryngeal edema and anaphylactic shock. Exploration of the mechanism of food allergy by taking advantage of animal models can not only overcome the difficulties caused by individual differences, but also avoid human trials that cannot be ethically conducted. After the models are developed and used, various tests can be performed. For example, serum can be collected 3 h after allergen challenge for the detection of IgE, IgG1, IgG2a, IL-4, IL-10, IFN-γ, etc., and intestinal mucosa tissue can be collected for HE staining to observe changes of eosinophils and mast cells. Spleen can be used for the preparation of a single-cell suspension that can be stimulated with specific allergens, followed by the measurement of IL-4, IL-10, IFN-γ, and other cell cytokines in the supernatant.

(iii) **Models of atopic dermatitis (AD)**

AD is a chronic and relapsing, pruritic inflammatory skin disorder, with complex interactions between genetic and environmental factors playing an important role in its development (Bieber 2008). However, the precise pathogenesis of AD remains to be fully elucidated. The clinical manifestations of AD include extremely itch skin, with scratching of the affected areas causing skin irritation, cracking, leaching of clear liquid, and even skin hardening, scaling, etc. A high incidence of AD occurs in infants and pre-school children. Dust mites and animal fur are important causes of AD. An animal model of AD is a very useful tool for evaluating the allergenicity of proteins with and without modification. BALB/c mice have often been used in animal models of AD induced by applying 2,4-dinitrochlorobenzene onto hairless dorsal skin.

13.3 Strategies for the Modification of Protein Allergenicity

After evaluation of protein allergenicity using bioinformatics and laboratory methods, allergenicity modification to reduce or even eliminate the allergenicity of the allergen proteins is necessary. Currently, the main strategy of protein allergenicity modification is to alter the coding sequences of the allergen proteins (Rabjohn et al. 2002). Certain steps are always involved in protein allergenicity modification. First, locating the high allergenicity sites on the allergen proteins using experimental and/or bioinformatics methods; second, predicting the specific amino acid sites that need to be modified using bioinformatics methods; and third, mutating the specific sites to reduce, or even completely eliminate, the allergenicity of the allergen proteins in order to ensure future security and safety in their application.

13.3.1 Epitope Prediction and Epitope Modification of Allergen Proteins

Allergen epitopes are the ultimate "culprit" in allergic reactions. Allergen epitopes can be either T cell epitopes or B cell epitopes depending on the different combinations of receptors. B cell epitopes can be linear epitopes or conformational epitopes, while T cell epitopes are generally linear epitopes. Experimental methods and forecasting methods are the usual methods used for epitope mapping (El-Manzalawy and Honavar 2010; Friedl-Hajek et al. 1999; Malherbe 2009; Newell et al. 2013). The former methods have many advantages but also have many shortcomings, while the latter methods provide new epitope mapping strategies, with simple, convenient features, and are widely used in research and modification of allergen protein epitopes. For specific principles and methods about predicting T and B cell epitopes, please refer to Chap. 14 in this book. This section will describe the prediction and modification of allergen protein T cell epitopes.

1. **Prediction of T cell epitopes using bioinformatics methods**

In recent years, a number of T cell epitopes prediction methods have been developed, of which the machine learning method is widely used due to its high efficiency and accuracy. Using machine learning algorithms to predict allergen epitopes includes data collection and processing, model building, and parameter optimization. Some common T cell epitope prediction softwares based on these algorithms are NetMHCII2.2 (http://www.cbs.dtu.dk/services/NetMHCII/) (Nielsen et al. 2007), Rankpep (http://bio.dfci.harvard.edu/Tools/rankpep.html) (Reche et al. 2002), and ProPred-I (http://www.imtech.res.in/raghava/propred1/) (Singh and Raghava 2003). The NetMHCII2.2 server predicts the binding of random peptides to HLA-DR, HLA-DQ, HLA-DP, and mouse MHC class II alleles using artificial neuron networks, which has high accuracy, and is widely used for the prediction of T cell epitopes.

2. **Using NetMHC-II for the prediction and modification of allergen protein T cell epitopes**
   (a) Obtain the sequence of an allergen from NCBI and predict the affinity between the allergen and MHC-II molecules using the online software NetMHC-II (http://www.cbs.dtu.dk/services/). If the affinity score is above 0.5, then the epitope is predicted to be a dominant epitope and needs modification.
   (b) According to the basic principle of allergenicity modification, select eight amino acids (AA) before and after the definitive AA to carry out random AA replacement followed by analysis of the affinity of the newly modified allergen with MHC-II.

13.3.2 Strategies of Allergen Protein Modification Using Laboratory Methods

At present, the primary method for allergen modification is to change the coding sequence of the allergen protein, thereby reducing the allergenicity of the target protein. Site-directed mutagenesis technology and error-prone PCR are the most commonly used methods to modify the allergenicity of the allergen

(Hakkaart et al. 1998; Tsai et al. 2003). Site-directed mutagenesis can introduce aimed changes to DNA fragments (including base addition, base deletion, and base substitution) by overlap extension using PCR (Nohr and Kristiansen 2003). Site-directed mutagenesis is usually used to modify the allergenicity of allergen proteins through amino acid substitutions at targeted sites. Error-prone PCR is performed under artificially controlled conditions whereby changing the reaction conditions can increase the frequency of error-prone incorporation of nucleotides to achieve direct modification of the target protein (Pritchard et al. 2005). Error-prone PCR is usually used to modify allergen proteins in cases where their structures and high allergenicity sites are unknown. The allergenicity of the allergen proteins can be reduced or even eliminated using amino acid site-directed mutagenesis technology and error-prone PCR technology, thus laying a foundation for safe application of the allergen proteins.

1. **Using site-directed mutagenesis to modify allergen proteins**

(a) **The principle of site-directed mutagenesis**

Site-directed mutagenesis is a method that is used to make specific and intentional changes (usually advantageous variations) to the target-DNA sequence (which may be genomic or plasmid DNA) including base addition, base deletion, and base substitution (Nohr and Kristiansen 2003). Site-directed mutagenesis can quickly and efficiently improve the traits of target proteins and is the common method for modifying the allergenicity of allergen proteins. When site-directed mutagenesis is used to modify the allergenicity of the allergen proteins, an oligodeoxynucleotide, which contains the pre-designed site, is synthesized as a primer. Then, pairing the mutant primer with the DNA template, the oligodeoxynucleotide is elongated by DNA polymerase to complete the DNA replication. Subsequently, the double-stranded DNA, which contains one strand of WT and one strand of mutant, is transformed into the host cells and the mutant is screened using sequence technology.

(b) **Methods of in vitro site-directed mutagenesis**

Site-directed mutagenesis technology was developed in the 1980s after combining gene cloning technology and DNA chemical synthesis methods. At present, in vitro site-directed mutagenesis methods are divided into three main types: single-strand phage for site-directed mutagenesis, cassette mutagenesis, and PCR-mediated in vitro site-directed mutagenesis (Carter et al. 1985; Reikofski and Tao 1992; Wells et al. 1985). Single-stranded phage for site-directed mutagenesis uses a single-strand M13 DNA as a template for mutagenesis and an oligonucleotide containing the desired mutation as a primer for extension. Consequently, heteroduplex DNA forms, consisting of one parental non-mutated DNA strand and a mutated DNA strand, which is then transformed into competent bacteria, and the corresponding mutant is screened using DNA sequence technology. Cassette mutagenesis, unlike other methods, does not require primer extension using DNA polymerase. In this method, a segment of DNA that contains the mutation of interest is synthesized and then inserted into a plasmid that is digested by restriction enzymes followed by subsequent ligation of the synthesized oligonucleotide containing the desired mutations to the plasmid. Single-stranded phage for site-directed mutagenesis needs to use a M13 phage vector and has low mutation efficiency, while cassette mutagenesis has close to 100 % mutant efficiency, but is limited by the availability of suitable restriction sites flanking the site that is to be mutated. The limitation of single-stranded phage for site-directed mutagenesis and cassette mutagenesis can be overcome using PCR-mediated in vitro site-directed mutagenesis, which combines PCR amplification technology and cloning transformation technology in a simple operation with high efficiency. PCR-mediated in vitro site-directed mutagenesis is always used to carry out site-directed mutation of proteins in vitro.

This section mainly focuses on the application of PCR-mediated site-directed mutagenesis in the modification of the allergenicity of allergen proteins.

PCR-Mediated In Vitro Site-Directed Mutagenesis

PCR-mediated site-directed mutagenesis is an in vitro method for creating specific mutations in a known gene. Exponential amplification by PCR generates a segment containing the desired mutation to be isolated from the original plasmid by electrophoresis, which can subsequently be inserted into the desired plasmid using recombinant DNA techniques. There are various methods to do this (Hemsley et al. 1989; Ho et al. 1989; Jeltsch and Lanio 2002; Landt et al. 1990; Weiner et al. 1994). Here, we describe a general strategy of PCR-mediated site-directed mutagenesis in vitro. The basic principle of PCR-mediated site-directed mutagenesis is to place the mutation site toward one end of the segment whereby the primers are designed to generate the mutations. This method requires just a single PCR step, but still has an inherent shortcoming of requiring long oligonucleotide primers with an appropriate restriction site near the mutation site(s). Alternatively, adopt three or four primers, two of which could be non-mutagenic primers that cover two suitable restriction sites, and produce a segment that can be digested and ligated into the target plasmid. These methods need multiple PCR steps so that the final segment to be ligated can contain the required mutations.

Improvements to PCR-Mediated Site-Directed Mutagenesis

In order to acquire the mutated genes, conventional PCR-mediated point mutation always requires three or four primer pairs and multiple PCR amplification steps. However, the mutated sequence can be acquired from one pair of primers and one cycle of PCR amplification, as outlined in Fig. 13.1 (Carey et al. 2013). First, one pair of primers, which contains the desired mutant site and has partially overlapping regions, is required for synthesis. The PCR products with complementary sticky ends can be generated using the synthesized mutated oligonucleotides, annealing the PCR product, and transforming into *E. coli*, whereby the sequence with the desired mutations can be acquired after sequencing picked colonies.

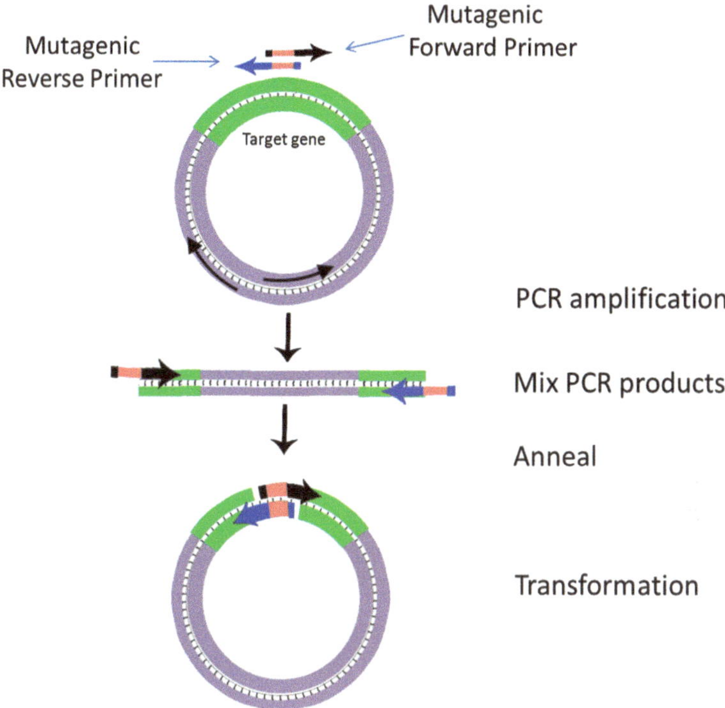

Fig. 13.1 The improvement of PCR-mediated site-directed mutagenesis (Carey et al. 2013)

The Application of PCR-Mediated In Vitro Site-Directed Mutagenesis in Protein Allergenicity Modification

PCR-mediated in vitro site-directed mutagenesis is a useful method for research on protein function such as activity sites, specificity, and allergenicity. In general, the following steps are always involved in allergenicity modification using PCR-mediated in vitro site-directed mutagenesis: (1) determining the specific sites to be modified; (2) designing mutant primers for PCR-mediated site-directed mutagenesis; (3) acquiring the mutated allergen sequence by amplifying the target allergen gene using the mutant primers; (4) annealing and transforming the mutant PCR products into host cells; and (5) verifying the mutant with the desired mutation by sequencing technology and evaluating the allergenicity of the mutant. In the past 10 years, many researchers have reported allergenicity modification using PCR-mediated in vitro site-directed mutagenesis. In summary, PCR-mediated site-directed mutagenesis is a simple, efficient, and cost-effective method for allergenicity modification. With the rapid development of hypoallergenic variants, PCR-mediated site-directed mutagenesis will become a powerful tool for the modification of allergenicity.

13.3.3 The Use of Error-Prone PCR in Protein Allergenicity Modification

1. **The principle of error-prone PCR**

Error-prone PCR is a method that can introduce random copying errors into any piece of DNA by utilizing an imperfect or sloppy reaction mixture (e.g., by adding Mn^{2+} or Mg^{2+} to the reaction mixture) (Pritchard et al. 2005). Generally, the replication of DNA by polymerase is extremely specific, with the difference in error-prone PCR being that the fidelity of the Taq DNA polymerase is modulated by alteration of the composition of the reaction buffer. When modifying allergen proteins using error-prone PCR, the coding sequence of the allergen proteins is used as a template and amplified under the error-prone conditions. The polymerase makes mismatches in the base paring during DNA synthesis that generates the introduction of errors into the newly synthesized DNA strand. By carefully controlling the reaction conditions, the frequency of the mismatched nucleotide bases can be regulated. This method has been proven useful for the generation of random mutations during the expression and screening process in a mutagenesis step.

2. **Reaction conditions for optimal error-prone PCR**

   (a) Using Taq DNA polymerase: It is critical that a Taq polymerase be used as it has naturally high error rate and does not have proofreading ability. The proofreading of nucleotide sequences is a feature that is found in many commercially available DNA polymerases (Biles and Connolly 2004; Keohavong and Thilly 1989; Peterson 1988); however, use of a proofreading DNA polymerase in error-prone PCR will result in the automatic correction of the mis-pairing nucleotides and any mutations that were incorporated during the replication will be lost.

   (b) Increasing Mg^{2+} concentration: In general, magnesium ions are an essential cofactor for the DNA polymerase in PCR reaction systems and its concentration must be optimized for each PCR reaction system (Cadwell and Joyce 1994; Ling et al. 1991). Many components of the system bind magnesium ions, including dNTP, DNA polymerase, template, and primers. Because the activity of DNA polymerase requires the presence of free magnesium ions, the magnesium ion concentration in the reaction system must exceed the total dNTP concentration. Typically, in the presence of 0.8 mM total dNTPs, 1.5 mM Mg^{2+} needs to be added to the PCR reaction system leaving approximately 0.7 mM free Mg^{2+} for the DNA polymerase. When the concentration of Mg^{2+} is below normal concentrations, the less the specificity of the DNA polymerase and the higher the magnesium ion concentration, the lower the specificity of the DNA polymerase. In order to elevate the error rate in the replication process, a higher concentration of magnesium ion (about 7 mM) should be added to the error-prone PCR reaction system.

(c) Adding Mn^{2+}: Generally, Mn^{2+} can also be added to the reaction system to promote the error rate (Lin-Goerke et al. 1997) in addition to increased Mg^{2+}. By varying the concentration of the Mn^{2+} added to the reaction system, the mutation rate can be varied from 1 to 5 nucleotides per 1 kb.
(d) Using unbalanced dNTP concentrations: dNTPs can be added at either equal or unbalanced amounts. An unbalanced dNTPs mixture can increase the misincorporation of nucleotides into the newly synthesized DNA strand (Lin-Goerke et al. 1997). Several methods have been used, including increasing the concentration of dGTP, increasing both dCTP and dTTP, or increasing all but dATP. The amount of certain nucleotides can also be decreased. In our experiments, an unbalanced dNTPs mixture that includes 0.2 mM each of dATP and dGTP and 1 mM each of dCTP and dTTP is suitable.
(e) Adding other substances to promote the error rate: Addition of alcohol and some triphosphate derivatives of nucleotides interfere with amplification and promote the rate of incorporation of mismatched bases (Fromant et al. 1995; Nishiya and Imanaka 1994; Shafikhani et al. 1997).
(f) Increasing the number of amplification cycles: The amplification step of PCR is commonly 35 cycles, and increasing the number of amplification cycles also promotes the incorporation of mismatched bases (Pritchard et al. 2005).

3. **The error-prone PCR procedure for allergen modification**

The method of error-prone PCR is similar to conventional PCR cloning of a target except that a much larger pool of transformants is needed to increase the diversity of the library. In general, the following steps are always involved in allergenicity modification using error-prone PCR (Hanson-Manful and Patrick 2013; Pritchard et al. 2005):

(a) Acquiring the randomly mutated allergen sequences by amplifying the target allergen gene under error-prone conditions;
(b) Digesting the PCR product with suitable restriction endonucleases;
(c) Ligating the digested PCR product into a suitable plasmid with DNA ligase;
(d) Transforming the ligated products into a host cell to acquire a mutant library;
(e) Screening mutants with reduced allergenicity by bioinformatics and experimental methods.

4. **The superiorities and characteristics of error-prone PCR in allergen modification**

Error-prone PCR has proved useful in directed modification of protein function in vitro, such as enzyme substrate specificity, enzyme activity, protein thermostability, and allergenicity (Huang et al. 2008; Liao et al. 2012; Stephens et al. 2009; Yu et al. 2009). A significant advantage for using error-prone PCR to modify the protein allergenicity is that it does not require detailed information on the allergen's structure or the precise prediction sites for amino acid substitution. Error-prone PCR involves producing enormous numbers of mutants of allergen mutant genes in order to modify the allergenicity of the allergen proteins. The success of allergenicity modification depends on the efficiency in creating mutagenesis libraries and screening them for the mutants with reduced allergenicity. The variety of the library is a decisive factor and a large number of techniques, such as increasing the Mg^{2+} concentration, adding Mn^{2+}, etc., have been developed to attain high-quality mutagenesis libraries (see above). The screening of the mutant libraries is also a crucial factor for successful allergenicity modification using error-prone PCR. Many high-throughput screening methods, such as allergen protein chips, dot-blot array, etc., have been developed for subsequent allergenicity screening. For more information on allergenicity screening after error-prone PCR, please refer to Chap. 8 in this book.

13.4 The Re-evaluation and Validation of the Allergenicity of Allergen Proteins After Modification

The allergenicity of the modified allergens must be re-evaluated and validated by laboratory methods. WBHRs, Mast Cell Challenge Assays, IgE Binding Assays, and animal models are the most common methods for evaluating the allergenicity of the allergen proteins. The first three methods, which evaluate the allergenicity of the allergen proteins in vitro, mainly assess the potential allergenicity from the perspective of serology and cytology. The last method, which assesses the allergenicity of the proteins by establishing allergic disease models followed by analysis of allergic symptoms and measurement of the specific IgE levels, is usually the allergenicity evaluation method used in vivo. Various allergic disease models have been established, such as animal models of asthma, AR, AD, and food allergic diseases, to evaluate the allergenicity of allergen proteins. When using animal models to evaluate the allergenicity of the target proteins, allergenicity strength is estimated according to changes of the specific antibodies, allergic disease-related cytokines and the symptoms of allergic reactions. Rats, mice, and guinea pigs are the common animals used for animal models for allergic diseases as they have similar immunologic response mechanisms as humans, can provide a wealth of related immunological information, and have a small size, short breeding cycle, and low cost. Therefore, using animal models to evaluate the allergenicity of the proteins has great advantages.

The products of error-prone PCR are a mixture of multiple mutants. Evaluation of the allergenicity of each mutant using laboratory experiments is difficult to implement, so preliminarily evaluation of the allergenicity of the mutants acquired from error-prone PCR is accomplished using bioinformatics methods such as homology comparing, cross reaction analyzing, and antigenic reactivity, in order to identify the mutants with reduced allergenicity that retain their original three-dimensional structure and biological activity. Unfortunately, screening of mutants using only bioinformatics evaluation methods cannot confidently identify those with completely reduced allergenicity and more stringent experimental evaluation methods are needed to assess the mutants.

At this time, there are no standardized methods for protein allergenicity assessment; however, evaluation of the allergenicity of the proteins can be achieved using existing technologies and methods. Although each method has its advantages and disadvantages, the allergenicity of the proteins can be evaluated effectively by combining a variety of evaluation methods.

References

Aalberse RC. Assessment of allergen cross-reactivity. Clin Mol Allergy. 2007;5:2.

Abril-Gil M, Garcia-Just A, Perez-Cano FJ, Franch A, Castell M. Development and characterization of an effective food allergy model in Brown Norway rats. PLoS One. 2015;10(4):e0125314.

Aldemir H, Bars R, Herouet-Guicheney C. Murine models for evaluating the allergenicity of novel proteins and foods. Regul Toxicol Pharmacol. 2009;54(3 Suppl):S52–7.

Bieber T. Atopic dermatitis. N Engl J Med. 2008;358(14):1483–94.

Biles BD, Connolly BA. Low-fidelity Pyrococcus furiosus DNA polymerase mutants useful in error-prone PCR. Nucleic Acids Res. 2004;32(22):e176.

Boyce JA, Assa'ad A, Burks AW, Jones SM, Sampson HA, Wood RA, Plaut M, Cooper SF, Fenton MJ, Arshad SH, et al. Guidelines for the diagnosis and management of food allergy in the United States: summary of the NIAID-Sponsored Expert Panel report. J Am Acad Dermatol. 2011;64(1):175–92.

Brusic V, Petrovsky N. Bioinformatics for characterisation of allergens, allergenicity and allergic crossreactivity. Trends Immunol. 2003;24(5):225–8.

Bublin M, Pfister M, Radauer C, Oberhuber C, Bulley S, Dewitt A, Lidholm J, Reese G, Vieths S, Breiteneder H, et al. Component-resolved diagnosis of kiwifruit allergy with purified natural and recombinant kiwifruit allergens. J Allergy Clin Immunol. 2010;125(3): 687–94, 94.e1.

Cadwell RC, Joyce GF. Mutagenic PCR. PCR Methods Appl. 1994;3(6):S136–40.

Carey MF, Peterson CL, Smale ST. PCR-mediated site-directed mutagenesis. Cold Spring Harb Protoc. 2013;2013(8):738–42.

Carter P, Bedouelle H, Winter G. Improved oligonucleotide site-directed mutagenesis using M13 vectors. Nucleic Acids Res. 1985;13(12):4431–43.

Chrysostomou C, Seker H. Prediction of protein allergenicity based on signal-processing bioinformatics approach. Conf Proc IEEE Eng Med Biol Soc. 2014; 2014: 808–11.

Dean JH, Cornacoff JB, Haley PJ, Hincks JR. The integration of immunotoxicology in drug discovery and development: Investigative and in vitro possibilities. Toxicol In Vitro. 1994;8(5):939–44.

Do CB, Katoh K. Protein multiple sequence alignment. Methods Mol Biol. 2008;484:379–413.

El-Manzalawy Y, Honavar V. Recent advances in B-cell epitope prediction methods. Immunome Res. 2010;6(Suppl 2):S2.

Friedl-Hajek R, Spangfort MD, Schou C, Breiteneder H, Yssel H, Joost van Neerven RJ. Identification of a highly promiscuous and an HLA allele-specific T-cell epitope in the birch major allergen Bet v 1: HLA restriction, epitope mapping and TCR sequence comparisons. Clin Exp Allergy. 1999;29(4):478–87.

Fromant M, Blanquet S, Plateau P. Direct random mutagenesis of gene-sized DNA fragments using polymerase chain reaction. Anal Biochem. 1995;224(1):347–53.

Furuta Y, Yoshioka M, Sugiyama T, Miyasaka K. Rapidly increasing allergic diseases. Hokkaido Igaku Zasshi. 2005;80(1):3–7.

Galbiati V, Mitjans M, Corsini E. Present and future of in vitro immunotoxicology in drug development. J Immunotoxicol. 2010;7(4):255–67.

Geier J, Uter W, Lessmann H, Schnuch A. Current contact allergens. Hautarzt. 2011;62(10):751–6.

Gibson J. Bioinformatics of protein allergenicity. Mol Nutr Food Res. 2006;50(7):591.

Gleich GJ, Jones RT. Measurement of IgE antibodies by the radioallergosorbent test. I. Technical considerations in the performance of the test. J Allergy Clin Immunol. 1975;55(5):334–45.

Goodman RE, Tetteh AO. Suggested improvements for the allergenicity assessment of genetically modified plants used in foods. Curr Allergy Asthma Rep. 2011;11(4):317–24.

Hakkaart GA, Aalberse RC, van Ree R. Epitope mapping of the house-dust-mite allergen Der p 2 by means of site-directed mutagenesis. Allergy. 1998;53(2):165–72.

Hanson-Manful P, Patrick WM. Construction and analysis of randomized protein-encoding libraries using error-prone PCR. Methods Mol Biol. 2013;996:251–67.

Hemsley A, Arnheim N, Toney MD, Cortopassi G, Galas DJ. A simple method for site-directed mutagenesis using the polymerase chain reaction. Nucleic Acids Res. 1989;17(16):6545–51.

Hileman RE, Silvanovich A, Goodman RE, Rice EA, Holleschak G, Astwood JD, Hefle SL. Bioinformatic methods for allergenicity assessment using a comprehensive allergen database. Int Arch Allergy Immunol. 2002;128(4):280–91.

Ho SN, Hunt HD, Horton RM, Pullen JK, Pease LR. Site-directed mutagenesis by overlap extension using the polymerase chain reaction. Gene. 1989;77(1):51–9.

Holm M, Andersen HB, Hetland TE, Dahl C, Hoffmann HJ, Junker S, Schiotz PO. Seven week culture of functional human mast cells from buffy coat preparations. J Immunol Methods. 2008;336(2):213–21.

Huang L, Li T, Zhou H, Qiu P, Wu J, and Liu L. Sinomenine potentiates degranulation of RBL-2H3 basophils via up-regulation of phospholipase A phosphorylation by Annexin A1 cleavage and ERK phosphorylation without influencing on calcium mobilization. Int Immunopharmacol. 2015a: (Epub ahead of print).

Huang Y, Cai Y, Yang J, Yan Y. Directed evolution of lipase of Bacillus pumilus YZ02 by error-prone PCR. Sheng Wu Gong Cheng Xue Bao. 2008;24(3):445–51.

Huang Q, Xu H, Yu Z, Gao P, Liu S. Inbred Chinese Wuzhishan (WZS) minipig model for soybean glycinin and beta-conglycinin allergy. J Agric Food Chem. 2010;58(8):5194–8.

Huang Q, Xu HB, Yu Z, Liu S, Gao P. In vivo digestive stability of soybean beta-conglycinin beta-subunit in WZS minipigs. Biomed Environ Sci. 2015;28(1):85–8.

Hviid L. Development of vaccines against Plasmodium falciparum malaria: taking lessons from naturally acquired protective immunity. Microbes Infect. 2007;9(6):772–6.

Ivanciuc O, Schein CH, Braun W. SDAP: database and computational tools for allergenic proteins. Nucleic Acids Res. 2003;31(1):359–62.

Jeltsch A, Lanio T. Site-directed mutagenesis by polymerase chain reaction. Methods Mol Biol. 2002;182:85–94.

Jensen BM, Falkencrone S, Skov PS. Measuring histamine and cytokine release from basophils and mast cells. Methods Mol Biol. 2014;1192:135–45.

Jeon WY, Shin IS, Shin HK, Lee MY. Samsoeum water extract attenuates allergic airway inflammation via modulation of Th1/Th2 cytokines and decrease of iNOS expression in asthmatic mice. BMC Complement Altern Med. 2015;15:47.

Keohavong P, Thilly WG. Fidelity of DNA polymerases in DNA amplification. Proc Natl Acad Sci USA. 1989;86(23):9253–7.

Kimber I, Dearman RJ, Penninks AH, Knippels LM, Buchanan RB, Hammerberg B, Jackson HA, Helm RM. Assessment of protein allergenicity on the basis of immune reactivity: animal models. Environ Health Perspect. 2003;111(8):1125–30.

King TP, Hoffman D, Lowenstein H, Marsh DG, Platts-Mills TA, Thomas W. Allergen nomenclature. WHO/IUIS Allergen Nomenclature Subcommittee. Int Arch Allergy Immunol. 1994;105(3):224–33.

Kinoshita T, Sawai N, Hidaka E, Yamashita T, Koike K. Interleukin-6 directly modulates stem cell factor-dependent development of human mast cells derived from CD34(+) cord blood cells. Blood. 1999;94(2):496–508.

Kirshenbaum AS, Petrik A, Walsh R, Kirby TL, Vepa S, Wangsa D, Ried T, Metcalfe DD. A ten-year retrospective analysis of the distribution, use and phenotypic characteristics of the LAD2 human mast cell line. Int Arch Allergy Immunol. 2014;164(4):265–70.

Kitagawa S, Zhang S, Harari Y, Castro GA. Relative allergenicity of cow's milk and cow's milk-based formulas in an animal model. Am J Med Sci. 1995;310(5):183–7.

Kuiper HA, Kleter GA, Noteborn HP, Kok EJ. Assessment of the food safety issues related to genetically modified foods. Plant J. 2001;27(6):503–28.

Ladics GS. Current codex guidelines for assessment of potential protein allergenicity. Food Chem Toxicol. 2008;46(Suppl 10):S20–3.

Ladics GS, Knippels LM, Penninks AH, Bannon GA, Goodman RE, Herouet-Guicheney C. Review of animal models designed to predict the potential allergenicity of novel proteins in genetically modified crops. Regul Toxicol Pharmacol. 2010;56(2):212–24.

Landt O, Grunert HP, Hahn U. A general method for rapid site-directed mutagenesis using the polymerase chain reaction. Gene. 1990;96(1):125–8.

Liao Y, Zeng M, Wu ZF, Chen H, Wang HN, Wu Q, Shan Z, Han XY. Improving phytase enzyme activity in a recombinant phyA mutant phytase from Aspergillus niger N25 by error-prone PCR. Appl Biochem Biotechnol. 2012;166(3):549–62.

Ling LL, Keohavong P, Dias C, Thilly WG. Optimization of the polymerase chain reaction with regard to fidelity: modified T7, Taq, and vent DNA polymerases. PCR Methods Appl. 1991;1(1):63–9.

Lin-Goerke JL, Robbins DJ, Burczak JD. PCR-based random mutagenesis using manganese and reduced dNTP concentration. Biotechniques. 1997;23(3):409–12.

Malherbe L. T-cell epitope mapping. Ann Allergy Asthma Immunol. 2009;103(1):76–9.

Mari A, Scala E. Allergome: a unifying platform. Arb Paul Ehrlich Inst Bundesamt Sera Impfstoffe Frankf A M. 2006;95:29–39 (discussion-40).

Marini A, Agosti M, Motta G, Mosca F. Effects of a dietary and environmental prevention programme on the incidence of allergic symptoms in high atopic risk infants: three years' follow-up. Acta Paediatr Suppl. 1996;414:1–21.

Martinez Barrio A, Soeria-Atmadja D, Nister A, Gustafsson MG, Hammerling U, Bongcam-Rudloff E. EVALLER: a web server for in silico assessment of potential protein allergenicity. Nucleic Acids Res. 2007;35(Web Server issue):W694–700.

Meng Z, Yan C, Deng Q, Dong X, Duan ZM, Gao DF, Niu XL. Oxidized low-density lipoprotein induces inflammatory responses in cultured human mast cells via Toll-like receptor 4. Cell Physiol Biochem. 2013;31(6):842–53.

Mobs C, Slotosch C, Loffler H, Jakob T, Hertl M, Pfutzner W. Birch pollen immunotherapy leads to differential induction of regulatory T cells and delayed helper T cell immune deviation. J Immunol. 2010;184(4):2194–203.

Mount DW. Using the basic local alignment search tool (BLAST). CSH Protoc. 2007;7:pdb–top17.

Muh HC, Tong JC, Tammi MT. AllerHunter: a SVM-pairwise system for assessment of allergenicity and allergic cross-reactivity in proteins. PLoS One. 2009;4(6):e5861.

Newell EW, Sigal N, Nair N, Kidd BA, Greenberg HB, Davis MM. Combinatorial tetramer staining and mass cytometry analysis facilitate T-cell epitope mapping and characterization. Nat Biotechnol. 2013;31(7):623–9.

Nielsen M, Lundegaard C, Lund O. Prediction of MHC class II binding affinity using SMM-align, a novel stabilization matrix alignment method. BMC Bioinform. 2007;8:238.

Nishiya Y, Imanaka T. Alteration of substrate specificity and optimum pH of sarcosine oxidase by random and site-directed mutagenesis. Appl Environ Microbiol. 1994;60(11):4213–5.

Nohr J, Kristiansen K. Site-directed mutagenesis. Methods Mol Biol. 2003;232:127–31.

Nolte H, Storm K, Schiotz PO. Diagnostic value of a glass fibre-based histamine analysis for allergy testing in children. Allergy. 1990;45(3):213–23.

Nordlee JA, Taylor SL, Townsend JA, Thomas LA, Bush RK. Identification of a Brazil-nut allergen in transgenic soybeans. N Engl J Med. 1996;334(11):688–92.

Pawankar R. Allergic diseases and asthma: a global public health concern and a call to action. World Allergy Organ J. 2014;7(1):12.

Payne V, Kam PC. Mast cell tryptase: a review of its physiology and clinical significance. Anaesthesia. 2004;59(7):695–703.

Penninks AH, Knippels LM. Determination of protein allergenicity: studies in rats. Toxicol Lett. 2001;120(1–3):171–80.

Peterson MG. DNA sequencing using Taq polymerase. Nucleic Acids Res. 1988;16(22):10915.

Piacentini GL, Vicentini L, Bodini A, Mazzi P, Peroni DG, Maffeis C, Boner AL. Allergenicity of a hydrolyzed rice infant formula in a guinea pig model. Ann Allergy Asthma Immunol. 2003;91(1):61–4.

Pritchard L, Corne D, Kell D, Rowland J, Winson M. A general model of error-prone PCR. J Theor Biol. 2005;234(4):497–509.

Rabjohn P, West CM, Connaughton C, Sampson HA, Helm RM, Burks AW, Bannon GA. Modification of peanut allergen Ara h 3: effects on IgE binding and T cell stimulation. Int Arch Allergy Immunol. 2002;128(1):15–23.

Reche PA, Glutting JP, Reinherz EL. Prediction of MHC class I binding peptides using profile motifs. Hum Immunol. 2002;63(9):701–9.

Reikofski J, Tao BY. Polymerase chain reaction (PCR) techniques for site-directed mutagenesis. Biotechnol Adv. 1992;10(4):535–47.

Rice AS, Cimino-Brown D, Eisenach JC, Kontinen VK, Lacroix-Fralish ML, Machin I, Mogil JS, Stohr T. Animal models and the prediction of efficacy in clinical trials of analgesic drugs: a critical appraisal and call for uniform reporting standards. Pain. 2008;139(2):243–7.

Richard G, Motohiro E, and Hugh S. AllergenOnline, a peer-reviewed protein sequence database for assessing the potential allergenicity of genetically modified organisms and novel food proteins. World Allergy Organ J. 2007. doi:10.1097/01.WOX.0000301456.01355.b0.

Saalbach I, Pickardt T, Machemehl F, Saalbach G, Schieder O, Muntz K. A chimeric gene encoding the methionine-rich 2S albumin of the Brazil nut (Bertholletia excelsa H.B.K.) is stably expressed and inherited in transgenic grain legumes. Mol Gen Genet. 1994;242(2):226–36.

Saha S, Raghava GP. AlgPred: prediction of allergenic proteins and mapping of IgE epitopes. Nucleic Acids Res. 2006; 34(Web Server issue):W202–9.

Schein CH, Ivanciuc O, Braun W. Bioinformatics approaches to classifying allergens and predicting cross-reactivity. Immunol Allergy Clin North Am. 2007;27(1):1–27.

Schmetzer O, Valentin P, Smorodchenko A, Domenis R, Gri G, Siebenhaar F, Metz M, Maurer M. A novel method to generate and culture human mast cells: Peripheral CD34+ stem cell-derived mast cells (PSCMCs). J Immunol Methods. 2014;413:62–8.

Scurlock AM, Burks AW. Peanut allergenicity. Ann Allergy Asthma Immunol. 2004;93(5 Suppl 3):S12–8.

Shafikhani S, Siegel RA, Ferrari E, Schellenberger V. Generation of large libraries of random mutants in *Bacillus subtilis* by PCR-based plasmid multimerization. Biotechniques. 1997;23(2):304–10.

Singh H, Raghava GP. ProPred1: prediction of promiscuous MHC Class-I binding sites. Bioinformatics. 2003;19(8):1009–14.

Stephens DE, Singh S, Permaul K. Error-prone PCR of a fungal xylanase for improvement of its alkaline and thermal stability. FEMS Microbiol Lett. 2009;293(1):42–7.

Sutovska M, Kocmalova M, Joskova M, Adamkov M, Franova S. The effect of long-term administered CRAC channels blocker on the functions of respiratory epithelium in guinea pig allergic asthma model. Gen Physiol Biophys. 2015;34(2):167–76.

Tandy JR, Mabry RL, Mabry CS. Correlation of modified radioallergosorbent test scores and skin test results. Otolaryngol Head Neck Surg. 1996;115(1):42–5.

Thomas K, Bannon G, Hefle S, Herouet C, Holsapple M, Ladics G, MacIntosh S, Privalle L. In silico methods for evaluating human allergenicity to novel proteins: International Bioinformatics Workshop Meeting Report. Toxicol Sci. 2005;88(2):307–10. 23–24 Feb 2005.

Thomas K, Aalbers M, Bannon GA, Bartels M, Dearman RJ, Esdaile DJ, Fu TJ, Glatt CM, Hadfield N, Hatzos C, et al. A multi-laboratory evaluation of a common in vitro pepsin digestion assay protocol used in assessing the safety of novel proteins. Regul Toxicol Pharmacol. 2004;39(2):87–98.

Thomas K, MacIntosh S, Bannon G, Herouet-Guicheney C, Holsapple M, Ladics G, McClain S, Vieths S, Woolhiser M, Privalle L. Scientific advancement of novel protein allergenicity evaluation: an overview of work from the HESI Protein Allergenicity Technical Committee (2000–2008). Food Chem Toxicol. 2009;47(6):1041–50.

Tsai WJ, Liu CH, Chen ST, Yang CY. Identification of the antigenic determinants of the American cockroach allergen Per a 1 by error-prone PCR. J Immunol Methods. 2003;276(1–2):163–74.

van Esch BC, Knipping K, Jeurink P, van der Heide S, Dubois AE, Willemsen LE, Garssen J, Knippels LM. In vivo and in vitro evaluation of the residual allergenicity of partially hydrolysed infant formulas. Toxicol Lett. 2011;201(3):264–9.

Vinje NE, Larsen S, Lovik M. A mouse model of lupin allergy. Clin Exp Allergy. 2009;39(8):1255–66.

Wan D, Wang X, Nakamura R, Alcocer MJ, Falcone FH. Use of humanized rat basophil leukemia (RBL) reporter systems for detection of allergen-specific IgE sensitization in human serum. Methods Mol Biol. 2014;1192:177–84.

Weiner MP, Costa GL, Schoettlin W, Cline J, Mathur E, Bauer JC. Site-directed mutagenesis of double-stranded DNA by the polymerase chain reaction. Gene. 1994;151(1–2):119–23.

Wells JA, Vasser M, Powers DB. Cassette mutagenesis: an efficient method for generation of multiple mutations at defined sites. Gene. 1985;34(2–3):315–23.

Wu H, Guo Y, Wang J, Zhang J, Wang S, Zou X, Tao A. The importance of allergen avoidance in high risk infants and sensitized patients: a meta-analysis study. Allergy Asthma Immunol Res. 2014;6(6):525–34.

Xia YC, Sun S, Kuek LE, Lopata AL, Hulett MD, Mackay GA. Human mast cell line-1 (HMC-1) cells transfected with FcepsilonRIalpha are sensitive to IgE/antigen-mediated stimulation demonstrating selectivity towards cytokine production. Int Immunopharmacol. 2011;11(8):1002–11.

Yu H, Li J, Zhang D, Yang Y, Jiang W, Yang S. Improving the thermostability of N-carbamyl-D-amino acid amidohydrolase by error-prone PCR. Appl Microbiol Biotechnol. 2009;82(2):279–85.

Zhang L, Huang Y, Zou Z, He Y, Chen X, Tao A. SORTALLER: predicting allergens using substantially optimized algorithm on allergen family featured peptides. Bioinformatics. 2012;28(16):2178–9.

Author's Biography

Dr. Zhaoyu Liu is working at the Guangdong Provincial Key Laboratory of Allergy and Clinical Immunology. Email: sysu1924@gmail.com.

Dr. Liu received his Ph.D. from Sun Yat-sen University while studying the interaction between host immunity and virus infection. He is currently working as a post-graduate student majoring in Food Safety Biology at Guangzhou Medical University. His main research interests focus on the innate immune system and its role in allergic disorders. He has published several scientific papers within internationally well-recognized journals and has participated in several projects of the National Natural Science Foundation.

Huifang Chen is currently working at the Guangdong Provincial Key Laboratory of Allergy and Clinical Immunology located in the Second Affiliated Hospital of Guangzhou Medical University. Email: 20051984chf163.com.

As a lab technician, Huifang Chen is skilled at cell culture, mast cell stimulation, animal experiments, flow cytometry, and confocal laser scanning microscopy. Her main research interests focus on the evaluation of the allergenicity of genetically modified allergens.

Professor Ailin Tao is a Professor at Guangzhou Medical University, Director of Guangdong Provincial Key Laboratory of Allergy and Clinical Immunology, Principal Investigator of the State Key Laboratory of Respiratory Disease, Deputy Director of the State Key Clinical Specialty in Allergy of the Second Affiliated Hospital of Guangzhou Medical University, Member of the State Committee for Transgenic Safety Assessment, Standing Committee Member of Allergy Branch of Guangdong Medical Association, and Member of Guangdong Provincial Committee for Transgenic Safety Assessment.

Prof. Ailin Tao earned his doctorate degree from the State Key Laboratory of Crop Genetic Improvement of Huazhong Agricultural University in 2002, followed by a postdoctoral training at Postdoctoral Station of Basic Medicine in Shantou University Medical College, majoring in allergen proteins. His most recent research has been on allergy bioinformatics, allergy and clinical immunology, and disease models such as allergic asthma, allergic rhinitis (AR), infection, and inflammation induced by allergy, inflammatory, and protracted diseases caused by antigens or superantigens. He has gained experience in the field of allergology including the mechanisms of immune tolerance, allergy triggering factors and chronic inflammation pathways, and allergenicity evaluation and modification for food and drugs. He proposed some new concepts including "Representative Major Allergens,"

"Allergenicity Attenuation" of immunotoxin and allergens, "broad-spectrum immunomodulator" as well as the theoretical hypothesis of "Balanced Stimulation by Whole Antigens." Prof. Tao's laboratory focuses on the diagnosis of allergic disease and the medical evaluation of food and drug allergenicity and its modification. Prof. Tao has now constructed a system for the prediction, quantitative assessment, and simultaneous modification of epitope allergenicity, which has been applied to more than 20 allergens, and he also developed a bioinformatics software program for allergen epitope prediction, SORTALLER (http://sortaller.gzhmu.edu.cn), which performed significantly better than other existing software, reaching a perfect balance of high specificity (98.4 %) and sensitivity (98.6 %) for discriminating allergenic proteins from several independent datasets of protein sequences of diverse sources. Furthermore, this program has a Matthews correlation coefficient as high as 0.970, a fast running speed, and can rapidly predict a set of amino acid sequences with a single click. The software has been frequently used by researchers from many institutions in China and over 30 countries worldwide, thus becoming the number one allergen epitope prediction software program. Prof. Tao has set up an allergen database ALLERGENIA (http://ALLERGENIA.gzhmu.edu.cn) that has several advantages over other databases such as a wide selection of non-redundant allergens, excellent astringency and accuracy, and friendly and usable analytical functions.

Chapter 14
Bioinformatics Methods to Predict Allergen Epitopes

Ying He and Ailin Tao

Abstract Epitopes are composed of contiguous or discontiguous specific amino acid residues on an antigen. Binding of these epitopes with their corresponding receptors on the surface of lymphocytes is a necessary step for an antigen to activate these cells and mount an immune response in the body. Therefore, precise determination of antigenic epitopes is very important in vaccine design research, modification of allergenicity, understanding immune tolerance mechanisms, and more. With the rapid development of bioinformatics, many free online servers with programs to predict B-cell or T-cell epitopes have emerged in recent years. These prediction methods are based on different modeling principles and algorithms, so their outcomes and accuracy varies. This chapter will introduce some typical programs for the prediction of B-cell and T-cell epitopes as well as their modeling principles and algorithms.

Keywords Bioinformatics · B-cell epitope · MHC binding · Prediction · T-cell epitope

Y. He · A. Tao (✉)
Guangdong Provincial Key Laboratory of Allergy and Clinical Immunology,
The State Key Clinical Specialty in Allergy, The State Key Laboratory of Respiratory Disease,
The Second Affiliated Hospital of Guangzhou Medical University,
250# Changgang Road East, Guangzhou, China
e-mail: Aerobiologiatao@163.com

Y. He
e-mail: heying0605@163.com

A. Tao and E. Raz (eds.), *Allergy Bioinformatics*, Translational Bioinformatics 8,
DOI 10.1007/978-94-017-7444-4_14

14.1 Introduction

Type I hypersensitivity can be divided into two stages, sensitization, and challenge. In the sensitization phase, an allergen induces B cells to produce IgE. With their Fc segment, IgE binds to the FcεRI on the surface of mast cells, basophils, and other target cells, and the body becomes sensitized. In the challenge stage, the same allergen is encountered again, specific IgE binds to the surface of the target cells, and anaphylaxis can occur. Mast cell and/or basophil degranulation causes an allergic reaction by releasing previously synthesized or newly synthesized inflammatory mediators such as histamine and leukotrienes. IgE plays a key role in this process and its production involves the differentiation of T_H2 cells and IgE class switching. The antigen-presenting cells process the allergen to antigenic peptides, forming a complex of antigenic peptide and MHC molecule that is recognized by the T-cell antigen receptor (TCR). The naive T cells then differentiate into T_H0 cells, which further differentiate into T_H2 cells under the influence of multiple factors, thereby promoting B-cell secretion of IgE. On the other hand, B-cell antigen receptors (BCR) also identify allergens, resulting in activation of STAT6, which can promote IgCε exon germline gene transcription and promote an IgE class switch (Linehan et al. 1998). Thus, for an allergen to act as an antigen to induce an immune response, it must first be processed and the antigenic peptides presented, with recognition by T cells and B cells being a key step. Currently, methods to predict T-cell epitopes are mainly achieved by predicting the binding affinity of antigen peptides to MHC molecules. However, rarely, bioinformatics tools are available to help identify how the TCR binds to the peptide-MHC complex and which amino acid residues on the peptide bind directly to the TCR.

The antigenic epitope refers to an antigen-specific molecule determined by specific chemical groups. Antigens bind through their antigenic epitopes to corresponding lymphocyte surface antigen receptors leading to the activation of the lymphocytes and eliciting an immune response. The character, number, and spatial configuration of the antigenic epitopes determine the specificity of the antigen, thus influencing the production of the corresponding antibodies or sensitized lymphocytes. Epitopes can be divided into B or T-cell epitopes according to which cell type they bind to. B-cell epitopes are linear fragments or have a conformation that can bind to the BCR or specific antibody. Linear antigen epitopes contain a contiguous amino acid sequence that is at the core of the interaction with the receptor or antibody. Most of the current predictions using bioinformatics are based on linear epitopes. However, 90 % of B-cell epitopes are conformational, not linear epitopes. Conformational epitopes contain some noncontiguous amino acids in the primary protein structure that become adjacent to each other in the folded tertiary structure. T-cell epitopes are linear epitopes that are recognized by a T-cell receptor (TCR). Using bioinformatics to predict antigenic epitopes will help us to study the modification of allergenicity, vaccine design, and immune tolerance mechanisms.

14.2 Prediction of Linear B-Cell Epitopes

14.2.1 Prediction Scheme

Methods to predict linear B-cell epitopes are based on the primary structure of the protein together with a comprehensive measure of physicochemical properties, structural features, and other characteristics of the protein antigen. The most commonly evaluated parameters include hydrophilicity, surface probability, antigenic index, flexibility, charge distribution scheme, and secondary structure.

- **Hydrophilicity**

Amino acid residues of the protein antigen are divided into hydrophilic residues and hydrophobic residues. Generally, hydrophobic residues are buried inside the protein, whereas hydrophilic residues are on the surface and are therefore closely linked to the antigen epitopes. Previous investigations have demonstrated that charged, polar, and hydrophilic amino acid side chains are common features of antigenic determinants. In 1981, a method to predict the antigenic determinants by analyzing amino acid sequences was proposed (Hopp and Woods 1981). In this prediction method, each amino acid residue is assigned a hydrophilicity value and then repetitively averaging these values in a hexapeptide along the peptide chain. The point of highest local average hydrophilicity would invariably lie within or be immediately adjacent to an antigenic determinant. Since the time this method was developed, the Kyte–Doolittle scale (Kyte and Doolittle 1982), HPLC scale (Parker et al. 1986), and other hydrophilic prediction schemes have been proposed.

- **Surface probability**

Surface probability refers to the possibility of amino acid residues contacting surface solvent molecules. It reflects the distribution of the outer layers and inside of the residues, which indirectly reflect the capacity for antibody binding. The Emini scale (Emini et al. 1985) is a typical method for predicting surface probability.

- **Flexibility**

Conformational protein antigens are not rigid and contain a polypeptide chain skeleton that has a certain degree of flexibility. The strong active amino acid residue is the site of a large flexibility locus. Since the antigen–antibody binding is a lock-and-key process and protein conformations change, "flexible" amino acid residues that are prone to twist and fold are most likely to become antigenic epitopes. Karplus and Schulz developed a predictive method for protein fragment activity based on the atomic temperature factor of 31 proteins with a known three-dimensional structure (Karplus and Schulz 1985).

- **Charge distribution**

The charge of specific antibodies produced in response to alkaline antigens tends to be acidic, whereas the charge of specific antibodies to acidic antigens tends to be alkaline (Gershoni et al. 1997).

- **Antigenic index**

Welling et al. studied 606 amino acids from 20 proteins of which the antigenic regions have been well researched in order to determine the average composition of an antigenic region (Welling et al. 1985). Each amino acid is described by the frequency that it occurs in the antigenic region which is then divided by the frequency of each amino acid in the total protein to generate a scale value. Therefore, this method is based on the percentage of each amino acid present in known antigenic determinants compared with the percentage of that amino acid in the total protein. Studies have shown that some antigenic regions predicted by Welling scale are consistent with the results from the hydrophilic method of Hopp and Woods (1981). But the disadvantage of this method is that the database is limited. Another antigenic index prediction method was proposed in 1988, which predicts the topological features of a protein directly from its primary amino acid sequence. The computer program generates values for surface accessibility parameters and combines these values with those obtained for regional backbone flexibility and predicted secondary structure (Jameson and Wolf 1988).

- **Secondary structure**

Secondary structure prediction programs show that a protruding β turn structure is more likely to become an epitope as it often appears on the surface of protein antigens and is easy for antibodies to bind to unlike α-helix and β-sheet structures that are generally not epitopes because they are not flexible and are difficult for antibodies to bind to. Protein secondary structure prediction methods include the Chou–Fasman (Chou and Fasman 1978), Garnier–Robson (Garnier–Robson 1978), Cohen (Colloc'h and Cohen 1991), and other methods, but the success rates of these methods for predicting secondary protein structures are below 65 %. Cohen has a high prediction rate for β turn with an accuracy rate as high as 95 % for the known types of protein folding (αα type, ββ type, and α/β type). In the case of a protein with unknown structure, prediction methods mentioned above for three types secondary structures are also available. In addition, the presence of a protruding β turn that is consistent in three prediction results is helpful in identifying epitopes.

The various methods for predicting correct epitopes are not that successful. Generally speaking, B-cell epitopes should be easily located or able to move to the cell surface in order to bind with their specific antibodies. In addition, a certain amount of flexibility is required because protein conformations will change when the antigen and antibody bind together. Typically, protein epitopes are present in the β turn. Therefore, the amino acid residues that are often present in the β turn also often appear in the antigenic epitope. Furthermore, the antigenic epitope generally consists of the amino acid residues with high hydrophilicity. However, current computer software programs cannot guarantee accurate predictions so further experiments are needed to verify.

14.2.2 Prediction Software

With the intensive study of the physical properties of amino acid residues, computer programs are now available, which predict B-cell linear epitopes including PEOPLE (Alix 1999), BEPITOPE (Odorico and Pellequer 2003), BcePred (Saha and Raghava 2004), and Protean of DNASTAR.

BcePred (http://www.imtech.res.in/raghava/bcepred/) is an online software program that predicts linear B-cell epitopes based on their physical and chemical properties (Saha and Raghava 2004). This dataset contains 1029 nonredundant B-cell epitope peptides from the Bcipep database, which were verified in that the same number of non-epitopic peptides was extracted from the SWISS-PROT database. The accuracy of this software based on the different prediction schemes was between 52.92 % and 57.53 %. After combining four properties (hydrophilicity, flexibility, polarity, and exposed surface) of the amino acid, the prediction accuracy increases to 58.70 % and the threshold value is 2.38. This server allows users to select any combination of amino acids and different physical and chemical characteristics including hydrophilicity, flexibility, accessibility, polarity, antigenic propensity, exposed surface and turns to predict the B-cell epitopes. The users can also choose any threshold value [−3 to +3] for the different parameters with the result being that when the threshold value is increased the specificity of the prediction can be improved but the sensitivity is decreased. The default threshold values for the different parameters are set based on the optimal sensitivity and specificity. BcePred prediction results can be presented in graphical or tabular form. The graphical format displays the score for every parameter of each amino acid residue along the protein backbone in different colors, helping users to easily observe the B-cell epitopes. Amino acid residues with values higher than the threshold value (default is 2.38) are considered to be B-cell epitopes.

14.3 Prediction of Conformational B-Cell Epitopes

Due to the limited number of known crystal structures of antigen–antibody complexes, the mechanism of antigen–antibody binding is not yet clear and, because of the difficulty in designing prediction algorithms as well as for other reasons, progress in predicting conformational B-cell epitopes has been relatively slow. In recent years, with the development of bioinformatics, some software programs that can predict conformational B-cell epitopes have been created.

The prediction of conformational B-cell epitopes usually involves the identification of surface and epitope residues, construction of benchmark datasets, algorithm design, and performance evaluation. Currently, prediction methods can be divided into four types: (i) machine learning methods; (ii) non-machine learning methods; (iii) methods based on phage display experiments; and (iv) protein–protein interface methods. Most of these methods are based on the protein structure, and some are based on amino acid sequence data or are combined with the phage display data. Some methods use training data that have been validated and some methods use nonvalidated data. These prediction methods generally need to take the three-dimensional structure of the antigen into consideration, and a fully automated protein structure homology-modeling server such as SWISS-MODEL is available for evaluating antigens with an unknown three-dimensional structure. The result of the epitope prediction includes the predicted epitopic residue, position and number of epitopic residues. Different methods have various forms of output. Some software programs give a score for each surface residue that reflects the probability of it being a protein epitope. Some software programs predict whether the surface patch, consisting of surface residues, is an epitope. Other software combines the above two functions. Some even provide a graphical visualization of three-dimensional structures to display the epitope mapping. Most of the software programs are able to predict both linear and conformational epitopes.

14.3.1 Machine Learning Method

In 2005, an online conformational B-cell epitope prediction software program CEP (http://bioinfo.ernet.in/cep.htm) was published based on the machine learning prediction method (Kulkarni-Kale et al. 2005). It is the first free online server that uses the Protein Data Bank (PDB) file of the antigen protein tertiary structure as input and a prediction of the conformational epitope is its output. It can also predict linear epitopes, and the results are visualized graphically. This algorithm employs a structure-based bioinformatics method and solvent accessibility of amino acids in a specialized manner. By calculating the relative surface area of each amino acid in the protein structure, this software is able to confirm which regions are sufficiently exposed to be an antigenic epitope. The antigenic residues with $\geq$ 25 percent accessible surface area are identified. Antigenic determinants are identified if at least three contiguous accessible residues are present. If the spatial distance between the antigenic residues is below a specified cut-off value, it is predicted to be a conformational epitope. The prediction accuracy of this software is as much as 75 %. This server was the first approach available for the prediction of conformational B-cell epitopes, which is of groundbreaking significance, but its disadvantage is that it has fewer feature selections compared with subsequently developed similar software.

ElliPro (Ponomarenko et al. 2008) and many other prediction software programs based on the machine learning method have now been created. Machine learning methods can reveal complex nonlinear relationships in the dataset and are suitable for this field with their huge amount of data, but there is noise and lack of a unified theory. In order to overcome the deficiencies that include poor prediction accuracy, these methods are usually combined with additional features, and sometimes data verified by experiments are used to make predictions. However, further improvements in the performance of the system depend on the growing number of benchmark datasets and other more valuable features yet to be found.

14.3.2 Non-machine Learning Method

Use of non-machine learning methods to predict B-cell epitopes refers to the use of geometric features of the protein antigenic structure, or the joint use of the geometric characteristics together with the physical and chemical characteristics, to make predictions directly through calculations and without the use of machine learning methods. Such methods are intuitively clear and trusted by some researchers. Representative software include DiscTope (Haste Andersen et al. 2006) and PEPITO (Sweredoski and Baldi 2008).

DiscTope (http://www.cbs.dtu.dk/service) is the first online server created just for the prediction of non-contiguous B-cell epitopes (Haste Andersen et al. 2006). The algorithm is currently amassing a dataset that includes the crystal structures of known antigen–antibody complex object data, conformational epitopes characterized by extensive statistical measurements, and analyses of the spatial characteristics and calculations related to the potential for surface location all of which contribute to the prediction of noncontiguous B-cell epitopes. DiscTope is trained on a compiled dataset of noncontiguous epitopes from 76 antigen–antibody complexes with X-ray diffraction structures. The DiscTope program allows for adjustments to the epitope threshold value in order to adjust the specificity and sensitivity of the prediction. DiscTope can detect 15.5 % of residues located in discontiguous epitopes with a specificity of 95 %, while the other linear epitope prediction method can only detect about 11 % of the residues with the same 95 % specificity. When software developers set a specificity of 90 % and a sensitivity of 24 % to predict the discontiguous epitopes of malaria protein apical membrane antigen 1 (AMA1), the results showed 43 of 311 residues in AMA1 were predicted as epitope residues.

14.3.3 Phage Display Method

After years of development and improvement, phage display technology provides an easy method for epitope research with high throughput, low costs, and a practical platform. The phage display prediction method involves finding mimetic peptides with antibody affinity obtained by antibody screening phage display peptide libraries, using them to obtain probe motifs, and then searching for the best matching sequence on the surface of the antigen's tertiary structure. Typical software programs include Findmap (Mumey et al. 2003), SiteLight (Halperin et al. 2003), 3DEX (Schreiber et al. 2005), Mapitope (Bublil et al. 2007), PepSurf (Mayrose et al. 2007b), Pep-3D-Search (Huang et al. 2008), MIMOX (Huang et al. 2006), MIMOP (Moreau et al. 2006), and Pepitope(Mayrose et al. 2007a).

The Pep-3D-Search (http://kyc.nenu.edu.cn/Pep3DSearch/) program is based on mimotope analysis to predict conformational B-cell epitopes (Huang et al. 2008). The algorithm flow is as follows: First, input a three-dimensional structure of the antigen (a PDB file) and a set of mimotopes or a motif. Pep-3D-Search identifies all surface residues of the query antigen and creates a graph of surface residues. Second, either of two modes can be used: the mimotope mode and motif mode. In the mimotope mode, the ACO (ant colony optimization) algorithm is employed to search for matching paths on the surface of the antigen. Each matching path is scored to the corresponding mimotope. Finally, a *P*-value calculation algorithm and DFS (depth-first search) algorithm are used to screen and cluster these paths and the candidate epitopes are revealed. In the motif mode, the motifs are located on the antigen surface using the ACO algorithm and the paths with a high score are directly output as candidate epitopes.

Phage display peptide library technology combined with computer modeling for the prediction of conformational epitopes has a more solid experimental foundation and therefore, predictions are more reliable. However, its limitation is obvious. It can only predict the epitopes of antigens for which a monoclonal antibody is available, thus, it is difficult to carry out more extensive applications for epitope prediction.

14.3.4 Protein–Protein Interface Method

The interaction between antigen and antibody is a type of protein–protein interaction, therefore, some prediction tools based on protein–protein interactions can also be used to predict B-cell epitopes. The predictions using these methods depend on the identification of the characteristics of amino acid residues in the protein complex interface. In recent years, there has been a rapid growth of data on the three-dimensional structures of proteins, and a number of protein–protein interface databases have been created for use. These databases can be used to analyze the characteristics of amino acid residues in the protein–protein interface, the features of sequence conservation, secondary structure, solvent accessibility, and side-chain conformational entropy that all make a prominent contribution in epitope prediction. The recently created servers for predicting protein–protein interfaces include the following: ClusPro (Comeau et al. 2004) (http://structure.bu.edu) and PatchDock (Schneidman-Duhovn et al. 2003) (http://bioinfo3d.cs.tau.ac.il/sources.html), both of which are based on protein–protein docking; PPI-PRED (Bradford and Westhead 2005) (http://bioinformatics.leeds.ac.uk/ppi-pred), which combines a support vector machine approach with surface patch analysis to predict protein–protein binding sites; PIER (Kufareva et al. 2007) (http://abagyan.ucsd.edu/PIER/), which is based on local statistical properties of the protein surface derived at the level of atomic groups for predicting interfaces.

Prediction of conformational B-cell epitopes is generally required to determine the tertiary structure of the antigen and computer programs are able to predict the tertiary structures for those antigens with unknown tertiary structures. The currently published tools for the prediction of conformational B-cell epitopes have some limitations, but computer prediction programs combined with biological experiments, or integrated with the induced fit mechanism of antigen–antibody recognition, will all contribute to improve the accuracy of future successful epitope prediction.

14.4 Prediction of T-Cell Epitopes

T cells usually do not directly recognize the epitopes of native protein antigens. In order for the antigen to be recognized by the TCR, it must first be processed by an antigen-presenting cell (APC). First, after entering the cell by endocytosis, the native antigen is degraded into antigen peptides that then bind to the MHC molecule and form a peptide-MHC complex, which expressed on the surface of the APC and recognized by TCR. Thus, the primary goal in predicting T-cell epitopes is to determine the binding affinity of antigenic peptides and MHC molecules. There are two types of MHC molecules involved in antigen presentation, class I and class II. MHC-I molecules bind an endogenous antigen peptide that is then expressed on the surface of $CD8^+$ cytotoxic T cells for recognition by TCR. In contrast, MHC-II molecules bind an exogenous antigen for recognition by the TCR on $CD4^+$ helper T cells. Thus, bioinformatics methods to predict T-cell epitopes include predictions for both cytotoxic T-cell (CTL) epitopes and helper T-cell (T_H) epitopes.

In the adaptive immune system, the binding of MHC-II molecules to antigenic peptides plays an important role in the immune defense response. Exogenous proteins are enclosed in an endosome after entering the APC by endocytosis. This endosome merges with the primary lysosome to form a phagolysosome, where the exogenous proteins are degraded into peptides. The processed peptides bind to MHC-II molecules and form complexes that are transported to the APC surface for recognition by T helper cells. Since allergens are generally exogenous antigens, the main way to predict allergenic T-cell epitopes is to predict the binding affinity of the antigenic peptide with MHC-II molecules, but this approach is not able to predict the binding affinity of the antigen peptide with TCR.

An empty MHC molecule contains a groove that the peptides fit into. The ends of the groove of MHC-I molecules are closed, so a protein fragment must have a relatively fixed length to fit this groove. However, unlike MHC-I molecules, the ends of the groove of MHC-II molecules are open, allowing for some extension of the amino terminal and carboxyl terminal of the peptide (Nielsen et al. 2010b). This increases the complexity of predicting peptide binding to MHC-II molecules, where changes in the length of T_H-cell epitopes are large, usually composed of 9–25 amino acid residues, and sometimes more than 30 amino acids residues.

In addition, genes encoding MHC molecules are extremely diverse. Human MHC molecules are termed human leukocyte antigen (HLA), and the HLA gene is located on human chromosome VI. The HLA class I genotype mainly includes HLA-A, HLA-B, and HLA-C. The HLA class II genotype mainly includes HLA-DP, HLA-DQ, and HLA-DR. So far, more than 3000 HLA II alleles have been found. When analyzing the MHC binding affinity with antigen peptide to predict T-cell epitopes, sufficient consideration should be given to the polymorphisms and geographical distribution differences of HLA alleles.

14.4.1 Modeling Methods and Software

Since the late 1990s, researchers have put a lot of effort into computer-based methods for predicting T-cell epitopes, and there have been many prediction programs based on bioinformatics, so choosing a prediction tool with the highest accuracy is very important. Li et al. (2010) selected 309 known epitope peptides, and comprehensively evaluated the accuracy, sensitivity, specificity, positive predictive value, and negative predictive value of seven popular epitope prediction software programs that include ARB Matrix(Bui et al. 2005), MHCPred (Guan et al. 2003), NetMHCII (Nielsen et al. 2007), NetMHCIIpan (Nielsen et al. 2010a), Propred (Singh and Raghava 2001), RANKPEP(Reche et al. 2004), and SYFPEITHI (Rammensee et al. 1999). The results show the existence of differences in the reliability of the prediction methods, of which NetMHCIIpan and NetMHCII have the highest accuracy (see Table 14.1). We can combine the results of these two software programs to determine the epitopes with relatively high reliability, which can provide important information for epitope modification and vaccine design.

Table 14.1 Statistical results of epitope prediction software (Li et al. 2010)

Software	Accuracy	Sensitivity	Specificity	Positive predictive values	Negative predictive values
ARB	0.77	0.80	0.71	0.87	0.37
RANKPEPE	0.74	0.68	0.89	0.94	0.34
SYFPET	0.84	0.86	0.84	0.93	0.40
Propred	0.81	0.75	0.96	0.98	0.37
NetMHCIIpan	0.88	0.83	0.85	0.93	0.40
NetMHCII	0.88	0.83	0.87	0.94	0.40
MHCPred	0.61	0.54	0.05	0.76	0.11

According to the different modeling methods that are based on peptide sequences, the prediction methods can be divided into the binding motif method, quantitative matrix method and machine learning method as described below:

- **Binding motif method**

This early modeling method is based on binding motifs, and its principle is that when a linear polypeptide binds to MHC molecules, only few amino acids contribute to the binding with their side chains and the tightest binding occurs at conserved amino acids located at both ends of the peptide. This binding site is called an "anchor site" and the corresponding amino acid residues at anchor sites are called "anchor residues." Some of these anchor residues together form a fixed pattern, commonly referred to as a "motif." This modeling method works by evaluating whether the peptide fragment has known motifs in order to determine whether there will be binding between peptide and MHC-II molecule. Thus, the more verified MHC-II binding motifs there are, the better the accuracy of T_H-cell epitope prediction.

SYFPEITHI (http://www.syfpeithi.de/) is a representative motif modeling method. Its database is based on T-cell epitopes and MHC ligands that have been verified in the literature and covers 7000 peptides that bind to MHC-I and MHC-II molecules (Rammensee et al. 1999). Each peptide has been cited in the literature and is directly linked to the EMBL and PubMed databases. Thus, the number of peptides is constantly updated. This prediction method is based on published motifs (pool sequencing and natural ligands) and takes into account the amino acids at anchor positions and auxiliary anchor positions, as well as other frequently occurring amino acids. A score for the amino acid residue is calculated from a set of criteria that looks at whether or not they are anchor residues, auxiliary anchor residues or preferred residues. Ideal anchor residues are given 10 points, unusual anchor residues 6–8 points, auxiliary anchor residues 4–6 and preferred residues 1–4 points. Amino acid residues that are considered to have a negative effect on the binding ability is given a score between −1 and −3. Recent studies on anchor residues have shown that, due to great variability in the binding groove of MHC-II molecules, the restriction of anchor residues is not as strong as that of MHC-I. In addition, some peptides that bind to MHC-II molecules are not required to exactly match the anchor site. Thus, using binding motif methods to predict T-cell epitopes is not very accurate and the reliability of SYFPEITHI prediction method is only about 50 %.

- **Quantitative matrix method**

The development of quantitative matrix modeling methods to identify MHC-II binding peptides heralded a new stage in epitope prediction. The principle of quantitative matrix modeling is that the contribution of each amino acid residue in a peptide to binding to MHC-II molecules is independent, and the binding values can be added linearly. In the quantitative matrix, a protein sequence is fragmented into 9-mer peptides in a sliding window, each amino acid in the 9-mer peptides is assigned a binding value based on the type of amino acid and its position, and then the values of each amino acid in the 9-mer peptide are added to obtain a total score. This score is compared with a threshold score to determine the likelihood that a peptide will bind to MHC molecules. In general, the prediction accuracy of the quantitative matrix approach is higher than the binding motif method.

TEPITOPE (http://www.imtech.res.in/raghava/propred/) is a typical quantitative matrix method that has been widely used to study a variety of subunit vaccines (Sturniolo et al. 1999). The MHC-II type alleles have enormous polymorphisms (there are more than 690 kinds of known HLA-DR alleles) to allow for binding to different peptides, so prediction of peptide molecule binding to MHC-II is complex. TEPITOPE covers more than 50 HLA-DR alleles. Using TEPITOPE as its foundation, Raghava et al. established the online server Propred that has an improved graphical display so that a variety of specific HLA-DR binding peptides can be displayed overlapping with each other at a glance, and these overlapping regions may be the heterozygous T-cell epitopes.

- **Machine learning method**

The machine learning method is an advanced method developed in recent years to predict helper T-cell epitopes and overcomes the disadvantages of the motif method and the matrix method. It can reveal the complicated nonlinear relationship in dataset with high specificity, accuracy, and adaptability, so it has been widely used. The main mode of the machine learning method is to first use machine learning approaches to determine the binding center, and then to establish prediction model based on the recognized binding center. A typical modeling method of machine learning was proposed in 1998 (Brusic et al. 1998), which was a MHC-II binding peptides prediction method based on an evolutionary algorithm (EA) and an artificial neural network (ANN). The main steps of this type of modeling are: (1) Binding data extraction—peptide sequences and their binding affinities are collected from a variety of sources; (2) Peptide alignment—an EA is used to generate alignment matrices that are then used to find and align putative nonamer cores of known binders; (3) ANN training and classification—the ANN training set comprising aligned binding nonamer cores and nonbinding nonamers is used to train ANNs to predict the binding affinity of query peptides. Machine learning algorithms include artificial neural networks, evolutionary algorithms, hidden Markov chains, Gibbs sampling, and support vector machines. NN-align and NetMHCIIpan are representative servers that are based on artificial neural networks to predict T-cell epitopes.

NetMHCIIpan 2.0 (http://www.cbs.dtu.dk/services/NetMHCIIpan-2.0/) is an improved pan-HLA-DR binding prediction method that uses concurrent alignments and a weight optimization training program (Nielsen et al. 2010a). This method is a pan-specific version of the earlier published allele-specific NN-align algorithm. It does not require pre-alignment of the input data, which means that it does not require a certain amount of data related to MHC molecules in the training data to generate an accurate result of specific alleles. Thus, this method is useful for making predictions for alleles with limited known binding data. Using large and diverse benchmark data to evaluate the predictive accuracy of NetMHCIIpan

2.0, the results showed that the method was significantly better than the earlier MHC-II prediction methods and was particularly improved for alleles with very limited binding data where the conventional methods often have poor prediction accuracy.

14.4.2 Application of Software Prediction

With advances in bioinformatics, a number of T-cell epitope prediction tools have been developed but the accuracy of these predictions must be experimentally verified. Ailin Tao et al. from Guangdong Provincial Key Laboratory of Allergy and Clinical Immunology have worked extensively in this field. They used the NetMHCII 2.2 online server to predict the T_H-cell epitopes for nearly 20 kinds of representative allergens including mites (Der f 1, Der f 2, Der f 3, Lep d 2, and Sar s 14), cat (Fel d 1), house (Equ c 1), fungi (Mal f 1), *Parietaria judaica* (Par j 1), birch (Bet v 1), ryegrass (Lol p 5b), oysters (Cra g 1), mackerel (Sco j 1), soybean (Gly m 4), peanut (Ara h 2 and Ara h 8), lentils (Len c 1), and so on. Then, in order to reduce their allergenicity, the strong binding sites of the allergen were modified by site-directed mutagenesis PCR to attenuate their MHC-II binding affinity. Finally, the allergenicity was evaluated using a human dendritic cell model as well as with sensitized mice models to verify the prediction accuracy of the epitope modification.

References

Alix A. Predictive estimation of protein linear epitopes by using the program PEOPLE. Vaccine. 1999;18(3–4):311–4.

Bradford JR, Westhead DR. Improved prediction of protein-protein binding sites using a support vector machines approach. Bioinformatics. 2005;21(8):1487–94.

Brusic V, Rudy G, Honeyman G, Hammer J, Harrison L. Prediction of MHC class II-binding peptides using an evolutionary algorithm and artificial neural network. Bioinformatics. 1998;14(2):121–30.

Bublil EM, Freund NT, Mayrose I, Penn O, Roitburd-Berman A, Rubinstein ND, Pupko T, Gershoni JM. Stepwise prediction of conformational discontinuous B-cell epitopes using the Mapitope algorithm. Proteins. 2007;68(1):294–304.

Bui H, Sidney J, Peters B, Sathiamurthy M, Sinichi A, Purton K, Mothe B, Chisari F, Watkins D, Sette A. Automated generation and evaluation of specific MHC binding predictive tools: ARB matrix applications. Immunogenetics. 2005;57(5):304–14.

Chou P, Fasman G. Prediction of the secondary structure of proteins from their amino acid sequence. Adv Enzymol Relat Areas Mol Biol. 1978;47:45 148.

Colloc'h N, Cohen FE. Beta-breakers: an aperiodic secondary structure. J Mol Biol. 1991;221(2):603–13.

Comeau SR, Gatchell DW, Vajda S, Camacho CJ. ClusPro: an automated docking and discrimination method for the prediction of protein complexes. Bioinformatics. 2004;20(1):45–50.

Emini E, Schleif W, Colonno R, Wimmer E. Antigenic conservation and divergence between the viral-specific proteins of poliovirus type 1 and various picornaviruses. Virology. 1985;140(1):13–20.

Garnier J, Osguthorpe D, Robson B. Analysis of the accuracy and implications of simple methods for predicting the secondary structure of globular proteins. J Mol Biol. 1978;120(1):97–120.

Gershoni JM, Stern B, Denisova G. Combinatorial libraries, epitope structure and the prediction of protein conformations. Immunol Today. 1997;18(3):108–10.

Guan P, Doytchinova IA, Zygouri C, Flower DR. MHCPred: a server for quantitative prediction of peptide-MHC binding. Nucleic Acids Res. 2003;31(13):3621–4.

Halperin I, Wolfson H, Nussinov R. SiteLight: binding-site prediction using phage display libraries. Protein Sci. 2003;12(7):1344–59.

Haste Andersen P, Nielsen M, Lund O. Prediction of residues in discontinuous B-cell epitopes using protein 3D structures. Protein Sci. 2006;15(11):2558–67.

Hopp TP, Woods KR. Prediction of protein antigenic determinants from amino acid sequences. PNAS. 1981;78(6):3824–8.

Huang J, Gutteridge A, Honda W, Kanehisa M. MIMOX: a web tool for phage display based epitope mapping. BMC Bioinform. 2006;7:451.

Huang Y, Bao Y, Guo S, Wang Y, Zhou C, Li Y. Pep-3D-Search: a method for B-cell epitope prediction based on mimotope analysis. BMC Bioinform. 2008;9:538.

Jameson BA, Wolf H. The antigenic index: a novel algorithm for predicting antigenic determinants. Comput. Appl. Biosci: CABIOS. 1988;4(1):181–6.

Karplus P, Schulz G. Prediction of chain flexibility in proteins. Naturwissenschaften. 1985;72(4):212–3.

Kufareva I, Budagyan L, Raush E, Totrov M, Abagyan R. PIER: protein interface recognition for structural proteomics. Proteins. 2007;67(2):400–17.

Kulkarni-Kale U, Bhosle S, Kolaskar AS. CEP: a conformational epitope prediction server. Nucleic acids research. 2005;33(Web Server issue):W168-71.

Kyte J, Doolittle R. A simple method for displaying the hydropathic character of a protein. J Mol Biol. 1982;157(1):105–32.

Li S, Liu C, Zou Z, Tao A. Comparative Assessment of Different Softwares for MHC - II Antigen Epitope Prediction (in Chinese with English Abstract). J Biomed Eng Res. 2010;29(2):128–32.

Linehan LA, Warren WD, Thompson PA, Grusby MJ, Berton MT. STAT6 is required for IL-4-Induced germline Ig gene transcription and switch recombination. J Immunol. 1998;161(1):302–10.

Mayrose I, Penn O, Erez E, Rubinstein ND, Shlomi T, Freund NT, Bublil EM, Ruppin E, Sharan R, Gershoni JM, et al. Pepitope: epitope mapping from affinity-selected peptides. Bioinformatics. 2007a;23(23):3244–6.

Mayrose I, Shlomi T, Rubinstein ND, Gershoni JM, Ruppin E, Sharan R, Pupko T. Epitope mapping using combinatorial phage-display libraries: a graph-based algorithm. Nucleic Acids Res. 2007b;35(1):69–78.

Moreau V, Granier C, Villard S, Laune D, Molina F. Discontinuous epitope prediction based on mimotope analysis. Bioinformatics. 2006;22(9):1088–95.

Mumey BM, Bailey BW, Kirkpatrick B, Jesaitis AJ, Angel T, Dratz EA. A new method for mapping discontinuous antibody epitopes to reveal structural features of proteins. J Comput Biol J Comput Mol Cell Biol. 2003;10(3–4):555–67.

Nielsen M, Lundegaard C, Lund O. Prediction of MHC class II binding affinity using SMM-align, a novel stabilization matrix alignment method. BMC Bioinform. 2007;8:238.

Nielsen M, Justesen S, Lund O, Lundegaard C, Buus S. NetMHCIIpan-2.0—Improved pan-specific HLA-DR predictions using a novel concurrent alignment and weight optimization training procedure. Immunome Res. 2010a;6:9.

Nielsen M, Lund O, Buus S, Lundegaard C. MHC class II epitope predictive algorithms. Immunology. 2010b;130(3):319–28.

Odorico M, Pellequer JL. BEPITOPE: predicting the location of continuous epitopes and patterns in proteins. J Mol Recogn JMR. 2003;16(1):20–2.

Parker JM, Guo D, Hodges RS. New hydrophilicity scale derived from high-performance liquid chromatography peptide retention data: correlation of predicted surface residues with antigenicity and X-ray-derived accessible sites. Biochemistry. 1986;25(19):5425–32.

Ponomarenko J, Bui HH, Li W, Fusseder N, Bourne PE, Sette A, Peters B. ElliPro: a new structure-based tool for the prediction of antibody epitopes. BMC Bioinform. 2008;9:514.

Rammensee H, Bachmann J, Emmerich N, Bachor O, Stevanovic S. SYFPEITHI: database for MHC ligands and peptide motifs. Immunogenetics. 1999;50(3–4):213–9.

Reche P, Glutting J, Zhang H, Reinherz E. Enhancement to the RANKPEP resource for the prediction of peptide binding to MHC molecules using profiles. Immunogenetics. 2004;56(6):405–19.

Saha S, Raghava G. BcePred: prediction of continuous B-Cell epitopes in antigenic sequences using physico-chemical properties. Lect Notes Comput Sci. 2004;3239:197–204.

Schneidman-Duhovn D, Inbar Y, Polak V, Shatsky M, Halperin I, Benyamini H, Barzilai A, Dror O, Haspel N, Nussinov R, et al. Taking geometry to its edge: fast unbound rigid (and hingebent) docking. Proteins. 2003;52(1):107–12.

Schreiber A, Humbert M, Benz A, Dietrich U. 3D-Epitope-Explorer (3DEX): localization of conformational epitopes within three-dimensional structures of proteins. J Comput Chem. 2005;26(9):879–87.

Singh H, Raghava GPS. ProPred: prediction of HLA-DR binding sites. Bioinformatics. 2001;17(12):1236–7.

Sturniolo T, Bono E, Ding J, Raddrizzani L, Tuereci O, Sahin U, Braxenthaler M, Gallazzi F, Protti M, Sinigaglia F, et al. Generation of tissue-specific and promiscuous HLA ligand databases using DNA microarrays and virtual HLA class II matrices. Nat Biotechnol. 1999;17(6):555–61.

Sweredoski MJ, Baldi P. PEPITO: improved discontinuous B-cell epitope prediction using multiple distance thresholds and half sphere exposure. Bioinformatics. 2008;24(12):1459–60.

Welling G, Weijer W, van der Zee R, Welling-Wester S. Prediction of sequential antigenic regions in proteins. FEBS Lett. 1985;188(2):215–8.

Author's Biography

Ying He is an Assistant Researcher working at the Guangdong Provincial Key Laboratory of Allergy and Clinical Immunology, The State Key Clinical Specialty in Allergy, The State Key Laboratory of Respiratory Disease, the Second Affiliated Hospital of Guangzhou Medical University. Email: heying0605@163.com.

Ying He received her Master's Degree from Jinan University with a major in Genetics in 2009 and working on the antifibrosis effect of the protein drugs FGFR2 and its mechanisms. Currently, she is working in allergy research. Her main research interests focus on the prediction of allergen epitopes using bioinformatics tools, the cloning and purification of recombinant allergens, and the evaluation of allergenicity and modification of recombinant proteins. She has participated in several research programs of National Natural Science Foundation of China.

Ailin Tao is a Professor at Guangzhou Medical University, Director of Guangdong Provincial Key Laboratory of Allergy and Clinical Immunology, Principal Investigator of the State Key Laboratory of Respiratory Disease, Deputy Director of the State Key Clinical Specialty in Allergy of the Second Affiliated Hospital of Guangzhou Medical University, Member of the State Committee for Transgenic Safety Assessment, Standing Committee Member of Allergy Branch of Guangdong Medical Association, Member of Guangdong Provincial Committee for Transgenic Safety Assessment and Master Tutor.

Professor Ailin Tao earned his doctorate degree from the State Key Laboratory of Crop Genetic Improvement of Huazhong Agricultural University in 2002, followed by a postdoctoral training at Postdoctoral Station of Basic Medicine in Shantou University Medical College, majoring in allergen proteins. His most recent research has been on allergy bioinformatics, allergy and clinical immunology, and disease models such as allergic asthma, allergic rhinitis, infection and inflammation induced by allergy, inflammatory, and protracted diseases caused by antigens or superantigens. He has gained experience in the field of allergology including the mechanisms of immune tolerance, allergy triggering factors, chronic inflammation pathways, and allergenicity evaluation and modification for food and drugs. He proposed some new concepts including "Representative Major Allergens," "Allergenicity Attenuation" of immunotoxin and allergens, "broad-spectrum immunomodulator" as well as the theoretical hypothesis of "Balanced Stimulation by Whole Antigens." Professor TAO's laboratory focuses on the diagnosis of allergic disease and the medical evaluation of food and drug allergenicity and its modification. Professor TAO has now constructed a system for the prediction, quantitative assessment, and simultaneous modification of epitope allergenicity, which has been applied to more than 20 allergens, and he also developed a bioinformatics software program for allergen epitope prediction, SORTALLER (http://sortaller.gzhmu.edu.cn), which performed significantly better than other existing software, reaching a perfect balance of high specificity (98.4 %) and sensitivity (98.6 %) for discriminating allergenic proteins from several independent datasets of protein sequences of diverse sources. Furthermore, this program has a Matthews correlation coefficient as high as 0.970, a fast running speed and can rapidly predict a set of amino acid sequences with a single click. The software has been frequently used by researchers from many institutions in China and over 30 countries worldwide, thus becoming the number one allergen epitope prediction software program. Professor TAO has set up an allergen database ALLERGENIA (http://ALLERGENIA.gzhmu.edu.cn) that has several advantages over other databases such as a wide selection of nonredundant allergens, excellent astringency and accuracy, and friendly and usable analytical functions.

Chapter 15
Allergen Database

Yuyi Huang and Ailin Tao

Abstract Over the past ten years, there has been steady growth in the identification of allergen sequences. Different research institutions have built allergen databases for the integration and screening of allergens, thus serving scientific research and medical applications. However, as fewer and fewer new allergens are being identified, it is time for the establishment of standards with the goal of perfecting allergen databases so that they can serve not only as a allergen query platform that allows more effective use of allergen information, but also be an excellent database, i.e., be comprehensive, non-redundant, stringent, accurate, and able to analyze allergen functions.

Keywords Allergen · Bioinformatics · Database · Data mining · Protein · Health

15.1 Introduction

With the advances in the technology and due to the economic effects, allergens have drawn more and more interest because allergic diseases greatly affect human health (Sicherer and Leung 2014, 2015). Allergy is a rapid onset, multisystem hypersensitivity reaction that may be caused by both immunological and nonimmunological mechanisms. Most allergic reactions are immunoglobulin (IgE) mediated (Kirkbright and Brown 2012), show clear genetic predispositions and individual differences, but generally will neither destroy histocytes nor cause tissue damage (Monitto et al. 2010; de Wang 2011). A substance that can induce an allergic

Y. Huang · A. Tao (✉)
Guangdong Provincial Key Laboratory of Allergy and Clinical Immunology,
The State Key Clinical Specialty in Allergy, The State Key Laboratory of Respiratory Disease,
The Second Affiliated Hospital of Guangzhou Medical University, 250# Changgang Road
East, Guangzhou 510260, Guangdong Province, People's Republic of China
e-mail: taoailin@gzhmu.edu.cn

Y. Huang
e-mail: chvipdata@126.com

A. Tao and E. Raz (eds.), *Allergy Bioinformatics*, Translational Bioinformatics 8,
DOI 10.1007/978-94-017-7444-4_15

reaction is called allergen and is an essential component of allergy (Soares-Weiser et al. 2014). According to the particular pathway that causes the allergic reaction, there are five types of allergens: (1) food allergens, like milk, egg, fish, beef, mutton, seafood, animal fat, foreign proteins, alcohol, sesame oil, green onion, ginger, garlic, and some vegetables, fruits, etc. (Meng juan et al. 2014); (2) inhalation allergens, such as pollen, catkins, dust, mites, animal dander, lampblack, paint, exhaust, coal gas, cigarettes, cold air, fog, etc. (Diamant et al. 2013); (3) contact allergens, like cold, heat, ultraviolet light, radiation, cosmetics, shampoo, detergent, hair dye, soap, chemical fiber products, plastic, metal jewelry, bacteria, mildew, viruses, parasites, etc. (Goossens 2015); (4) drug allergens, like drugs, antimicrobial and anti-inflammatory drugs, penicillin, streptomycin, xenogeneic serum, etc. (5) self-allergens, i.e., autoantigens, such as mental stress, working stress, microbial infections, ionizing radiation, burns, and other biological, physical, and chemical factors including trauma or infection can cause changes in the structure or composition of autoantigens that results in the formation of allergens (Hopp et al. 2014).

Identifying and collecting all known allergens and creating an allergen database will provide a convenient query platform for medical researchers, doctors, and the general public that will provide complete information about known allergens. At present, several original and relatively complete international allergen databases have been created.

The existing allergen databases include: (1) the WHO/IUIS database (Radauer et al. 2014) (World Health Organization and International Union of Immunological Societies) that was created by the World Health Organization and International Union of Immunological Societies in 1984. It focuses on the systematic nomenclature of allergens according to their characteristics, structure, function, molecular biologic features, and bioinformatics data. The WHO/IUIS database only includes officially certified allergens. Newly submitted allergens are investigated by an Executive Committee, and the process of adding new allergens to this database requires extensive evidence as demonstrated in published papers and their validity must be further verified. Only when an allergen meets all requirements will it be formally named and included in the allergen database; (2) the Allergome database (Mari and Scala 2006), an online website, was created to provide relevant information about all aspects of an allergen. It includes allergens that cause IgE-dependent allergic diseases such as asthma, allergic dermatitis, allergic conjunctivitis, allergic rhinitis, and urticaria. Moreover, because allergy researchers desire that all allergenic molecules be supported by explicit documented evidence, every allergen in the Allergome database describes the allergen source (e.g., animal or plant, dandruff, fruit, pollen, seeds, spores, venom, or the human body) and its route of exposure (e.g., contact, ingestion, inhalation, or injection). The relevant information about each allergen is attached in a specific page. In addition, all allergens in the WHO/IUIS database have been incorporated into the Allergome database, including their relevant information and systematic nomenclature. The screening protocol of the search function in the Allergome database is the same as the protocol used by the WHO/IUIS database, thus, all data in the Allergome database also come from officially certified allergens. In addition to all the allergens included in the WHO/IUIS database, the Allergome database also contains a number of allergens that have had their immunological mechanism, clinical data, and other documented evidence thoroughly investigated. The Allergome database devotes itself to serving scientists working in allergy and immunology research. Its web interface and comprehensive data also provide a very convenient and efficient way for the clinician to obtain knowledge of allergens; (3) the AllergenOnline database is a platform that provides a list of manually examined allergens that are searched using certified serial numbers. This database focuses on assessing the safety of a protein that has undergone gene mutation or other food processing methods. Proteins have been assessed by additional experiments such as the serum fixation test, stimulation of basophils for histamine release, and cross-reactivity in vivo. This database is updated annually and provides a simple but efficient tool for safety assessment; (4) the ALLERGENIA database aims to promote the sharing of knowledge. All resources in this database are open to all researchers as well as to the public without reservation. Just a click on the "Browse Database" button displays all contents in the database, including accession number of an allergen and information about whether the allergen has been included in the UniProt protein database, the Allergome database or IUIS international allergens database, related information regarding IgE reactivity, and any published support. In addition, with just a click users can link to the interface of the data source, where more detailed content can be acquired. In order to provide the most availability, nonredundancy, astringency, accuracy, and functional analysis for researchers and the public, every sequence in the ALLERGENIA database has undergone strict double blind verification. The ALLERGENIA database is a tool for scientific research that promises accuracy and efficiency. It makes it possible for researchers to focus on the problem itself instead of wasting time searching for the relevant data. Moreover, ALLERGENIA, a public knowledge database, provides a feasible way for our institute, which is based in Guangzhou, and is dedicated to popularizing experimental science, to guide the public away from pseudoscience and lead them to reliable knowledge. Other influential allergen databases are named in Table 15.1.

15.2 Overview of Databases

Never perfect, every database has its own merits and drawbacks. The WHO/IUIS database is an internationally recognized database and includes officially certified allergens. There is no doubt about its authority. Therefore, many scientific researchers choose it as their certified allergen standard. Furthermore, nearly all other allergen databases are founded on this database. For example, the search function in the Allergome database is identical with that in the WHO/IUIS database. But since the WHO/IUIS database is very careful in investigating new allergens and/or due to inadequate funding for the necessary experiments, the process of adding new allergens is very slow. At present, the creation of genetically

Table 15.1 Allergen databases

Name	Website	# Allergens	Organization
Defra	http://allergen.fera.defra.gov.uk	640	Food and Envir. Res Agency
AllergenOnline	http://www.allergenonline.com	1706	University of Nebraska
SDAP (Ivanciuc et al. 2003)	http://fermi.utmb.edu/SDAP/index.html	1526	Univ. of Texas Medical Branch
WHO/IUIS (Radauer et al. 2014)	http://www.allergen.org	771	WHO/IUIS Allergen Nomenclature Sub-committee
Allallergy	http://allallergy.net	6329	Food and Preservative Allergy and Intolerance Database
Allergome (Mari and Scala 2006)	http://www.allergome.org	2915	Allergy Data Laboratories (ADL)
ADFS (Nakamura et al. 2005)	http://allergen.nihs.go.jp/ADFS	1777	NIHS, Japan
AllerMatch (Fiers et al. 2004)	http://www.allermatch.org	1080	Wageningen University
UniProtKB/Swiss-Prot (Boutet et al. 2007)	http://www.uniprot.org	435	EMBL-EBI, PIR
ALLERGENIA	http://ALLERGENIA.gzhmu.edu.cn	2150	GdPKL of Allergy and Immunology

modified products is growing explosively and suspect allergens caused by severe environmental pollution have increased significantly. Therefore, new protein information is very limited in this database.

Allergome, a database that is based on the allergens in the WHO/IUIS allergen database, takes advantage of the searching capabilities of the Internet and can automatically find biological molecules that are IgE sensitive or induce IgE-mediated diseases. Thanks to new computer-based programs, this database is regularly updated and its contents are plentiful. However, results obtained from computer programs inevitably contain some errors. Allergen databases, due to their importance to personal health and safety, should be more rigorous than other types of databases. Therefore, the judgment rule that “no omissions are allowed even at the expense of false judgment” is too inclusive.

The AllergenOnline database is a balance of the above two databases, that is to say, it includes all known allergens and ensures their careful assessment. Information about an allergen is regularly collected from the NCBI protein database and then investigated by a panel. Most importantly, a unified reviewing procedure is in place that all experts abide to when they independently investigate any particular protein. Finally, the panel votes to decide whether an allergen should be included. In this way, the accuracy as well as the growth ability of a database can be ensured. However, the database depends too strongly on preliminary artificial screening, and various defects, such as invalid sequences, repetitive sequences, and error sequences, can exist in different databases. A more obvious problem is that, huge amounts of sequences with high similarity point to the same allergen, providing several difficulties for the application of data.

In conclusion, an excellent database should be: (1) **comprehensive**, meaning it should include all known sequences of all allergens so that it can function as and meet the requirements of a database. Only when the database contains sequences from comprehensive resources will the users choose it as a regular tool; (2) **non-redundant**, meaning it does not include a large number of redundant sequences i.e., the same sequences with different names, the same sequences with different accession numbers, or integrated sequences and their fragments coexisting in the database. Using redundant sequences to expand the database is unwise. It can mislead users to possibly repeat positive or negative results. Importantly, it can weaken the accuracy and credibility of the software program that is utilized by some users to search the allergen database; (3) **stringent**, meaning that the allergen database should be a useful tool to solve practical problems and not just be an archive of allergens. Similarity of some sequences can reach 97 % and even as high as 99 %. These sequences have limited usefulness for practical applications. Time that is needed to acquire valid information from the database should be reduced. Repeat sequences should be avoided. In other words, the database should provide a practical tool for users to fully solve problems; (4) **accurate**, meaning of all its data should be investigated repeatedly and confirmed by experiment history. This is the most important standard and is at the core of a reliable database. If a database cannot promise its accuracy, any other information will be useless. This is particularly crucial for a medical database, because information regarding the sequences it contains will directly affect personal safety and medical decisions. The right of security is the most significant right for our citizens. Therefore, in order to embody the social responsibility required of a medical database, accuracy of the database is paramount; and (5) **analyzing performance**, meaning the operability and practicability of the database conveniently allows for the acquisition of comprehensive information by the user. Without this, a database will just look like a gold mine. The users, as miners, must spend most of their time excavating data. And this will be even more difficult for the general public who may not have basic computer knowledge. A functional analysis that is too complicated results in a database that looks like a super-excavator ahead of its time, which is hard to manipulate. In other words, the database itself should do the data mining and offer the gold directly to the users. This will reduce the costs in time and money and maximize the benefits.

15.2.1 WHO/IUIS

The WHO/IUIS allergy database (Radauer et al. 2014), an internationally recognized database, contains the officially confirmed allergens. Many institutions use this database as the gold standard. Because of its authority, almost all other allergen databases are based on it. But there are some irregular entries and invalid data in this database, which impacts its usefulness. By 2014, this database contained a total of 794 allergen indexes, 133 of which only had titles or were invalid. Of 1028 sequences, 941 sequences are marked with an UNIPROT accession number (906 actual effective sequences) and 56 sequences are labeled with NCBI accession number rather than UNIPROT accession number (13 actual effective sequences). After removing the repetitive sequences from the 919 actual effective ones, 911 sequences remain.

15.2.2 Allergome

The Allergome database (Mari and Scala 2006) is based on the WHO/IUIS database. Using published literature and the computer, this database automatically finds biological molecules that react with IgE or induce IgE-mediated disease. Because a computer program automatically searches new literature, this database is updated regularly, thus, its content is rich but it also has more invalid sequences. This database, updated at 10:15 +1 GMT on October 9, 2014, contains a total of 3885 protein allergens sequences and 4561 displayable login entries. As a large number of single entries each correspond to multiple sequences, 6306 sequences can be downloaded through these entries. After removing duplicate sequences, there are 5916 remaining sequences. When 99 % of the similar sequences are removed, 4630 sequences remain and when 97 % of similar sequences are removed, 3738 sequences remain.

15.2.3 AllergenOnline

The AllergenOnline database (Goodman et al. 2014) is updated annually and has a unified approval process that ensures its validity and expandability. According to version 14 updated on January 20, 2014, a total of 1706 allergens are indexed and this includes 1666 nonredundant sequences. After removing 99 % of similar sequences, there are 1404 remaining sequences and after removing 97 % of similar sequences, there are 1112 remaining sequences.

15.3 Introduction to the ALLERGENIA Database

15.3.1 Background

An increasing number of protein molecules have been identified as either allergens or nonallergens during the past 20 years (Sicherer and Leung 2014, 2015). Currently, there is no consensus among different Web-based databases because they have individually gathered different numbers of so-called "authentic" allergens and, in some cases, have even incorrectly included some nonallergens as "allergens." Collecting, organizing, and displaying veritable data that is reported in the scientific literature and having them interactively serve to draw unanimous conclusions is becoming a major concern of many investigators who rely on this knowledge to evaluate the allergenicity of their starting materials.

The allergen database ALLERGENIA has obvious advantages with accessibility, nonredundancy, astringency, accuracy, and functional analysis.

15.3.2 General Information

The ALLERGENIA database was developed by the Guangdong Provincial Key Laboratory of Allergy and Clinical Immunology, the State Key Laboratory of Respiratory Disease, the Second Affiliated Hospital of Guangzhou Medical University. At this time, FARRP (allergenonline), IUIS, and Allergome, etc., are the most widely used allergen website databases. But these websites have faults and/or limitations. Our ALLERGENIA database substantially outperforms these comprehensive allergen databases and by far exceeds the more narrow databases such as the InformAll Allergenic Food Database. The following four important aspects are featured in the ALLERGENIA database: (1) the highest number veritable allergens are included with the authenticity of each allergen sequence individually verified; (2) no false allergens (nonallergens) are included; (3) only nonredundant allergenic sequences are listed; and (4) the interactive interface is user friendly for a wide range of applications. After extensive and thorough investigation, we gathered 2108 authentic allergenic protein sequences and eliminated any redundant sequences.

15.3.3 BLAST Search

Gene modification as a technology has its advantages and disadvantages. On one hand, it can amplify genes of excellent agronomic traits in a crop to help resolve the conflict between resource shortages, environmental deterioration, and population growth; on the other hand, it may bring novel genes into crops that may also be allergens, thus affecting allergic people. Therefore, it is necessary to evaluate the allergenicity of candidate genes before gene modification and employ a "brave exploration but cautious promotion" approach for the breeding of gene-modified crops. At present, all genetically modified crops that have passed through safety assessment are considered to be safe (Wal 2015). All newly expressed proteins in recombinant-DNA plants that could be present in the final food should be assessed for their potential to cause allergic reactions. Genes derived from known allergenic sources should be assumed to encode an allergen unless scientific evidence demonstrates otherwise. Several bacteria, such as *Staphylococcus aureus*, *Aspergillus niger*, *Aspergillus fumigatus*, etc., are important allergens (FAO/WHO 2003, 2008), but instead are being neglected by some experts. All gene transformations should comply with the Codex Principle in order to verify the safety of the transgenic food using a programmed process of assessments before commercialization (FAO/WHO 2003). Production and utilization should be suspended immediately if any allergic cases related to a GMO food are found. Transgenic foods should never be rushed to market or be imported from abroad without definite information in regard to allergenic potential to the people of the (importing) countries.

According to the Codex Alimentarius Commission Guidelines (FAO/WHO 2008), if a gene fragment comes from some known allergenic source, it should be considered as an allergen. That is to say, we should assume that this type of gene fragment is harmful before conducting experiments to exclude this assumption. When testing allergenicity for new GMOs, we need to compare the results of allergenicity assessment on a case-by-case basis. If the coding amino acid sequence of one candidate transgene is highly identical to that of a mite allergen, while the amino acid sequence of another candidate gene only resembles that of a peanut allergen, the former has more risk potential than the latter. Furthermore, if the amino acid sequence of a candidate gene only has large identity with the allergen from rice, it will not pose any safety issues to the people feeding on rice.

Therefore, we developed an allergenicity evaluation standardization system, which is divided into three levels. The first standard is called Rice Level, i.e., the acceptable lowest limit or least hazardous level. This standard determines that a new GMO is safe only when the allergenicity of its transformed gene(s) and products is similar to or lower than that of rice, even when the transformed genes and their encoding amino acid sequences are similar to the most potent allergens derived from rice amino acid sequences. The second standard is called Peanut/Soybean Level, i.e., a hazard level that has warning exclusion criteria. This standard deems that a GMO product may be hazardous to the humans when the allergenicity of the newly introduced protein encoded by the transformed gene is similar to the most potent allergens derived from peanut/soybean, especially if the sequence(s) is highly similar to the most potent allergens from peanut/soybean on contiguous and identical fragment(s). This standard states that the transforming candidate genes should have allergenicity assessment and modification. The third standard is called Mite/Olive level, i.e., the highest allergenicity level and candidate for mandatory exclusion. This standard determines that these GMO products would be hazardous to humans, and that the encoding gene should not be transformed when the allergenicity of the encoded protein by the future-transformed gene is similar to the most potent allergens derived from mite/olive (pollen), especially if the sequence(s) is highly similar to the most potent allergens from mite/olive (pollen) on contiguous and identical fragment(s). Here we propose a three-level standardization system for allergenicity evaluation as follows (Fig. 15.1).

Based on the above considerations, the ALLERGENIA database provides various search and comparison methods so as to offer customers comprehensive

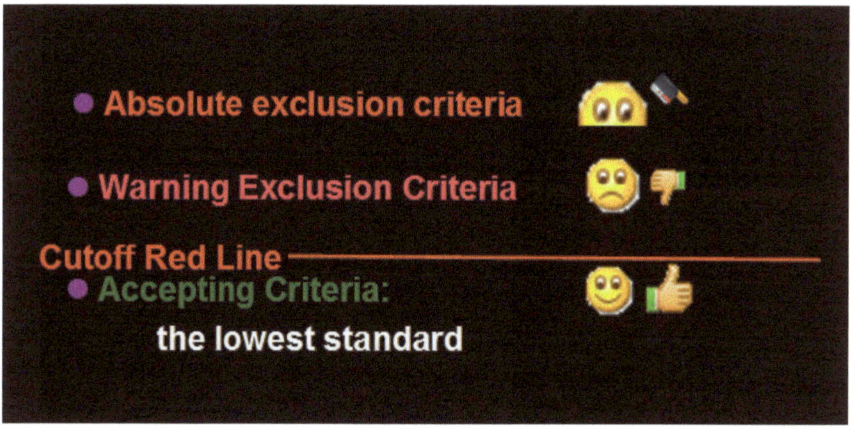

Fig. 15.1 A system for the standardization of allergen evaluation divided into three levels

results. Customers can draw their own conclusions using their own standards if desired or carry out subsequent verification work. This database provides two ways of BLAST, i.e., routine BLAST and BLAST based on FAO/WHO rules. The routine BLAST, this method does not include short identical fragments in the results because its goal is to check whether there sequences that are identical or similar among the entries listed in the queried database. Figure 15.2 shows an example of a routine BLAST. The BLAST method based on FAO/WHO rules searches for identity with contiguous amino acids or a minimum of 35 % sequence identity over a window of 80 amino acids with an allergen as defined by FAO/WHO rules. Through this special search, results obtained from the ALLERGENIA website display different identity levels in five columns: Column 1, a minimum of 35 % sequence identities over a window of 80 amino acids of each query protein with an allergen; Column 2, identities less than 35 % over sliding windows of 80 amino acid segments of each query protein, but with at least one segment identity of eight contiguous amino acids; Column 3, identities less than 35 % for sliding windows of 80 amino acid segments of each query protein, but with a sequence of only seven exact match amino acids; Column 4, identities less than 35 % for sliding windows of 80 amino acid segments of each query protein, but only with six exactly matching amino acids; Column 5, identities less than 35% for sliding windows of 80 amino acid segments of each query protein, but with no more than five exactly matching amino acids. This design was chosen based on a recent study demonstrating that shared identity with only five contiguous amino acids can also lead to cross-reactivity between nonhomologous allergens. Figure 15.3 shows an example of the BLAST comparison method using FAO/WHO rules.

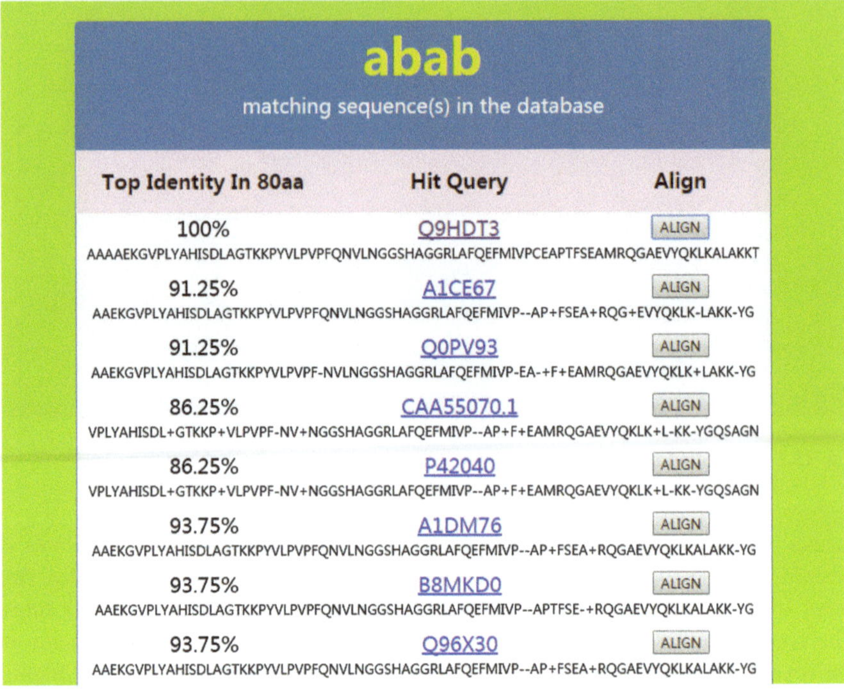

Fig. 15.2 Partial results of a sequence "abab" demonstrating a routine BLAST for

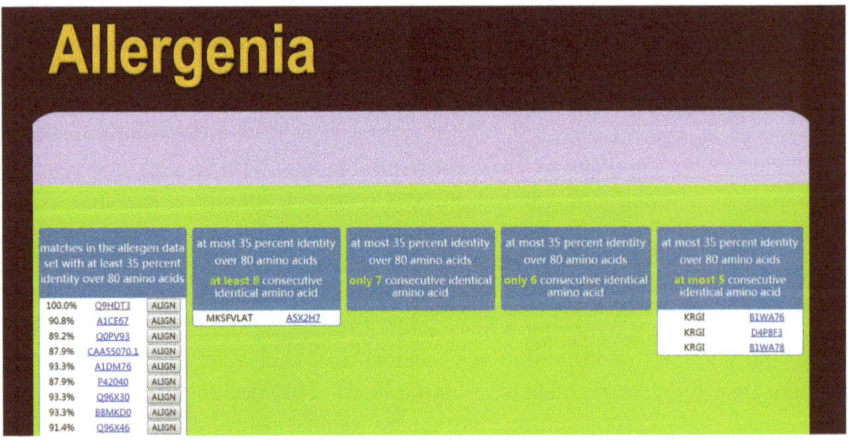

Fig. 15.3 An example a sequence "abab" demonstrating the BLAST comparison method using FAO/WHO rules and the ALLERGENIA database

15.3.4 Browse Database

When browsing the ALLERGENIA database, the page displayed shows all the allergen data gathered in the database for any particular allergen. The "ID" column shows the protein accession number in UniProt or in NCBI format. The allergen facts column includes five items: "UniProt," "Allergome," "IUIS," "IgE," and "Reference." For each item, "y" indicates that the objective protein(s) is/are collected as allergens in certain databases, such as UniProt, Allergome, IUIS, or that the protein possesses IgE-binding ability or that other references definitely regard the protein as an allergen, whereas "n" indicates that the corresponding database does not identify this protein as an allergen, or that the protein lacks proof of IgE-binding or other supporting references. It is important that all allergens display complete information in each of the five categories. Faults and/or limitations relating to the existing websites can be clearly viewed, thus allowing researchers to rely on the most authentic allergen database.

15.3.5 Condition Search

Condition searches in the ALLERGENIA database include five options: "UniProt," "Allergome," "IUIS," "IgE," or "Reference," which all contain allergen facts. In each module, when the "Existent" condition is selected, the allergens for which there are known facts are retrieved and displayed, whereas when the "Non-Existent" condition is selected, allergens that lack individual facts are retrieved and displayed. These searches allow the allergen facts to be viewed in a three-dimensional way and also show the differences among different databases for each individual allergen.

References

Boutet E, Lieberherr D, Tognolli M, Schneider M, Bairoch A. UniProtKB/Swiss-prot. Methods Mol Biol. 2007;406:89–112.

de Wang Y. Genetic predisposition for atopy and allergic rhinitis in the Singapore Chinese population. Asia Pac Allergy. 2011;1(3):152–6.

Diamant Z, Gauvreau GM, Cockcroft DW, Boulet LP, Sterk PJ, de Jongh FH, Dahlen B. Inhaled allergen bronchoprovocation tests. J Allergy Clin Immunol. 2013;132(5):1045–55.

FAO/WHO. Guideline for the conduct of food safety assessment of foods produced using recombinant DNA microorganisms (CAC/GL 46-2003). Foods derived from modern biotechnology (Second ed.). 2003. p. 35–57.

FAO/WHO. Guideline for the conduct of food safety assessment of foods derived from recombinant DNA animals (CAC/GL 68-2008). Foods derived from modern biotechnology (Second ed.). 2008. p. 57–76.

Fiers MW, Kleter GA, Nijland H, Peijnenburg AA, Nap JP, van Ham RC. Allermatch, a webtool for the prediction of potential allergenicity according to current FAO/WHO codex alimentarius guidelines. BMC Bioinform. 2004;5:133.

Goodman R, van Ree R, Vieths S, Ferreira F, Ebisawa M, Sampson HA, et al. Criteria used to categorise proteins as allergens for inclusion in allergenonline.org: a curated database for risk assessment. Clin Transl Allergy. 2014;4(Suppl 2): P12.

Goossens A. New cosmetic contact allergens. Cosmetics. 2015;2(1):22–32.

Hopp AK, Rupp A, Lukacs-Kornek V. Self-antigen presentation by dendritic cells in autoimmunity. Front Immunol. 2014;5:55.

Ivanciuc O, Schein CH, Braun W. SDAP: database and computational tools for allergenic proteins. Nucleic Acids Res. 2003;31(1):359–62.

Kirkbright SJ, Brown SG. Anaphylaxis–recognition and management. Aust Fam Physician. 2012;41(6):366–70.

Mari A, Scala E. Allergome: a unifying platform. Arb Paul Ehrlich Inst Bundesamt Sera Impfstoffe Frankf A M. 2006;95:29–39.

Meng juan W, Jiang hua L, Lin yu G, Jia jie L, Zhi gang W. Food allergen labeling management: a review of the European experience and its inspiration for China. Food Sci. 2014;35(1):261–5.

Monitto CL, Hamilton RG, Levey E, Jedlicka AE, Dziedzic A, Gearhart JP, Boyadjiev SA, Brown RH. Genetic predisposition to natural rubber latex allergy differs between health care workers and high-risk patients. Anesth Analg. 2010;110(5):1310–7.

Nakamura R, Teshima R, Takagi K, Sawada J. Development of Allergen Database for Food Safety (ADFS): an integrated database to search allergens and predict allergenicity. Kokuritsu Iyakuhin Shokuhin Eisei Kenkyusho Hokoku. 2005;(123):32–6.

Radauer C, Nandy A, Ferreira F, Goodman RE, Larsen JN, Lidholm J, Pomes A, Raulf-Heimsoth M, Rozynek P, Thomas WR, et al. Update of the WHO/IUIS allergen nomenclature database based on analysis of allergen sequences. Allergy. 2014;69(4):413–9.

Sicherer SH, Leung DY. Advances in allergic skin disease, anaphylaxis, and hypersensitivity reactions to foods, drugs, and insects in 2013. J Allergy Clin Immunol. 2014;133(2):324–34.

Sicherer SH, Leung DY. Advances in allergic skin disease, anaphylaxis, and hypersensitivity reactions to foods, drugs, and insects in 2014. J Allergy Clin Immunol. 2015;135(2):357–67.

Soares-Weiser K, Takwoingi Y, Panesar SS, Muraro A, Werfel T, Hoffmann-Sommergruber K, Roberts G, Halken S, Poulsen L, van Ree R, et al. The diagnosis of food allergy: a systematic review and meta-analysis. Allergy. 2014;69(1):76–86.

Wal JM. 8-Assessing and managing allergenicity of genetically modified (GM) foods. Handbook of Food Allergen Detection and Control. 2015: 161–78.

Author's Biography

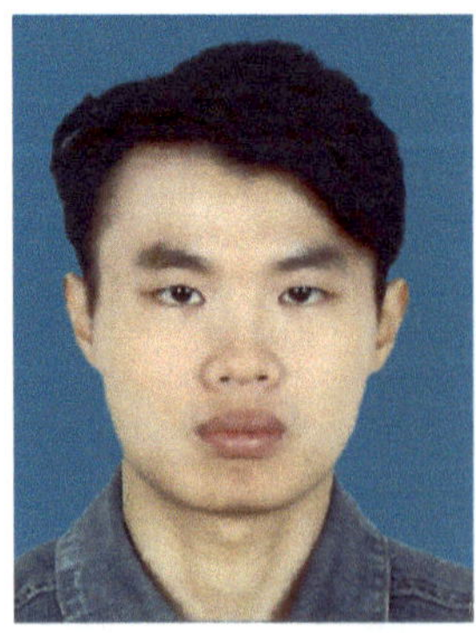

Yuyi Huang is an Assistant Researcher working at the Guangdong Provincial Key Laboratory of Allergy and Clinical Immunology, the State Key Laboratory of Respiratory Disease, and the State Key Clinical Specialty in Allergy of the Second Affiliated Hospital of Guangzhou Medical University. Email: chvipdata@126.com.

Dr. Huang received his Master's Degree in Immunology and Molecular Biology from Guangzhou Medical University in 2010 and is currently working at Guangzhou Medical University. His experience is in alleviation of the allergenic potential of allergens with a main focus on allergen bioinformatics. He has developed an allergen database ALLERGENIA and an allergen predicting software SORTALLER.

Ailin Tao is a Professor at Guangzhou Medical University, Director of Guangdong Provincial Key Laboratory of Allergy and Clinical Immunology, Principal Investigator of the State Key Laboratory of Respiratory Disease, Deputy Director of the State Key Clinical Specialty in Allergy of the Second Affiliated Hospital of Guangzhou Medical University, Member of the State Committee for Transgenic Safety Assessment, Standing Committee Member of Allergy Branch of Guangdong Medical Association, Member of Guangdong Provincial Committee for Transgenic Safety Assessment and Master Tutor.

Professor Ailin TAO earned his doctorate degree from the State Key Laboratory of Crop Genetic Improvement of Huazhong Agricultural University in 2002, followed by a post-doctoral training at Postdoctoral Station of Basic Medicine in Shantou University Medical College, majoring in allergen proteins. His most recent research has been on allergy bioinformatics, allergy and clinical immunology and disease models such as allergic asthma, allergic rhinitis, infection and inflammation induced by allergy, inflammatory and protracted diseases caused by antigens or superantigens. He has gained experience in the field of allergology including the mechanisms of immune tolerance, allergy triggering factors, and chronic inflammation pathways and allergenicity evaluation and modification for food and drugs. He proposed some new concepts including “Representative Major Allergens,” “Allergenicity Attenuation” of immunotoxin and allergens, “broad-spectrum immunomodulator” as well as the theoretical hypothesis of “Balanced Stimulation by Whole Antigens.” Professor TAO’s laboratory focuses on the diagnosis of allergic disease and the medical evaluation of food and drug allergenicity and its modification. Professor TAO has now constructed a system for the prediction, quantitative assessment, and simultaneous modification of epitope allergenicity, which has been applied to more than 20 allergens, and he also developed a bioinformatics software program for allergen epitope prediction, SORTALLER (http://sortaller.gzhmu.edu.cn), which performed significantly better than other existing software, reaching a perfect balance of high specificity (98.4 %) and sensitivity (98.6 %) for discriminating allergenic proteins from several independent datasets of protein sequences of diverse sources. Furthermore, this program has a Matthews correlation coefficient as high as 0.970, a fast running speed and can rapidly predict a set of amino acid sequences with a single click. The software has been frequently used by researchers from many institutions in China and over 30 countries worldwide, thus becoming the number one allergen epitope prediction software program. Professor TAO has set up an allergen database ALLERGENIA (http://ALLERGENIA.gzhmu.edu.cn) that has several advantages over other databases such as a wide selection of nonredundant allergens, excellent astringency and accuracy, and friendly and usable analytical functions.

Zeitfracht Medien GmbH
Ferdinand-Jühlke-Straße 7
99095 Erfurt, Deutschland
produktsicherheit@kollbri360.de